Thermal Evaluation of Indoor Climate and Energy Storage in Buildings

There is a need to reduce energy consumption for space cooling and heating via energy efficient solutions/technologies for implementation in the buildings. Thermal energy storage regulates indoor temperature, shifting the peak load to the off-peak hours and reducing the energy need for space cooling and heating. This book presents the most recent advances related to the thermal energy storage system design and integration in buildings. Additionally, modelling, application, synthetization, and characterization of energy efficient building materials are also considered.

Features:

- Provides a deep understanding of thermal energy storage technology and summarizes its utility and feasibility that can be commercially implemented worldwide
- Covers recent advancements related to thermal energy storage system design and integration in buildings
- Discusses modelling, application, synthetization, and characterization of energy-efficient building materials
- Details novel and emerging heat storage materials and their application to energy and environmental processes
- Highlights the need for future research on building comfort in cooling, heating, and ventilation through a green energy perspective

This book is aimed at researchers and graduate students in mechanical, renewable energy, and HVAC engineering.

Thermal Evaluation of Indoor Climate and Energy Storage in Buildings

Edited by Shailendra Kumar Shukla

CRC Press
Taylor & Francis Group
Boca Raton London New York

CRC Press is an imprint of the
Taylor & Francis Group, an **informa** business

Designed cover image: Shutterstock

First edition published 2025
by CRC Press
2385 NW Executive Center Drive, Suite 320, Boca Raton FL 33431

and by CRC Press
4 Park Square, Milton Park, Abingdon, Oxon, OX14 4RN

CRC Press is an imprint of Taylor & Francis Group, LLC

ISBN: 978-1-032-52777-2 (hbk)
ISBN: 978-1-032-54212-6 (pbk)
ISBN: 978-1-003-41569-5 (ebk)

DOI: 10.1201/9781003415695

Typeset in Times
by Apex CoVantage, LLC

The electronic version of this book was funded to publish Open Access through Taylor & Francis' Pledge to Open, a collaborative funding open access books initiative. The full list of pledging institutions can be found on the Taylor & Francis Pledge to Open webpage.

Contents

About the Editor

Shailendra Kumar Shukla is Professor in the Department of Mechanical Engineering, Indian Institute of Technology (BHU). He is also Founder Coordinator of the Centre for Energy and Resources Development (CERD), IIT (BHU), and Coordinator of the CST UP Incubation Centre for Grass-Root Innovators at the Mechanical Engineering Department, IIT (BHU), Varanasi. He has developed research collaborations with different institutes in the UK and Japan in the field of thermal engineering and renewable energy technology. He has published more than 100 research papers in high-impact SCI/Scopus journals, five books, and many book chapters in books published by reputed publishers. He has been placed on the boards of editors and reviewers of highly reputed journals. As a writer and researcher, he is best known for his ability to put theoretical things in a practical way.

Recently he has been conferred FRSC, fellow of the Royal Society UK. He has been awarded a Japanese Society for Promotion of Sciences JSPS Fellow Award 2014–15 for Visiting GIFU University, Japan. A recipient of the UGC-TEC Consortium Agreement Award 2010 in Sustainable Energy by University Grants Commission, New Delhi, India, he is also former Pro-Vice Chancellor of Ranchi University.

Contributors

Ali A. F. Al-Hamadani
Mechanical Engineering Department
Wasit Universiy
Wasit, Iraq

Banyal, Ankush
Rajiv Gandhi Government Engineering College
Kangra, India

Chaube, Akhilesh K.
Krishi Vigyan Kendra, Singrauli
Jawaharlal Nehru Krishi Vishwavidyalaya, Jabalpur
India

Chauhan, Vikash Kumar
Department of Mechanical Engineering
Indian Institute of Technology (BHU)
Varanasi, India

Gautam, Abhishek
Department of Mechanical Engineering
Indian Institute of Technology, Bombay
Mumbai, India

Ghosh, Rajgourab
Amity Institute of Biotechnology
Amity University, India

Jani, D. B.
Government Engineering College Dahod
Ahmedabad, Gujarat, India

Jeon, Yongseok
Department of Mechanical Engineering
Korea Maritime and Ocean University
Busan, Republic of Korea

Kaul, Sanjay
Engineering Technology, School of Business and Technology
Fitchburg State University
Fitchburg, USA

Kaushik, Amit
Biotechnology
Amity Institute of Biotechnology
Noida, U.P., India

Khandal, R. K.
Former Vice Chancellor, Uttar Pradesh Technical University
India

Kumar, Sonu
Koneru Lakshmaiah Education Foundation
KL Deemed to be University, Vaddeswaram
Guntur, Andhra Pradesh, India

Kumar, Rajan
Dr. B R Ambedkar National Institute of Technology
Jalandhar, India

Kumar, Prabhanshu
Amity Institute of Biotechnology
Amity University
Noida, India

Mishra, Manish
Indian Institute of Technology
Roorkee, India

Mishra, Abhishek
Invertis University
Bareilly, U.P., India

Mishra, Shubheksh
Samrat Ashok Technological Institute
Vidisha, M.P., India

Mishra, Anshuman
Institute of Advanced Materials, IAAM
Gammalkilsvägen 18 Ulrika 59053, Sweden

Nindhia, Tjokorda Gde Tirta
Engineering Faculty
Udayana University
Bali, Indonesia

Panat, Ashish
S. N. D. T. Women's University
Mumbai, India

Pandey, Sonika
Institute of Basic Science
Bhopal, M.P., India

Pathak, Saurabh
Government Polytechnic Gazipur
U.P. India

Roy, Biswajit
CSIR-Centre for Cellular and Molecular Biology
Hyderabad, India

Sahoo, P. K.
Botswana International University of Science and Technology (BIUST)
Botswana, South Africa

Sahu, Govind
Department of Mechanical Engineering
Government Engineering Collage Raipur
Chhattisgarh, India

Saini, R. P.
Department of Hydro and Renewable Energy
Indian Institute of Technology Roorkee
Roorkee, India

Shukla, Shailendra Kumar
Department of Mechanical Engineering
Indian Institute of Technology (BHU)
Varanasi, India

Singh, P. P.
Invertis University
Bareilly, U.P., India

Singh, Rajeev
Department of Environmental Science
Jamia Millia Islamia
New Delhi, India

Singh, D. K.
Dr. B R Ambedkar National Institute of Technology
Jalandhar, India

Suresh, C.
Department of Mechanical Engineering
Korea Maritime and Ocean University
Busan, Republic of Korea

Tripathi, Sunil Kumar
CSIR-Centre for Cellular and Molecular Biology
Hyderabad, India

Tripathi, Ajay
Department of Mechanical Engineering
Government Engineering Collage Raipur
Chhattisgarh, India

Tripathi, Bhartendu Mani
Department of Mechanical Engineering
Indian Institute of Technology (BHU)
Varanasi, India

Upadhyay, Chanakya
Institute of Basic Science
Bhopal, India

Verma, Samakshi
Rama University, Mandhana
Kanpur, India

Preface

Welcome to the fascinating domain of thermal evaluation of indoor climate and energy storage in buildings. The demand for sustainable and energy-efficient buildings has never been greater in today's rapidly changing world. In this book, we embark on a journey to explore the intricate relationship between advanced building materials, energy storage systems, and bioclimatic building technologies and their impact on indoor thermal comfort and local climate change. From high-performance insulation materials to innovative facade systems, these materials revolutionize building construction, ensuring better energy conservation and reduced environmental impact. This book explores cutting-edge energy storage technologies, ranging from thermal storage systems to battery-based solutions, that are instrumental in achieving energy self-sufficiency and reducing the reliance on external energy grids. Ventilated walls and double-skin facades are innovative building envelope solutions that have gained traction due to their significant impact on indoor climate and energy performance. Bioclimatic building technologies take inspiration from nature to create sustainable and climate-responsive designs. By utilizing passive design strategies such as orientation, shading devices, and natural ventilation, these technologies harmonize buildings with their environment, maximizing occupant comfort while minimizing energy demands. Responsive buildings with smart sensors and automated systems are at the forefront of intelligent and energy-efficient design. These structures optimize heating, cooling, and lighting systems by monitoring and adapting to environmental conditions in real time, thereby improving energy efficiency and occupant comfort. Passive and active exploitation of renewable energy resources is a cornerstone of sustainable building design. By incorporating solar panels, wind turbines, geothermal systems, and other renewable energy technologies, buildings can generate clean energy and significantly reduce their carbon footprint. This book investigates integrating renewable energy systems in buildings, providing insights into their benefits, challenges, and best practices. By adopting holistic approaches to urban planning, including efficient transportation, waste management, and infrastructure design, we can create urban environments that optimize resource usage and promote a high quality of life for inhabitants. At the heart of all sustainable building practices lie occupants' well-being and thermal comfort. Buildings can create healthy and productive indoor environments by considering indoor air quality, natural light, thermal insulation, and ergonomic design. We examine the relationship between the built environment and human well-being, emphasizing the importance of thermal comfort and occupant satisfaction.

I hope this book serves as a valuable resource for architects, engineers, researchers, and students who are passionate about advancing sustainable building practices. By understanding the intricate interplay between advanced building materials, energy storage systems, and indoor thermal evaluation, it will pave the way for a more sustainable and energy-efficient built environment.

1 Advanced Building Materials

Shailendra Kumar Shukla and Sanjay Kaul

1.1 INTRODUCTION

1.1.1 Brief History of Building Materials

Building supplies have been essential throughout human history. Different construction materials' growth and accessibility have molded the built environment and impacted how people live, work, and interact with their surroundings. Ancient humans constructed their dwellings and other constructions from natural resources, including stone, clay, wood, and animal skins. These resources were plentiful and easily accessible, and the region's topography and climate determined how they should be used. For instance, lumber was widely utilized for construction in locations with many kinds of wood, but stone and mud were employed in areas with little access to wood. New construction materials were discovered and produced as civilizations evolved and grew. For instance, the Romans used concrete considerably while building their famous aqueducts, bridges, and other grand constructions. Brick and stone were more often used throughout the Middle Ages, which allowed for the construction of cathedrals, castles, and other impressive structures still standing today. New building materials like steel, glass, and concrete revolutionized the construction sector in the modern age, enabling the construction of more significant and intricate buildings. New building materials also made building structures more quickly and effectively feasible, facilitating the rapid growth of cities and metropolitan regions. Beyond their actual usage in construction, building materials are significant. Different materials have symbolic, cultural, and social significance corresponding to society's values and ideals. For instance, although the use of glass and steel in contemporary skyscrapers represents the principles of development and innovation, wood in traditional Japanese architecture is directly tied to the nation's traditional aesthetics and love of natural beauty. Creating and applying building materials have profoundly influenced human history and the built environment. They have a cultural, social, and symbolic significance in shaping how people live, work, and engage with their environment.

1.2 DIFFERENT FACTORS FOR CONSIDERATION IN THE SELECTION OF BUILDING MATERIALS

The selection of building materials is a crucial step in the construction of any project, and several factors should be considered. Some of the essential considerations are as follows.

DOI: 10.1201/9781003415695-1

1.2.1 Climatic Conditions

The weathering effect significantly influences building materials' effectiveness and performance. You will find quite different climatic conditions depending on where you go in the world. For instance, tropical areas experience higher levels of atmospheric moisture, while certain regions see temperature fluctuations that negatively or positively impact the durability of building materials. In the same way that different materials have been developed for different types of places, various construction methods are used worldwide. Therefore, choosing the appropriate material for a specific area becomes very crucial. A building's primary purpose is to offer protection from the environment outside. Therefore, a building should be well insulated against outside heat, cold, rain, or other climatic variables. These days, construction materials' performance is impacted by unpredictable environmental conditions around the world. A building material should therefore be able to resist weathering effects to withstand wet and dry environments without breaking down or losing its mechanical characteristics. In addition, if the building material doesn't offer appropriate insulation from the weather outside, more energy is needed to maintain the indoor atmosphere.

1.2.2 Strength and Durability

The materials must be strong and durable enough to sustain the loads and pressures they will endure over the building's lifespan. Different types of structures require different levels of strength and durability. For example, high-rise buildings, bridges, and industrial facilities typically require materials with exceptional strength to handle heavy loads and withstand forces, while residential buildings may have different requirements. Building codes and industry standards provide guidelines for minimum strength and durability requirements. It is essential to ensure that selected materials meet or exceed these standards and undergo proper testing to validate their performance. By considering strength and durability parameters, builders and architects can choose materials that provide structural stability, longevity, and safety for the intended application, resulting in well-designed and reliable structures.

1.2.3 Thermal Capabilities and Availability

Thermal capabilities refer to a material's ability to resist heat transfer, regulate temperature, and provide insulation properties. These capabilities are crucial in maintaining indoor comfort, reducing energy consumption, and promoting energy efficiency in buildings. To guarantee that the building maintains a pleasant temperature and that energy consumption is minimal, the materials used should have strong thermal insulation capabilities. Some materials have high thermal mass, which can absorb, store, and release heat slowly. This property can help regulate temperature variations and reduce the need for mechanical heating or cooling. Materials like concrete, stone, or rammed earth are examples of those with high thermal mass. The commodities should be freely accessible in the local market to avoid delays and additional costs associated with obtaining resources from distant places. Utilizing locally available materials can reduce transportation costs, carbon footprint, and construction time. It also supports the local economy and promotes sustainable practices. Assessing the reliability and stability of the supply chain is essential to avoid delays or disruptions in construction. Understanding materials' availability and potential limitations can help plan construction schedules effectively.

1.2.4 Moisture and Fire Resistance

Moisture resistance refers to a material's ability to withstand exposure to moisture without undergoing significant degradation or damage. Moisture can penetrate buildings through sources like rain, humidity, or plumbing leaks, and if not properly addressed, it can lead to structural issues, mold growth, and compromised indoor air quality. Materials with good moisture resistance help maintain the integrity of the building envelope, prevent water infiltration, and minimize the risk of moisture-related problems. To avoid damage and deterioration from moisture penetration, the materials used should be water- and moisture-resistant. In some cases, it is essential for materials to allow moisture vapor to escape from the building to avoid trapped moisture and associated problems. Fire resistance is the ability of a material to withstand or slow down the spread of fire, thereby providing vital time for occupants to evacuate and reducing property damage. Fire-resistant materials play a crucial role in improving the overall fire safety of a building. To reduce the danger of fire-related damage or harm, the materials used should be fire resistant.

1.2.5 Maintenance and Cost Effectiveness

Maintenance refers to the activities required to keep a building in good condition, prevent deterioration, and ensure the proper functioning of its components. Materials that require minimal maintenance offer several benefits, like good durability and longevity. They can withstand environmental stresses, resist wear and tear, and maintain their performance over an extended period. The materials should be simple, and any necessary repairs or maintenance should be quick and affordable. Regular maintenance is often necessary to preserve the appearance of certain materials. Low-maintenance materials can retain their visual appeal with minimal effort, ensuring that the building maintains its desired aesthetic over time. Cost effectiveness refers to achieving the desired functionality and performance of a building while optimizing expenses. The upfront cost of materials is an important consideration. It is crucial to balance the initial investment with the long-term benefits and durability of the chosen materials. Higher-quality and more durable materials may have a higher upfront cost but can provide cost savings in the long run. Evaluating the total cost of ownership over the lifespan of the building is essential. This includes the initial cost and the maintenance, repairs, and potential replacements or upgrades required over time. Materials with long-term performance and minimal upkeep can contribute to cost effectiveness.

1.2.6 Sustainability and Aesthetics

Both during manufacture and after usage in the building, the materials used should be sustainable and have no negative influence on the environment. The materials should be aesthetically pleasing and blend well with the building's overall design.

1.3 ANCIENT BUILDING MATERIALS

Our future is greatly influenced by our history. The employment of outdated technologies by our own progenitors is very symbolic. Originally, there was an ecological balance maintained between humans and the natural world. They held that in order to create a stunning image that protected the natural beauty of the surroundings, construction and nature should coexist. There are presently 3650 nationally significant, acknowledged sites and monuments associated with ancient culture in India. This local study emphasises Bengal's traditional building techniques and underappreciated historical beauty. India is well-known throughout the world for its contributions to and diversity within culture. The construction process and structural stability that ensure a heritage building's survival even in the face of natural disasters, calamities, and negligence stand out as common features.

While the World Heritage Commission, the Archaeological Survey of India, and the State Heritage Commission protect certain buildings, it is distressing to learn that more than 1 lakh buildings, precincts, and sites remain nameless and unprotected.[1] This highlights a specific area and its architectural design, which was renowned for its simplicity and grandeur while employing materials that could be available locally. Some of the commonly used materials are classified in Figure 1.1.

1.3.1 Stone

Natural stone has been utilized in construction for a very long time. It may be utilized for roofing, flooring, and walls. Due to its tenacity, sturdiness, and aesthetic appeal, stone has been a popular natural building material for thousands of years. Granite, limestone, sandstone, marble, slate, and other stone forms may be utilized in building projects. Some modern and ancient stone based sculpture are shown in Figure 1.2 and Figure 1.3.

Stone may be used in buildings in various ways, such as cladding, flooring, roofing, and structural components like columns, beams, and arches. It is a flexible

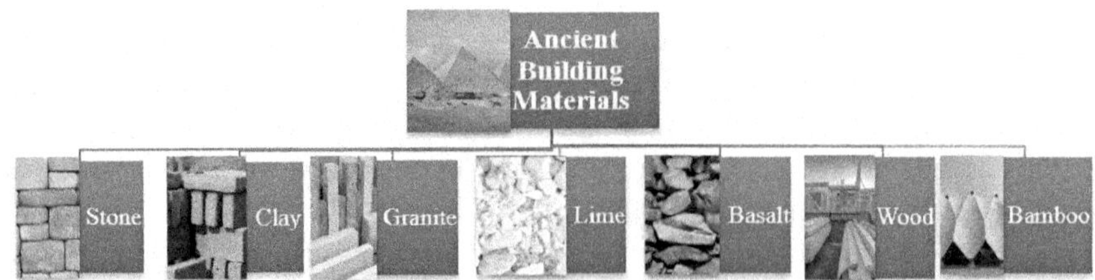

FIGURE 1.1 Classification of traditionally used building materials.

FIGURE 1.2 Stone-based structures in Konya cultural heritage: (a) Kızılören Caravanserai, (b) Kuruçeşme caravanserai.[2]

material that can be polished, cut, and molded in various ways to provide a variety of aesthetic effects. Stone's durability as a construction material is one of its key benefits. It may persist for millennia without degrading and is resistant to weathering, erosion, and fire. Additionally, it has superior thermal mass, which helps save energy use and assists in controlling indoor temperature. The visual attractiveness of stone is another benefit. It can offer a natural, rustic appearance or a more polished and elegant aesthetic that fits well with the surroundings. Additionally, stone comes in a wide range of hues, patterns, and textures, giving architects and designers the freedom to construct distinctive buildings.

However, there are several drawbacks to employing stone for construction. Stone may be expensive to ship and install since it is heavy and brutal. The cost may also increase since it takes specialized tools and training to correctly cut and shape the stone. Additionally, over time and especially if they are exposed to extreme climatic conditions, some varieties of stone may be more prone to weathering or breaking. For a stone to keep its durability and look, frequent care procedures like washing and sealing may be necessary. Stone is still a common and highly appreciated building material because of its numerous benefits and capacity to increase any construction project's value and beauty, despite these shortcomings.

FIGURE 1.3 Ancient stone-based sculpture.[3]

1.3.2 Clay

Clay has been utilized as a natural building material for thousands of years due to its availability, low cost, and adaptability. It is a fine-grained soil comprising silicon, oxygen, and aluminum minerals. Bricks, tiles, plaster, and adobe buildings are just a few construction projects that may employ clay. Because of their durability and fire resistance, clay bricks and tiles are frequently employed in buildings. They also offer superior insulation and are energy efficient. Clay is a cheap and accessible building material that is widely accessible and readily available. Additionally, it is simple to work with and can be molded and sculpted to match various design specifications. Due to clay's superior thermal insulation properties, a building's interior temperature may be controlled, decreasing energy use. Thanks to its intense heat retention qualities, it can keep a structure cool in hot regions and warm in cold ones. Clay's resilience to fire is another benefit. Because of their excellent fire resistance, clay bricks and tiles can help contain a fire if it breaks out. Because of this, clay is the perfect substance for creating fire-resistant constructions like chimneys, ovens, and fireplaces.

Nevertheless, there are several drawbacks to employing clay for construction. It is unsuitable for supporting big loads or constructions subject to seismic activity since it is relatively weak under tension. Clay is prone to erosion and weathering, eventually leading to degradation. Additionally, using clay as a building material necessitates the development of specialized abilities and expertise. The final product's strength and durability can be significantly influenced by the type of clay used and the techniques employed for processing, molding, and burning. Despite these disadvantages, clay is a well-liked and often used construction material, particularly in areas where it is easily accessible and reasonably priced. It may assist in building energy-efficient and fire-resistant structures because it is a sustainable and ecologically beneficial material.

1.3.3 Granite

Granite is a natural building material formed from volcanic activity and has been used in construction for centuries. It is a dense, hard, and durable material with many construction advantages. One of the primary advantages of granite as a building material is its strength and durability. Granite is one of the hardest and most durable natural stones and can withstand heavy loads and harsh environmental conditions. It has excellent resistance to wear, abrasion, and weathering, making it ideal for applications requiring high strength and durability. Granite is a robust and long-lasting construction material that resists weathering and moisture. Its compressive strength has been measured at more than 200 MPa.[4] Granite also has excellent thermal properties, which can help regulate the temperature inside a building and reduce energy consumption. It is also an excellent sound insulator and can reduce noise transmission, making it ideal for applications requiring sound insulation. Another advantage of granite is its aesthetic appeal. It has a unique and distinctive appearance with various colors, textures, and patterns. It can be used for various applications in construction, including cladding, flooring, countertops, and landscaping, and it can provide a natural and elegant look to any project.

However, there are also some disadvantages to using granite as a building material. It is a relatively heavy material, which can add to the cost of transportation and installation. It also requires specialized skills and equipment to cut and shape the stone to fit the desired design, which can further increase the cost. In addition, using granite in construction requires careful consideration of its environmental impact. Granite mining and processing can significantly impact the environment, including air and water pollution, soil erosion, and habitat destruction. Ensuring that granite is sourced and processed using sustainable and environmentally friendly practices is essential. Despite these drawbacks, granite remains a popular and valued building material due to its many advantages and ability to provide strength, durability, and aesthetic appeal to any construction project.

1.3.4 Lime

A common building element, lime is used for rendering, plastering, and as a mortar binding agent. It is long lasting, permeable, and mold and mildew resistant. People have used lime as a traditional building material for thousands of years. It is produced by burning shells or limestone at high temperatures to produce a powder that, when combined with water, may be used to make a paste. Lime provides many benefits, including longevity, sustainability, and adaptability.

Lime's durability as a construction material is one of its main benefits. It is perfect for mortars, plasters, and renders because of its outstanding adhesive characteristics. Additionally, it can absorb and release moisture, enabling it to adjust to variations in temperature and humidity and lowering the likelihood of cracks and other types of damage. Due to its natural composition and minimal carbon impact, lime is also a sustainable building material. It is an environmentally favorable option for building because it is also recyclable and biodegradable. Another advantage of lime is its flexibility. It may be used for brickwork, plastering, flooring, and roofing, among other construction-related applications. It may also be blended with other elements, like sand and fibers, to create more robust construction materials. Lime also has strong insulation qualities, which can aid in lowering energy bills and expenditures associated with heating and cooling systems. It can protect structures from fire to a certain extent and is also fire resistant.

The use of lime as a construction material has significant drawbacks, however. It may be pricey, and the proper installation calls for professional labor. To preserve its durability, it also needs frequent upkeep and repairs. Due to its many benefits and capacity to provide longevity, sustainability, and adaptability to any construction project, lime continues to be a well-liked and highly valued building material despite these shortcomings.

1.3.5 Basalt

A naturally occurring construction material called basalt is created when volcanic lava flows. It is a heavy, sturdy material that has been employed in building for a long time. Basalt may be found in many different forms, including basalt fiber, basalt rebar, and basalt rock, each of which has specific qualities and uses in building. Strength and durability are two of basalt's main benefits as a construction material. One of the hardest natural stones, basalt offers exceptional resistance to deterioration from wear, abrasion, and the elements. It is the perfect choice for applications that need exceptional strength and durability since it can handle large loads and extreme climatic conditions. Additionally, basalt offers exceptional insulating qualities for both heat and sound. It can offer insulation against sound to lessen noise transmission, assist in controlling the temperature within a structure, and use less energy. The visual attractiveness of basalt is another benefit. Its fine-grained texture and dark grey-to-black color give it a distinct and unusual look. It has several uses in construction, including cladding, flooring, and landscaping, and can provide any project with a natural and modern appearance. The tensile strength of basalt fibers in their fibrous state ranges from 1700 MPa to 4800 MPa, and they also have a high melting point of up to 1450°C.[5] Basalt fibers are utilized as insulating materials in the aerospace, automotive, and electrical sectors because they are nontoxic, fireproof, exhibit resistance to nuclear and ultraviolet (UV) radiations, and sustain a relatively large range of temperatures ranging from –260°C to 750°C. Basalt fibers are added to cement mortar to strengthen the building structure's flexural and compressive properties.

However, employing basalt as a construction material has significant drawbacks as well. Because of its weight, installation and shipping expenses may increase. The cost may also increase since it takes specialized tools and training to correctly cut and shape the stone. The environmental impact of basalt use in buildings must also be carefully considered. Environment-harming effects of basalt mining and processing include soil erosion, air and water pollution, and habitat degradation. It is crucial to confirm that basalt is obtained and processed in a sustainable and eco-friendly manner. Despite these disadvantages, basalt is a well-liked and highly valued building material because of its many benefits and capacity to provide any construction project strength, durability, and aesthetic appeal.

1.3.6 Wood

Wood is a multipurpose material employed in building for ages. It is resilient, portable, and renewable. It is suitable for cladding, flooring, and framing. Since ancient times, wood has been utilized in construction as a popular and adaptable building material. It offers many benefits, like cost, simplicity of building, sustainability, and natural beauty. The inherent beauty of wood is one of its main benefits as a construction material. It is kind and welcoming and may give any project a natural and organic vibe. Wood is available in various hues, grains, and textures, which may give a building's design more depth and personality. The sustainability of wood is another benefit. It is an eco-friendly material for building since it is a renewable resource that can be cultivated and collected sustainably. Since trees absorb carbon dioxide as they grow, using wood can also help a building's carbon footprint reduction. Wood is also relatively straightforward; it can be molded, cut, and assembled using essential tools and methods. Due to its small weight, it can facilitate and reduce the cost of building and shipping. Additionally, wood may be manufactured off-site to speed up construction and save expenses. Additionally, wood has strong insulation qualities, which may aid in lowering energy expenses and heating and cooling requirements. It is excellent for use in applications that require sound insulation since it is also a superb sound insulator and can aid in limiting noise transmission.

Nevertheless, there are several drawbacks to employing wood for construction. Insect, rot, and fire damage may impact wood's resilience and longevity. To safeguard against these threats, it needs routine upkeep and care. Additionally, wood is unsuitable for some purposes, such as high-rise structures or structures in seismically active regions. Other materials, such as concrete or steel, could be a better option. Despite these disadvantages, wood continues to be a well-liked and highly regarded building material because of its many benefits and capacity to provide natural beauty, sustainability, and affordability to any construction project.

1.3.7 Bamboo

Since ancient times, bamboo has been utilized as a building material because of its quick growth and sustainability. Its benefits include sustainability, toughness, durability, and adaptability. The sustainability of bamboo as a construction material is one of its main benefits. It is a plant that overgrows and may be harvested in as little as three years, making it a significant renewable resource. In addition to being biodegradable and having a minimal carbon impact, it makes for an environmentally beneficial building material. The resilience and strength of bamboo are further benefits. It is a superior building material for buildings that need strength and stability since it has a higher tensile strength than steel and more excellent compression resistance than concrete. Insects, dampness, and rot cannot harm it, extending its lifespan and lowering maintenance expenses. Another adaptable material that may be utilized in a variety of construction projects is bamboo. It may be used for flooring, cladding, roofing, and structural components like beams and columns. Its minimal weight and ease of use can speed up construction and cut expenses. Additionally, bamboo has effective thermal and acoustic insulation qualities that can aid in lowering energy use and noise transmission. Additionally, it functions as a natural air cleaner and can enhance indoor air quality.

Nevertheless, there are several drawbacks to employing bamboo for construction. It must be treated with fire-retardant chemicals to increase its fire resistance since it is flammable. Additionally, it is inappropriate for some applications, such as high-rise structures or structures in seismically active regions. Due to its many benefits and capacity to provide sustainability, durability, strength, and adaptability to any construction project, bamboo continues to be a well-liked and highly valued building material despite these limitations.

1.3.8 Thatch

A classic building material used for millennia in construction is thatch. In order to create a roofing material, dried grass, straw, or reeds are piled and weaved together. Thatch offers several benefits, including sustainability, insulating capabilities, and a rustic appearance. Thatch's sustainability as a construction material is one of its main benefits. It is a sustainable building alternative because it is constructed from abundant and renewable natural elements. In addition to being biodegradable and having a minimal carbon impact, it is an environmentally beneficial substitute for synthetic roofing materials. Additionally, thatch offers superior insulating qualities that may assist in controlling temperature and saving energy use. Buildings can be kept warm in the winter and cool in the summer, which helps save energy expenditures for heating and cooling. Thatch has a natural look, which is an additional benefit. It may give a structure more personality and charm, with a distinct and rustic aspect. Thatched roofs are a flexible option for architects and builders since they can be modified to meet the style and architecture of the structure.

However, using thatch in buildings has several disadvantages. It requires routine maintenance and repairs to maintain its longevity because it is susceptible to weather and insect damage. It could also need to be replaced because it is less durable than other roofing materials. Additionally, not all climates or building types may be suited for thatch. Since it cannot endure extremely severe weather, it is more frequently utilized in regions with mild or moderate temperatures. Thatch is still a common and highly appreciated building material because of its many benefits and capacity to give any construction project sustainability, insulation, and natural beauty.

1.3.9 Adobe

Adobe construction material comprises mud, straw, and occasionally other unprocessed organic resources like clay or sand. After being shaped into bricks or blocks, the mixture is then left to cure in the sun. Since ancient times, adobe has been used for construction, especially in regions with arid climates like the Southwest of the United States, Central and South America, the Middle East, and Africa. The fact that adobe is a relatively energy-efficient construction material is one of its benefits. Buildings made of adobe have strong, well-insulated walls, making the inside comfortable all year round. In addition to being widely accessible and reasonably priced, adobe is a preferred option in many regions of the world. In many regions of the world, traditional construction materials are still often employed, particularly in rural areas where contemporary materials are neither easily accessible nor reasonably priced. They have withstood the test of time and are prized for their sturdiness, sustainability, and few adverse environmental effects.

1.4 CONVENTIONAL BUILDING MATERIALS

Conventional building materials refer to the widely used and traditional materials used in construction for many years. These materials have a long history of use and are commonly found in various types of buildings and structures. Some of the conventional building materials and their properties are shown in the Table 1.1

TABLE 1.1
Different Properties of Conventionally Used Construction Materials[6–8]

Building material	Density (kg/m^3)	Thermal conductivity (W/(m×K))	Compressive strength (MPa)	Tensile strength (MPa)
Concrete	2000	1.3–2.25	15–30	2–5
Steel	7500–8000	45	350–1000	450–750
Brick	1600–1750	0.5–1.0	30–35	2.8
Glass	2580	0.5–1.38	1000	7
Wood	600–900	0.1–0.2	30–60	70–140

1.4.1 Concrete-Based Materials

A combination of cement, water, and aggregates like sand, gravel, or crushed stone makes the construction material concrete. In order to create a strong and long-lasting structure, the mixture is poured into a mold or formwork and allowed to solidify and cure. Concrete is utilized in many construction projects, including bridges, buildings, and pavements. Concrete's strength and durability as a construction material are two benefits. Concrete is a common material for buildings that must survive challenging circumstances since it is fire, water, and weather resistant. Compared to other building materials like steel or wood, it is also reasonably affordable. Concrete's adaptability is another benefit. It may be molded into various sizes and forms to satisfy various design needs. Steel bars or fibers can also be used as reinforcement to boost the strength and load-bearing capacity of concrete. However, utilizing concrete has certain drawbacks as well. The influence on the environment is among the main issues. Cement manufacture, an essential ingredient in concrete, generates a sizable portion of the carbon dioxide emissions linked to climate change. Additionally, the exploitation of aggregates may harm ecosystems and natural habitats. Concrete's overall cost and environmental effect might be increased by the substantial energy and resource requirements for its production and transportation. Additionally, over time and especially if they are subject to freeze–thaw cycles or other environmental stressors, concrete constructions can be vulnerable to cracking and other deterioration. Concrete may be a valuable and adaptable building material. However, to ensure its long-term sustainability and durability, it must be carefully considered concerning its environmental impact and maintenance requirements.

1.4.2 Steel-Reinforced Concrete

Concrete and reinforcing steel are combined to create reinforced concrete (RC), a composite material. Concrete's tensile strength, which is usually substantially lower than its compressive strength, is increased by adding steel. The finished product is robust, long-lasting, and damage resistant, making it perfect for various construction applications. Steel bars, wire mesh, and fibers are just a few examples of the various reinforcing options for RC. Before the concrete is poured, steel bars are usually inserted and set inside the concrete formwork in a grid pattern. The wire is used to join the bars, creating a cage-like structure buried in the concrete. There are various advantages to steel reinforcement in RC. First, it strengthens the concrete, enabling it to support more weight and endure more stress, which makes it perfect for use in construction projects like bridges, skyscrapers, and dams. Second, the steel reinforcement aids in preventing concrete from cracking.[9] The steel bars assist in more uniformly dispersing the stress when the concrete is subjected to tensile stresses, which lowers the likelihood of cracking and stops any existing fractures from spreading. An image of reinforced concrete is shown in Figure 1.4.

Finally, the steel reinforcement contributes to the concrete's increased durability. The reinforcement can aid in extending the structure's service life and lower the need for expensive repairs and replacements by minimizing cracking and increasing the strength of the concrete. Reinforced concrete is a very effective and frequently

FIGURE 1.4 Illustrative image of reinforcement concrete.

utilized building material with several advantages over conventional concrete. Its versatility and affordability make it a popular choice for builders and architects worldwide, and its strength, durability, and damage resistance make it perfect for a broad range of applications.

1.4.3 Synthetic Fiber-Reinforced Concrete

Concrete reinforced with synthetic fibers such as polyester, nylon, or polypropylene, as shown in Figure 1.5, to increase strength and durability is known as synthetic fiber-reinforced concrete (SFRC). The fibers are added to the concrete mixture during the mixing phase and evenly dispersed throughout the concrete matrix. Concrete's performance may be significantly enhanced by the use of synthetic fibers in a variety of methods. First, a prevalent issue with concrete constructions, cracking may be managed with fibers. The fibers disperse the stresses inside the concrete more equally, lowering the possibility of fractures developing and spreading. This may increase the concrete structure's overall durability and useful life. These fibers are included in the concrete matrix like that of other concrete components. Although the volume proportion of fibers in conventional FRC is less than 2%, it can reach 15% in select high-performance concrete buildings.[10] Second, SFRC can increase the concrete's hardness and impact resistance. When the concrete is exposed to impact or dynamic stresses, the fibers reinforce, absorbing and dispersing energy. The use of SFRC in high-stress applications, including bridge decks, airport runways, and industrial floors, is excellent as a result. Third, SFRC can make concrete more fire resistant. When subjected to high temperatures, the synthetic fibers melt and disintegrate, creating a network of channels that let gases escape. It aids in avoiding the pressure buildup that may otherwise result in concrete exploding or spalling during a fire. SFRC is a flexible and affordable solution to enhance concrete buildings' toughness, performance, and longevity. It is beneficial when consideration of cracking, impact, and fire resistance is crucial.

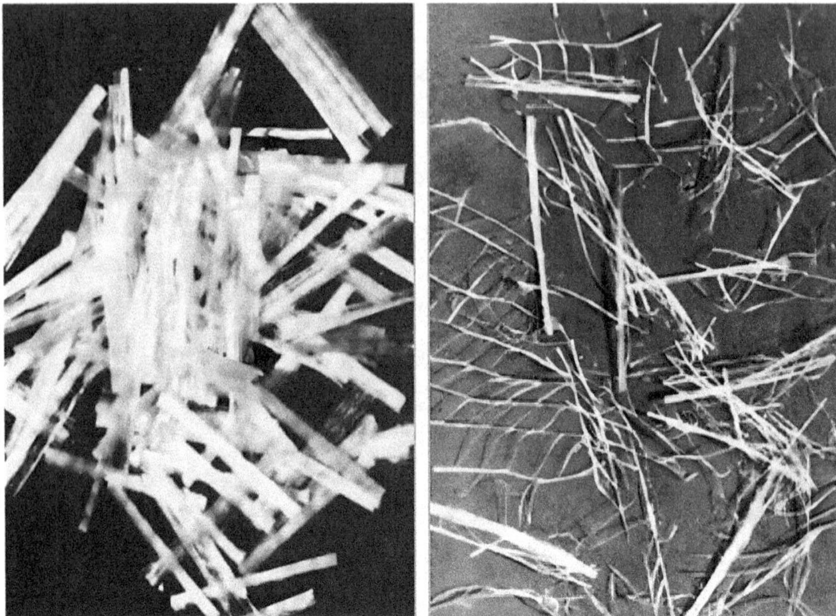

FIGURE 1.5 Synthetic fiber for reinforcement in concrete.[11]

1.4.4 Glass Fiber-Reinforced Concrete

Glass fibers are added to concrete to increase its strength and durability, creating glass fiber-reinforced concrete (GFRC). Glass strands are woven into a fabric-like substance to create the glass fibers. The fibers are then bound together and attached to the concrete by impregnating the fabric with a polymer or cementitious matrix. GFRC is superior to conventional concrete in several ways. It is easier to work with and install since it is lighter and more flexible and can lessen the need for expensive machinery and the time and expense of building. Second, GFRC is ideal for challenging situations due to its exceptional resistance to corrosion and wear. The concrete is strengthened and stabilized more by the glass fibers, reducing the likelihood of cracking and other damage. Third, GFRC offers greater versatility in architectural and design applications since it can be molded into several forms and patterns. In recent years, the usage of GFRC in architectural panels, facades, and ornamental components has grown in popularity because it enables the creation of intricate shapes and patterns that are either impractical or impossible to produce with conventional concrete. Last but not least, GFRC is a green and sustainable building material. Because the glass fibers used in GFRC are created from recycled resources, the manufacture of concrete has a lower environmental effect. Glass fibers improve a concrete structure's surface integrity and homogeneity by reducing bleeding and the risk of fracture development in the concrete mixture. Glass fiber-reinforced concrete's compressive, flexural, and split tensile strengths were all shown to increase by 20% after 28 days of curing.[12] When the fiber percentage was increased from 0.0%

to 1.2% by weight, the compressive strength of alkali-resistant glass fiber-reinforced concrete rose by 15.12%.[13,14] Additionally, splitting tensile and flexural strength improved by 60.78% and 50.20%, respectively, with the same percentage increase in fiber content.

Additionally, GFRC has a more negligible carbon impact than conventional concrete since it uses less energy to make and transport it. GFRC is a flexible and cutting-edge construction material with several benefits over conventional concrete. Its resilience, adaptability, and strength make it perfect for various architectural and design applications, and its environmental responsibility makes it a wise option for the industry's future.

1.4.5 Carbon Fiber-Reinforced Concrete

Concrete that the addition of carbon fiber has strengthened is called carbon fiber-reinforced concrete (CFRC). Carbon atoms are bound together in a crystalline structure to form carbon fibers, which are very strong and light in weight. Concrete's performance may be enhanced in several ways by including carbon fibers. The carbon fibers improve the concrete's tensile strength, improving its resistance to bending and other forms of stress. It is crucial because concrete is prone to breaking when subjected to bending forces and is generally stronger in compression than tension. Second, CFRC has excellent corrosion and fatigue resistance, making it perfect for use in severe locations or in applications where the structure would be exposed to frequent loading and unloading cycles. Using carbon fibers further strengthens the concrete, reducing the risk of cracking and other damage. Third, CFRC is lightweight and incredibly strong, perfect for large-scale construction projects like bridges and high-rise skyscrapers.[15] Because carbon fibers are lightweight, less concrete is needed to produce the same degree of strength and durability, which lowers the cost of materials and transportation. The ideal range of carbon fibers in concrete mixtures is thought to be between 0.2% and 0.3% by volume, and evenly dispersed 20 mm long carbon fibers at 0.35% by volume in CFRC beams have shown the maximum energy absorption, which is around 2.3 times higher than that of non-reinforced concrete beams. Utilizing reclaimed carbon fibers from carbon fiber-reinforced plastic bicycle frames and removing the resin using microwave-assisted pyrolysis (MAP) technology has improved the impact resistance, compressive strength, and flexural strength of CFRC samples.[16] Compared to concrete samples reinforced with regular carbon fibers, the maximum compressive strength of CFRC increased by 48.8% when recycled carbon fibers were included at 10% by weight.[17]

Finally, CFRC offers greater versatility in architectural and design applications since it can be molded into several forms and patterns. Because it enables complicated shapes and patterns that would be challenging or impossible to accomplish with conventional concrete, the use of CFRC in architectural panels, facades, and ornamental components has grown in popularity in recent years. CFRC is a flexible, cutting-edge building material with several advantages over conventional concrete. Its flexibility and design diversity make it an appealing option for architects and designers, and its strength, durability, and lightweight nature make it excellent for a wide range of building applications.

1.4.6 Steel

Steel is the primary material used to produce steel-based constructions. Due to its power, toughness, and adaptability, this kind of structure has grown in popularity recently. Large open spaces may be built using steel, a very robust and long-lasting material, without load-bearing walls or columns. The strength and endurance of steel-based structures are two of their key advantages. Steel is perfect for usage in various locations since it is resistant to many different harms, including fire, dampness, and pests. Steel can sustain huge weights and endure intense pressures because of its excellent resistance to bending and deformation. Depending on the specific requirements of the uses, carbon and a few other elements are added to iron to create steel, an iron alloy. Different kinds of steel alloys may be produced by altering the carbon content of the steel. High-carbon steel is tough, strong, and brittle, containing 0.5% to 2.0% carbon by weight. Pre-stressed concrete wires, high-strength wire ropes, springs, and suspension cables for bridges are all frequently made from this type of steel alloy. Conversely, low-carbon steel may be easily formed into thin sheets or strips because of its softness, malleability, ductility, and high toughness.[18]

The adaptability of steel-based structures is another benefit. Greater versatility in architectural and design applications is made possible by the ability to produce steel in various forms and sizes. This adaptability also enables the construction of extensive, open areas without the requirement for load-bearing walls or columns, increasing useable floor space and additional design options. Buildings made of steel are likewise very environmentally friendly and sustainable. Steel is a highly recyclable material, and recycled steel is used to construct many steel-based buildings. Steel is also very energy efficient, which may aid in lowering energy expenses and expenditures associated with heating and cooling systems. Lastly, steel-based structures are frequently more affordable to build than conventional ones, as shown in Figure 1.6.

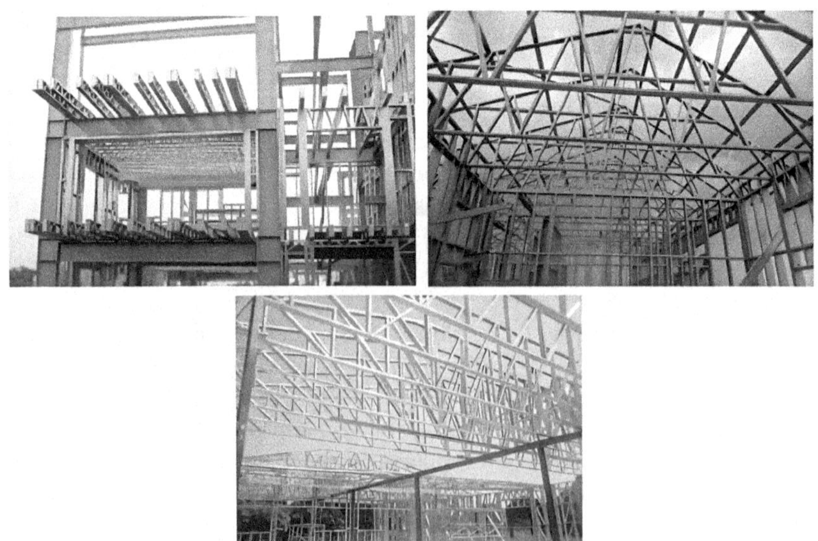

FIGURE 1.6 Basic construction of steel structure and truss detail.[19]

Construction is often completed quicker, and the lack of load-bearing walls or columns may lead to lower construction costs. Steel is also quite strong, which means that maintenance and repairs will be unnecessary throughout a building's lifespan. Steel-based buildings have several advantages over conventional building techniques. Steel is a desirable alternative for various construction applications, from residential residences to commercial structures and industrial facilities, because of its strength, durability, adaptability, sustainability, and affordability.

1.4.7 Brick

Brick is a classic building material that has existed for a long time. It is constructed of clay or another natural material molded into a rectangular shape and heated to a high temperature to transform into a potent, long-lasting substance. For its strength, tenacity, and aesthetic appeal, brick is highly prized and employed in various construction applications. Brick's strength and longevity are two of its key features. Brick is excellent for usage in various locations since it is very resistant to harm from fire, dampness, and pests. Brick can sustain huge weights and endure intense pressures since it is also very resistant to bending and distortion. The thermal mass characteristics of brick are another benefit. Brick can absorb and store heat from the environment because of its large thermal mass, which helps control indoor temperatures and lower energy use and can cut energy expenses for heating and cooling while enhancing interior comfort. Additionally, brick is adaptable and may be utilized in various architectural and design contexts. It may be formed into many sizes and forms, enabling more design versatility. Additionally, brick is available in various hues and textures, making it simple to complement other construction materials or produce striking and appealing patterns. Moreover, brick is a very environmentally friendly and sustainable building material. It is manufactured from natural resources, and many brick producers employ environmentally friendly processes and supplies. Brick is also highly resilient and, with proper care, may survive for decades or centuries, decreasing the need for replacement and minimizing waste. Brick is a flexible, strong, and environmentally friendly building material with several advantages over other materials. It is an appealing solution for various construction applications, from residential residences to commercial structures and industrial facilities thanks to its strength, thermal mass qualities, adaptability, and eco-friendliness. Nowadays in the construction of structures, concrete bricks are also occasionally used. Various attempts have recently been made to create bricks from waste materials such as fly ash, as shown in Figure 1.7, in addition to sawdust, slag, sewage sludge, plastic trash, and cigarette butts.[20]

FIGURE 1.7 Sample of fly ash-based brick.[21]

1.4.8 Glass

One of the earliest materials manufactured by humans' dates back to 500 BC. The same raw elements needed to make ceramics are also needed to make glass, including soda, lime, and silica. Even so, these raw materials are mixed in various compositions and treated differently for glass manufacturing than for ceramics. At higher temperatures, the fusion of siliceous material with alkalis results in the creation of glass. The primary sources of silica for making perfect glass are sand, quartz, and flintstone, albeit melting these minerals requires much thermal energy. While glass made entirely of quartz can be produced at temperatures as high as 2300°C, pure quartz must first melt at a temperature of 1710° C. Soda, natron, potash, or occasionally shards of recycled glass are added as alkalis to lower the melting temperatures.[22,23] The temperature required to make glass drops to 1500°C when soda is added. Lime enters the picture to balance the solubility and improve the durability of glass because adding alkalis to the process makes the mixture soluble in water. Glass acquires its color from certain impurities in essential ingredients. Therefore, the opacity, transparency, and color of glass can be altered by adjusting the amount of a coloring agent in the glass composition. Depending on the various geographic regions from which raw materials for glass are purchased, the impurities change. Due to its extensive use in various building styles and its long history in construction, glass is regarded as a standard building material. It is a versatile material with unique qualities and advantages. Some of the essential characteristics of glass as a typical building material are that it is renowned for its transparency, which permits natural light to enter structures and produces a setting that appears open and expansive. It promotes well-being and fosters a sense of connection to the outside world. Glass allows for the transmission of natural light into interior spaces, which reduces the demand for artificial lighting during the day.

Additionally, it enables passive solar heat gain, which aids in regulating indoor temperatures and lowering energy use. Glass is valued for its aesthetic attributes, which give buildings a contemporary and chic appearance. It can be employed in various architectural designs, from sleek and modern structures to traditional and historical ones.

Glass is used in various structural applications, including glass facades, curtain walls, and floors. While allowing for wide-ranging views and daylight, it can offer structural stability. Insulated glass units (IGUs) filled with insulating gas and comprising several glass panes provide thermal insulation. IGUs improve thermal insulation by lowering heat transfer and raising a building's energy efficiency. Glass can aid in sound insulation, lowering the amount of noise transmitted between interior and outside spaces. Laminated glass, which has numerous layers and an interlayer, offers improved sound insulation. Tempered or laminated glass is frequently used to increase security and safety. It contains an interlayer that holds the glass together even when cracked, preventing it from falling apart. In contrast, tempered glass is made to break into little, less dangerous fragments when it is broken. Glass is a potent substance that resists UV rays, corrosion, and the elements.[24] Its long-term performance can be ensured with proper upkeep and care. Compared to other building materials, glass surfaces are comparatively simple to keep clean and maintain. Their transparency and appearance can be maintained with regular cleaning. Glass is a recyclable substance, making it a sustainable option. It may be reused and recycled to make new glass items, which lowers the need for raw resources. Glass has been a common building material for ages, and technological developments have increased its uses and enhanced its performance. Due to the utilitarian and aesthetically pleasing aspects it brings to structures, it is still frequently utilized in the construction sector.

1.4.9 Wood

One of the most common and commonly used building materials, wood, has a long history of use in architecture. It has been used in various building types and architectural styles and offers many benefits. An essential characteristic of wood as a typical building material is that it is a resource that comes from trees that is regenerative. Unlike to non-renewable materials, responsibly harvested wood and timber products can be obtained from sustainably managed forests. Wood has advantageous structural characteristics such as strength, durability, and load-bearing capability. It can be applied to building frameworks, beams, columns, and other load-bearing components. Wood naturally insulates, which aids in temperature control and lessens heat transfer. Compared to materials like steel or concrete, it offers higher thermal performance. Wood has a warm, natural aspect, giving structures character and aesthetic appeal. It may produce a warm and welcoming ambiance in both interior and exterior uses. Wood is also a flexible material that is easily cut, molded, and formed, creating unique and creative patterns. It provides versatility in terms of architectural styles and can be employed in traditional, modern, or rustic designs. Wood is relatively lightweight compared to many other building materials, making it simpler to handle and transport. It may result in shorter construction schedules and lower labor expenses. Wood has outstanding natural sound absorption and dampening capabilities. It is appropriate for flooring, wall panels, and ceilings since it can help create a peaceful and comfortable indoor environment. Wood that has been appropriately cared for and treated may be solid and long lasting. Timber preservation and protective coatings can improve its resistance to decay, pests, and weathering. Wood has the remarkable capacity to retain carbon dioxide (CO_2) that is taken in from the atmosphere as trees develop. By lowering greenhouse gas emissions, this helps to combat climate change. Wood is a material that is both recyclable and degradable. At the end of its life cycle, it can be recycled, repurposed, or used again, reducing waste and environmental harm. Wood continues to be the material of choice for various construction uses, including residential, commercial, and industrial projects. Wood can offer environmentally friendly, visually beautiful, and structurally sound solutions in the built environment with the right design, building, and maintenance techniques.

1.5 ADVANCED BUILDING MATERIALS

Advanced building materials are cutting-edge, high-performance substances utilized in construction that provide better qualities than conventional building substances. These materials are often created to increase a building's overall performance, durability, sustainability, and energy efficiency. A few instances of cutting-edge building materials are classified in Figure 1.8.

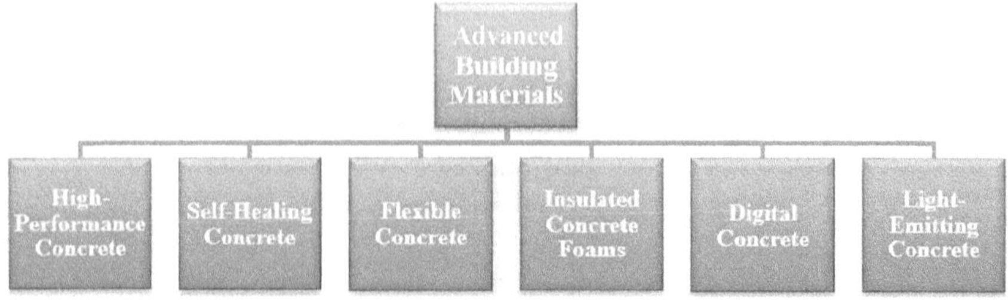

FIGURE 1.8 Classification of advanced building materials.

1.5.1 High-Performance Concrete

High-performance concrete (HPC) is a modern building material with better qualities than regular concrete. It is designed to offer improved strength, durability, and resistance to different environmental variables. Compared to regular concrete, high-performance concrete very often includes silica fume. High-performance concrete usually incorporates fly ash, ground granulated blast furnace slag (slag), or both components. In high-performance concrete, the aggregate's maximum size is typically 10 to 14 mm, meaning a smaller maximum size than ordinary concrete. This is because of two factors.[25] First, a smaller maximum size reduces the differential stresses that could cause microcracking at the aggregate–cement paste interface. Second, smaller aggregate particles are more robust than bigger ones because the most significant faults in the rock are removed during comminution, which controls strength. Another thing to note about the ingredients is that a superplasticizer is required because of the mixture's low water-to-cement ratio and silica fume. During the past decades, high-performance concrete has gradually grown, mainly due to creating concrete with increasing strengths: 80, 90, 100, 120 MPa, and occasionally even higher. Nowadays, 140 MPa can be routinely produced in various places worldwide.[26] But high-strength concrete and high-performance concrete are different. Some of the salient features and advantages of employing high-performance concrete are.

1. Strength and durability: HPC is appropriate for projects that demand a higher load-bearing capacity because of its much higher compressive strength than conventional concrete. Its capacity to bear enormous loads and withstand breaking increases the structure's overall toughness and lifetime.
2. Reduced permeability: Compared to traditional concrete, HPC has a lower permeability, making it less susceptible to water infiltration. It lessens the possibility of embedded steel reinforcement corroding and increases the structure's endurance, especially in settings exposed to dampness, chemicals, or severe weather.
3. Enhanced resistance to chemical attack: HPC is designed to have a high resistance to attacks from chemicals, including acids, sulfates, and chlorides. As a result, it is appropriate for buildings situated in urban or coastal locations where the risk of exposure to corrosive substances exists.

4. Increased freeze–thaw resistance: HPC has excellent freeze–thaw resistance. It lessens the chance that freezing conditions may cause damage due to the expansion of water inside the concrete. This characteristic is essential in cold locations with frequent freeze–thaw cycles.
5. Improved workability: HPC can maintain good workability despite having more strength, simplifying placement and consolidation during construction. With proper handling, it may be pumped and put in intricate formations to obtain a flawless surface finish.
6. Design flexibility: Compared to conventional concrete, HPC offers greater design flexibility. It enables architects and engineers to develop cutting-edge and visually beautiful structures by making thin and lighter structural elements.
7. Sustainability: HPC can help with environmentally friendly building techniques. It frequently includes additional cementitious materials like fly ash or slag, which minimize the carbon footprint and the amount of cement needed. Additionally, due to its improved resilience, structures may survive longer and require fewer repairs or reconstructions.
8. Rapid construction: HPC can promote speedier construction operations due to its increased early-age strength development. It enables building timelines to be hastened, saving time and money.

High-rise buildings, bridges, dams, tunnels, maritime constructions, and infrastructure projects where toughness and durability are essential frequently utilize high-performance concrete. Due to its excellent qualities, it is a fundamental component of strong and durable constructions.

1.5.2 Ultra-High-Performance Concrete

An advanced building material with extraordinary strength, durability, and performance qualities is ultra-high-performance concrete (UHPC). This particular kind of concrete is carefully mixed with fine particles, fibers, and chemical admixtures to produce concrete with exceptional mechanical and structural qualities. UHPC has a substantially higher compressive strength than regular concrete, generally reaching 150 MPa (megapascals). With less material used, slim, lightweight constructions may be built thanks to this outstanding strength. UHPC also exhibits exceptional tensile strength, frequently attained by mixing steel or synthetic fibers. This increased tensile strength aids in reducing cracking and enhancing the material's overall toughness and resilience. Tiny powders give UHPC a high particle-packing density, creating a dense and impenetrable microstructure. The material's resistance to abrasion, chemical assault, and freeze–thaw cycles is influenced by its density.[27] UHPC is highly resistant to environmental variables like chloride intrusion, carbonation, and chemical exposure because of its dense microstructure and reduced porosity. The service life of structures is increased due to their durability, which also lowers maintenance needs. Because of its dense matrix, UHPC has a low permeability to liquids and gases, making it ideal for applications requiring water tightness such as water tanks, tunnels, and bridge decks. UHPC has high strength as well as excellent ductility. Structures can handle dynamic loads and seismic occurrences better because they can experience considerable deformations before failing. The high strength and structural performance of UHPC allow for the construction of smaller and lighter parts, increasing design freedom and opening up more architectural options.

Due to UHPC's great strength and durability, these vital components, such as bridge decking, girders, and connectors, can last longer and handle more weight. Architectural facades can be made with UHPC panels and cladding for their light weight, sturdiness, and aesthetic appeal. To increase the load-bearing capacity and prolong the service life of deteriorated concrete structures, including columns, beams, and slabs, UHPC is used in their repair and strengthening.[28] UHPC makes it possible to create precast concrete components with complex designs, robust connections, and increased durability. Even though UHPC has several benefits, there are a few things to consider. Because UHPC requires specialized materials and strict manufacturing specifications, it is often more expensive than regular concrete. UHPC calls for precise control over the mix design and production process to guarantee consistent performance. Equipment and specialized skills may be needed while working with UHPC to ensure optimal mixing, placement, and curing to maximize performance.

1.5.3 Self-Healing Concrete

Self-healing concrete is a type of concrete that can repair cracks and damage on its own, reducing the need for regular maintenance and repairs. Self-healing concrete is an innovative technology that can be used in green buildings to reduce the need for maintenance and repairs and to extend the life of the building. A model of self-healing concrete is shown in Figure 1.9. Self-healing concrete was developed as a novel method of self-repairing concrete cracks.[29] Yet several creative methods for self-healing cementitious materials have been devised and suggested. This is referred to as a life cycle assessment (LCA). The LCA's goal is quantifying the environmental impact reduction that could be achieved by using the proposed self-healing concrete instead of more traditional concrete. Some direct benefits of concrete self-healing include a reduction in the rate of deterioration, the extension of service life, and a reduction in repair frequency and cost over the life cycle of the concrete infrastructure.[30] Certain unique components (such as fibers or capsules), which include some adhesive substances, are dispensed into the concrete mix to generate self-healing concrete. When fibers or capsules break, the liquid that is within immediately seals the fracture and repairs it. Since concrete has a relatively low tensile strength, cracks are a regular occurrence. These gaps make concrete less durable since they make it simple for gases and liquids that could contain toxic compounds to travel. Concrete itself may be harmed, as well as the steel reinforcement bars, if microcracks spread and eventually reach the reinforcement.[31] Thus, it's crucial to keep crack width under control and to get them repaired as soon as feasible.

Concrete buildings' service lives could be extended by self-healing fissures, which would also increase the material's durability and sustainability. Another approach involves incorporating microorganisms into the concrete mixture. These microorganisms can produce calcium carbonate when they come into contact with water, which can fill in cracks and prevent further damage. Self-healing can be classified in to two categories.

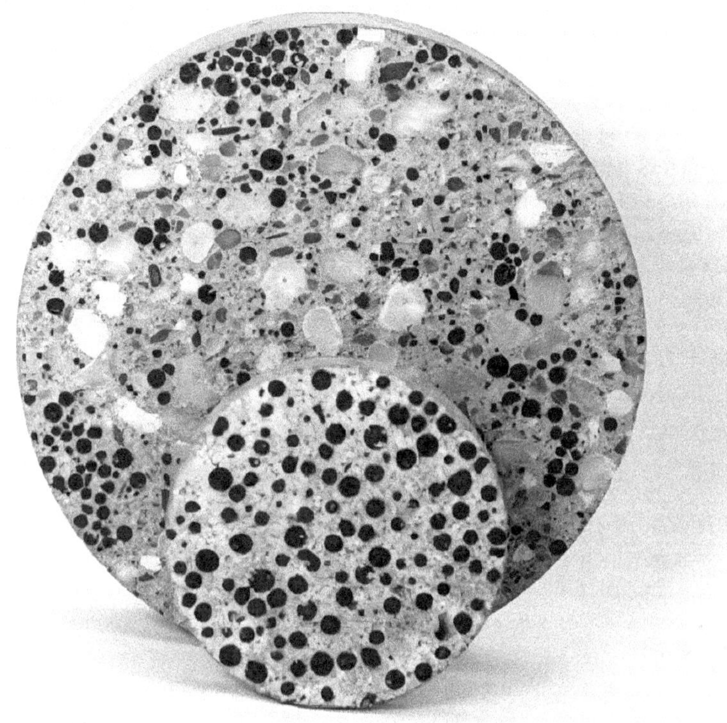

FIGURE 1.9 Model of self-healing concrete.[32]

1.5.3.1 Autogenous Self-Healing Concrete

This is an example of self-healing concrete that demonstrates concrete's innate ability to mend or heal cracks in the presence of water. The autogenous crack healing in concrete is accomplished by further hydrating un-hydrated cement grains, precipitating calcium carbonate crystals on the surface of cracks in the presence of water and CO_2, chipping concrete particles inside crack surfaces, and consolidating debris or impurities present in the ingress water to seal the crack. The autogenous healing process is recognized at a very early stage of creating concrete buildings. It restores the strength of the concrete while also repairing the fissures. If the pH of fresh water is close to 7 and there is a lot of carbon dioxide in the air, it is possible to seal 200 m wide cracks. Fibers, shrinkable polymers like polyethylene terephthalate (PET), and crystalline admixtures like SCM (superplasticizers and crystalline admixtures) an all be used to improve.

1.5.3.2 Autonomous Self-Healing Concrete

This concrete exhibits self-healing properties, utilizing the inherent capability of concrete to repair cracks when exposed to water. The autogenous damage healing process involves several mechanisms: further hydration of un-hydrated cement grains, precipitation of calcium carbonate crystals on crack surfaces in the presence of water and CO_2, fragmentation of concrete particles within cracks, and consolidation of debris or impurities in the water to seal the cracks. The autogenous healing mechanism has been recognized since the early days of concrete construction, effectively filling cracks and preserving concrete durability, particularly in freshwater environments with a pH close to 7. In environments rich in carbon dioxide, cracks up to 200 micrometers wide can also be sealed. Enhancements to autogenous healing include the addition of fibers, shrinkable polymers like polyethylene terephthalate (PET), or crystalline admixtures such as SCM. However, research suggests that incorporating GGBS or fly ash reduces calcium ion production, thereby diminishing the concrete's self-healing capacity.

1.5.3.3 Advantage of Self-Healing Concrete

1. Improve durability: Self-healing concrete contains microcapsules of healing agents that can be released to repair cracks and other damage in the concrete. This can extend the life of the building and reduce the need for repairs and maintenance.
2. Reduced environmental impact: Self-healing concrete can reduce the need for new concrete, which can significantly reduce the environmental impact of construction. Additionally, it can reduce the need for transportation and disposal of construction materials, which can further reduce the environmental impact.
3. Cost-effective: While self-healing concrete may have a higher initial cost than traditional concrete, it can provide long-term cost savings through reduced maintenance and repairs. This can make it a cost-effective option for green buildings.
4. Energy efficiency and safety: Self-healing concrete can also improve energy efficiency by reducing the need for heating and cooling to maintain comfortable indoor temperatures. This can lead to lower energy bills and reduced greenhouse gas emissions. Self-healing concrete can improve the safety of the building by preventing damage from structural failures or natural disasters such as earthquakes or hurricanes.
5. Design flexibility: Self-healing concrete can be used to create a wide range of building designs, including custom shapes and sizes. This provides architects and builders with more design flexibility and can result in unique and aesthetically pleasing building designs.

1.5.4 Flexible Concrete

Over the past few years, academics have concentrated on creating sustainable concrete. Fly ash, slag, and recycled aggregates are some examples of recycled resources included in sustainable concrete, which helps lessen the carbon footprint of the concrete production process. A type of concrete with greater flexibility and elasticity than regular concrete is known as flexible concrete, bendable concrete, or flexible cementitious composite. Deformation won't cause it to break or lose its structural integrity because it is built to endure it. Cementitious materials, aggregates, and additives are the main components of flexible concrete. The precise composition may change depending on the desired level of flexibility and the intended application. Fibers such as steel or polymer fibers are frequently added to the mix to increase the concrete's tensile strength and flexural capacity. Due to its great flexibility, flexible concrete can experience severe deformation without cracking.[34] Fibers that span fissures and distribute stress throughout the material are incorporated to achieve this. The fibers allow the concrete to bend and flex without breaking by absorbing energy and preventing fracture progression. Flexible concrete's resistance to cracking is one of its main benefits. The presence of fibers aids in limiting crack initiation and halting crack spread. This feature is beneficial when the concrete is subjected to dynamic loads, such as in earthquake-prone areas or structures that experience vibrations. Pavements, bridge decks, precast components, architectural facades, and repairs are just a few uses for flexible concrete. Due to its flexibility, it can adjust to various structural movements and loads, lowering the possibility of damage or failure brought on by outside forces. Flexibility in concrete is intended to offer exceptional resilience and lifespan. The material's resistance to shrinkage, cracking, and environmental conditions, including freeze–thaw cycles, is improved by adding fibers and other additives. As a result, the concrete construction becomes more robust and lasts longer. It also provides multiple construction advantages. Due to its flexibility, expansion joints are less necessary, streamlining construction procedures and lowering maintenance needs. It can be shaped into many forms, giving architects flexibility and design freedom. Although flexible concrete offers performance benefits, it is crucial to consider the sustainability issues related to the materials employed. The sustainability of flexible concrete can be improved by using environmentally friendly cementitious components, such as additional cementitious materials or substitute binders with lower carbon footprints and recycled aggregates.[35] It's important to remember that the study and use of flexible concrete is still a new topic. The qualities and application scope of this novel building material are continually being enhanced by ongoing research and developments in material science.

1.5.5 Insulated Concrete Forms

Insulated concrete forms are blocks made from expanded polystyrene foam that are stacked together and filled with concrete to create a high-performance, energy-efficient building envelope. Figure 1.10 shows an illustrative image of insulated concrete forms. Insulated concrete forms are a sustainable building technology that can be used in green buildings to provide superior insulation, energy efficiency, and durability. Insulated concrete forms are designed to provide a high level of insulation to a building's walls and can be used for both residential and commercial construction projects.[36] They are also known for their durability, energy efficiency, and ease of use. Insulated concrete forms come in a variety of sizes and shapes and can be customized to fit the specific needs of a construction project. They are typically made from expanded polystyrene foam (EPS) or extruded polystyrene foam (XPS), and the foam is reinforced with steel or plastic webbing to add strength to the walls. CFs offer several advantages over traditional building materials.[37] They provide a high level of

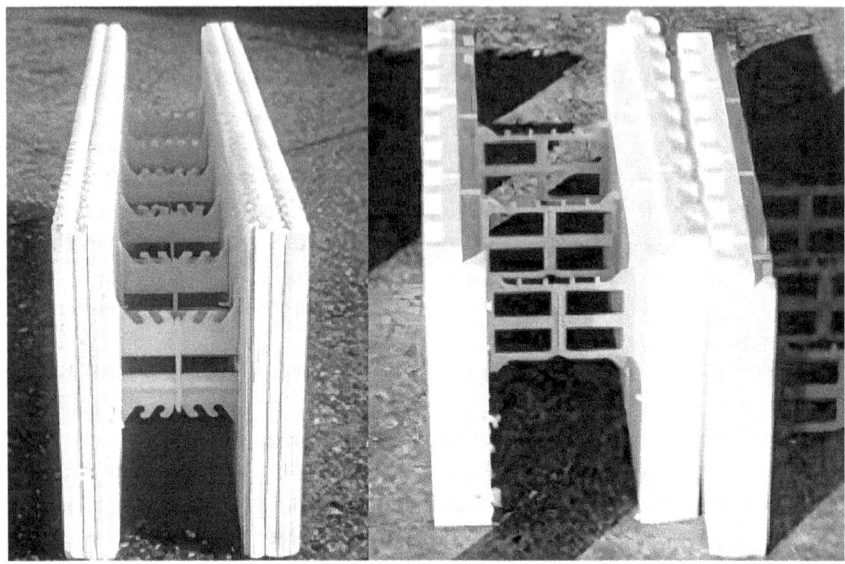

FIGURE 1.10 Illustrative image of insulated concrete forms.[39]

insulation, reducing energy costs and improving indoor comfort. They are also fire resistant and have excellent soundproofing properties. Additionally, because insulated concrete forms are lightweight and easy to install, construction times are typically shorter and labor costs are reduced. Insulated concrete forms provide strong and durable building structures that are resistant to damage from wind, water, and fire.[38] This can reduce the need for repairs and maintenance and extend the life of the building.

Advantages of Insulated Concrete Forms

1. Superior insulation: Insulated concrete forms are made from insulating foam blocks that are stacked and filled with concrete. This creates a building envelope that provides superior insulation and reduces energy consumption for heating and cooling.
2. Energy efficiency: Insulated concrete forms create airtight building envelopes that reduce the amount of air leakage and energy loss compared to traditional building methods. This results in lower energy bills, reduced greenhouse gas emissions, and improved indoor air quality.
3. Cost effectiveness: While insulated concrete forms may have a higher initial cost than traditional building methods, they can provide long-term cost savings through reduced energy consumption, lower maintenance costs, and increased durability.
4. Sustainable materials: Insulated concrete forms are made from environmentally friendly materials, including foam insulation made from recycled materials and concrete made from sustainable materials, such as fly ash.

5. Design flexibility: Insulated concrete forms can be used to create a wide range of building designs, including custom shapes and curved walls. This provides architects and builders with more design flexibility and can result in unique and aesthetically pleasing building designs.

Insulated concrete forms are a sustainable building technology that can be used in green buildings to provide superior insulation, energy efficiency, and durability. By incorporating insulated concrete forms into building designs, architects and builders can create sustainable buildings that are environmentally responsible and cost effective, with long-term benefits for both the environment and occupants.

1.5.6 Digital Concrete

A new idea, "digital concrete," combines conventional concrete with digital technologies, embedded sensors, and data collection. Various digital components must be integrated into the physical structure to track the performance of the concrete structure, gather data, and enable real-time analysis. Digital concrete includes strain gauges, accelerometers, and temperature and moisture[40] sensors within the concrete mixture or into precast concrete components. These sensors continuously monitor the structural load, temperature swings, humidity levels, and strain distribution within the concrete. Digital concrete's integrated sensors allow for real-time environmental conditions, structural performance, and behavior monitoring.[40] Engineers and stakeholders will have access to the most recent data on the structural health and integrity of the concrete thanks to the wireless transmission of this data to a central system for examination. Digital concrete allows for continuous structural health monitoring, identifying and evaluating potential problems such as fractures, deformations, and stress concentrations. This proactive technique enables early structural fault detection, allowing for prompt maintenance or repairs to prevent catastrophic failures. Digital concrete makes it easier to optimize the performance of concrete by gathering data on variables like strain, temperature, and moisture. Based on real-time feedback, engineers can use data analysis to optimize concrete mix designs and curing procedures and increase material performance. It also offers valuable information about the aging and deterioration of concrete structures. Engineers can minimize downtime and maximize the structure's lifespan by predicting maintenance needs through analysis of the gathered data and creating proactive maintenance plans. It can create accurate structural models, and simulations can be made using digital concrete data. Engineers can then analyze the effects of environmental conditions, simulate various load situations, and optimize the structural design for increased performance and safety. Digital concrete's real-time monitoring and early detection features help keep construction sites safer. Identifying potential hazards and weaknesses reduces the likelihood of mishaps or structural breakdowns.

Digital concrete offers valuable information during all stages of a structure's existence, from construction to operation and maintenance. Based on the concrete's performance history, this data-driven method offers better decision making about repairs, renovations, and future design concerns. It's crucial to remember that the field of digital concrete is still developing, and its current level of mainstream usage could be higher. However, the creation and application of digital concrete in construction projects are driven by ongoing research and improvements in sensor technology, data analytics, and Internet of Things (IoT) applications.

1.5.7 Light-Emitting Concrete

Researchers from Michoacán University in San Nicolás de Hidalgo, Mexico, have created a unique cement that collects solar energy during the day and emits it at night. This luminescent cement might be used to illuminate buildings, cycling routes, and roadways. Figure 1.11 (a) and (b) shows light-emitting cement used as pathways and in building applications, respectively. Cement is often an opaque substance that prevents light from penetrating its core. It is really a blend of powders that, when mixed with water, dissolve as an effervescent tablet and begin to solidify into a highly robust gel. In addition, this procedure produces crystal flakes in the gel that deflect sunlight.[41] Due to this, researchers concentrated on changing the cement's microstructure in order to remove all crystals and turn it completely into a gel by adding a mixture of polymers with luminescent properties. These polymers can absorb solar energy and release it as light for about 12 hours at night into the environment.[42,43]

Advantage of light-emitting cement

1. Improved visibility: The luminescent properties of the cement can help improve visibility in low-light conditions, making it easier to see walkways, stairs, and other important areas.
2. Energy efficiency: Light-emitting cement doesn't require electricity to produce light, so it can help save on energy costs.
3. Low maintenance: The cement doesn't require any additional maintenance beyond standard concrete upkeep.

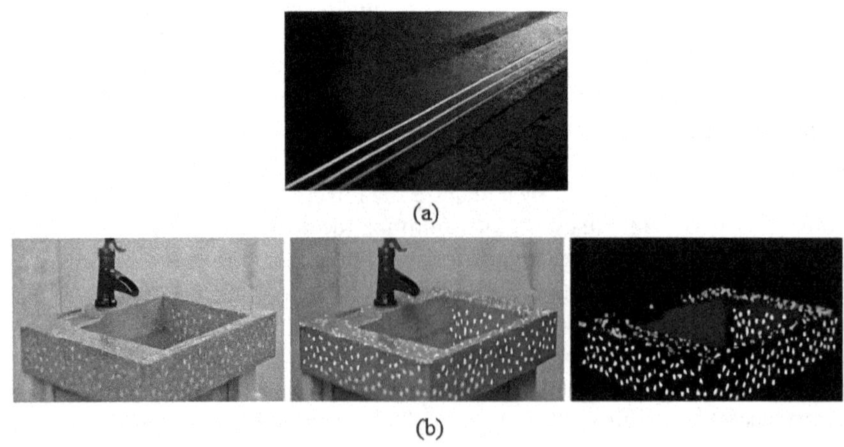

FIGURE 1.11 (a) Light-emitting cement used as pathways and (b) used in building.[44]

4. **Aesthetic appeal and safety:** Light-emitting cement can be used for decorative purposes, creating unique and eye-catching designs. Light-emitting cement can improve safety in public areas such as sidewalks, parks, and bike paths by increasing visibility and reducing the risk of accidents.
5. **Environmental benefits:** Some types of light-emitting cement are made from recycled materials, which can help reduce waste and promote sustainability.

Overall, light-emitting cement has the potential to offer a range of benefits, from improving safety to enhancing the aesthetic appeal of buildings and public spaces.

1.5.8 Bio-Cementation

"Bio-cementation" refers to using specific microbes to solidify or "cement" loose soil or sand particles together. It is also known as microbial-induced calcite precipitation (MICP). It has drawn interest as a cutting-edge method for enhancing building materials' stability and mechanical qualities. Bacteria, mainly species of *Sporosarcina*, *Bacillus*, or other urease-producing microbes, are used in the bio-cementation process. These microorganisms can produce the enzyme urease, which hydrolyzes urea into carbon dioxide and ammonia. The ammonia that is created elevates the pH of the environment in the presence of calcium ions, causing calcium carbonate ($CaCO_3$) to precipitate. The created calcium carbonate serves as a binder, gluing the soil particles together to form a solid mass.[45] The bio-cementation method has several benefits over conventional cement. For example, strength and durability are increased because the pores between soil particles are filled with calcium carbonate, which leads to better cohesion and interlocking. It also improves the material's stability and strength. Bio-cementation can significantly improve the ability of weak or loose soils to support loads, making them appropriate for use in buildings. As soil particles solidify due to calcium carbonate precipitation, the material's permeability is decreased, making it less permeable to water infiltration and erosive forces. Because bio-cementation uses naturally occurring bacteria instead of cementitious products with a high carbon footprint, it is considered an environmentally benign process.

However, there are several restrictions and difficulties with bio-cementation. The bio-cementation method works best with sandy soils and may have some difficulty with other soil types. The soil's composition and environmental factors must be favorable for bacteria to flourish and calcium carbonate to precipitate. It can take weeks or months for the bacteria to acquire the necessary level of soil consolidation during the bio-cementation process. Something other than this might be appropriate for construction jobs with tight deadlines. It can be challenging to ensure that bacteria and calcium carbonate precipitation are distributed uniformly throughout the soil mass. Quality control procedures must be implemented to track and confirm the efficiency of the bio-cementation process. Research on bio-cementation is still in its early stages, and new studies are being conducted to improve the method and overcome its drawbacks. It might offer long-term options for improving building materials and stabilizing soil in various technical applications.

1.5.9 Recycled Materials

Using recycled materials, such as recycled steel or reclaimed wood, can help reduce the environmental impact of building materials. The use of recycled materials in green buildings is a sustainable practice that can significantly reduce the environmental impact of the building industry. Here are some ways that recycled materials can be incorporated into green buildings.

1.5.9.1 Recycled Steel

Steel is a common building material, and using recycled steel reduces the environmental impact of steel production. Recycled steel can be used as a building material in a variety of applications, including structural framing, roofing, and cladding. Steel is a strong, durable, and versatile material that is commonly used in construction, and recycling steel can help to reduce the environmental impact of the construction industry. Recycling steel involves melting down scrap steel and remolding it into new steel products. The process of recycling steel requires less energy and produces fewer greenhouse gas emissions than producing new steel from raw materials. In addition, recycling steel can help to conserve natural resources, reduce landfill waste, and extend the lifespan of existing steel products. In construction, recycled steel can be used in a variety of ways. For example, it can be used for structural framing in buildings, such as columns, beams, and trusses. Recycled steel can also be used for roofing and cladding, as well as in the manufacture of building components such as stairs, handrails, and elevators. In addition to the environmental benefits of using recycled steel in construction, there are also economic benefits. Using recycled steel can be more cost-effective than using new steel, as the cost of recycled steel is typically lower than the cost of new steel. In addition, using recycled steel can help to reduce the risk of price fluctuations associated with new steel. However, there are also some challenges associated with using recycled steel in construction. For example, the quality and consistency of recycled steel can vary depending on the source and the recycling process. This can pose challenges for builders and manufacturers who need to ensure that the steel meets the required standards for strength and durability. Overall, the use of recycled steel as a building material can provide environmental and economic benefits. By using recycled steel, the construction industry can help to reduce greenhouse gas emissions, conserve natural resources, and reduce landfill waste. However, careful consideration must be given to the quality and consistency of recycled steel to ensure that it meets the required standards for strength and durability in construction applications

1.5.9.2 Recycled Concrete

Concrete is another common building material that can be made with recycled materials such as crushed concrete or fly ash. Recycled concrete can be used as a building material in a variety of construction applications. Concrete is a widely used building material, and recycling concrete can help to reduce the environmental impact of the construction industry. Recycling concrete involves crushing and grading old concrete to produce a material that can be used as an aggregate in new concrete. The process of recycling concrete requires less energy and produces fewer greenhouse gas emissions than producing new concrete from raw materials. In addition, recycling concrete can help to reduce landfill waste and conserve natural resources. In construction, recycled concrete can be used in a variety of ways. For example, it can be used as an aggregate in new concrete, as well as in the manufacture of concrete blocks and other precast concrete products. Recycled concrete can also be used as a base material for roads and pathways and as a filler material in construction projects. There are several benefits associated with using recycled concrete in construction. For example, using recycled concrete can help to reduce the environmental impact of the construction industry, as it can help to reduce greenhouse gas emissions and conserve natural resources. In addition, using recycled concrete can be more cost-effective than using new concrete, as the cost of recycled concrete is typically lower than the cost of new concrete. However, there are also some challenges associated with using recycled concrete in construction. For example, the quality and consistency of recycled concrete can vary depending on the source and the recycling process. This can pose challenges for builders and manufacturers who need to ensure that the concrete meets the required standards for strength and durability. Overall, the use of recycled concrete as a building material can provide environmental and economic benefits. By using recycled concrete, the construction industry can help to reduce greenhouse gas emissions, conserve natural resources, and reduce landfill waste. However, careful consideration must be given to the quality and consistency of recycled concrete to ensure that it meets the required standards for strength and durability in construction applications.

1.5.9.3 Reclaimed Wood

Reclaimed wood is a popular choice for recycled materials in building and construction applications. It is obtained from old buildings, barns, and other structures that have been dismantled or demolished. Reclaimed wood can provide a unique look to a building project while also offering environmental benefits. Using reclaimed wood in construction can help to reduce the demand for new timber, which can help to conserve natural resources and reduce the environmental impact of the construction industry. Additionally, reclaiming and reusing old wood can help to prevent it from ending up in landfills, which can help to reduce waste and greenhouse gas emissions. Reclaimed wood can be used in a variety of ways in construction, including as structural framing, flooring, wall cladding, and decorative features. When using reclaimed wood, it is important to ensure that it is properly prepared and treated to ensure its durability and structural integrity. There are several benefits to using reclaimed wood in construction. For example, it can offer a unique aesthetic that cannot be replicated with new materials. Reclaimed wood also has a character and warmth that can add value to a building project. In addition, using reclaimed wood can be more cost-effective than using new timber, as the cost of reclaimed wood is typically lower than the cost of new timber. However, there are also some challenges associated with using reclaimed wood in construction. For example, the quality and condition of the wood can vary depending on the source and the previous use of the wood. This can pose challenges for builders and manufacturers who need to ensure that the wood meets the required standards for strength and durability. Overall, the use of reclaimed wood as a building material can provide environmental and aesthetic benefits. By using reclaimed wood, the construction industry can help to reduce the demand for new timber, conserve natural resources, and reduce waste. However, careful consideration must be given to the quality and condition of reclaimed wood to ensure that it meets the required standards for strength and durability in construction applications.

1.5.9.4 Recycled Glass

Recycled glass is a versatile and sustainable material that can be used in building and construction applications. Glass recycling involves melting down used glass and forming it into new products, which can help to conserve natural resources and reduce waste. In construction, recycled glass can be used in a variety of ways. For example, it can be used as an aggregate in concrete and asphalt, as well as in the manufacture of glass blocks, countertops, and tiles. Recycled glass can also be used as a decorative element in landscaping and as a filler material in construction projects. There are several benefits associated with using recycled glass in construction. For example, using recycled glass can help to reduce the demand for new raw materials, which can help to conserve natural resources and reduce the environmental impact of the construction industry. Additionally, using recycled glass can help to reduce landfill waste and greenhouse gas emissions. Recycled glass can offer a unique and attractive aesthetic to a building project, with its texture and color adding visual interest. Recycled glass can also be cost effective, as the cost of recycled glass is typically lower than the cost of new glass. However, there are some challenges associated with using recycled glass in construction. For example, the quality and consistency of recycled glass can vary depending on the source and the recycling process. This can pose challenges for builders and manufacturers who need to ensure that the glass meets the required standards for strength and durability. Overall, the use of recycled glass as a building material can provide environmental and aesthetic benefits. By using recycled glass, the construction industry can help to conserve natural resources, reduce waste, and reduce greenhouse gas emissions. However, careful consideration must be given to the quality and consistency of recycled glass to ensure that it meets the required standards for strength and durability in construction applications.

1.5.9.5 Recycled Insulation

Recycled insulation refers to insulation materials that are made from recycled materials. These materials can include recycled glass, recycled cotton, recycled denim, recycled plastic bottles, and other materials that have been diverted from the waste stream. Using recycled insulation in buildings can provide several benefits. First, it helps to reduce the amount of waste that goes into landfills. Second, it can help to reduce the environmental impact of manufacturing new insulation materials from virgin resources. Third, recycled insulation can provide a more sustainable and healthier indoor environment by reducing the use of materials that contain harmful chemicals. Recycled insulation can be used in a variety of applications, including wall cavities, attics, floors, and roofs. Depending on the type of material used, recycled insulation can have different R-values, which refers to the insulation's ability to resist heat flow. For example, recycled denim insulation is made from post-industrial and post-consumer denim that has been diverted from the waste stream. It is a natural and sustainable insulation material that is free from harmful chemicals and is safe for people and the environment. Recycled denim insulation has a high R-value, which makes it an effective insulator for walls, attics, and floors. Another example of recycled insulation is cellulose insulation, which is made from recycled paper products such as newspaper, cardboard, and other materials. It is an eco-friendly and sustainable insulation material that is treated with fire retardant to prevent fire hazards. Cellulose insulation has a high R-value and is suitable for use in walls, attics, and floors. Overall, using recycled insulation in buildings can provide environmental, economic, and health benefits. It can help to reduce waste, conserve resources, and provide a more sustainable indoor environment

1.5.9.6 Recycled Plastic

Recycled plastic is a sustainable and versatile material that can be used in building and construction applications. Plastic recycling involves collecting, cleaning, and processing used plastic products, which can then be used to make new products, including building materials. In construction, recycled plastic can be used in a variety of ways. For example, it can be used as a substitute for traditional materials like wood and concrete in the form of plastic lumber, decking, and fencing. Recycled plastic can also be used in roofing materials, insulation, and as a water-resistant barrier. There are several benefits associated with using recycled plastic in construction. For example, using recycled plastic can help to reduce the demand for new raw materials, which can help to conserve natural resources and reduce the environmental impact of the construction industry. Additionally, using recycled plastic can help to reduce landfill waste and greenhouse gas emissions. Recycled plastic can offer a range of benefits to construction projects, such as resistance to moisture, insects, and decay, as well as durability over time. Recycled plastic building materials can also be cost effective in the long term, as they typically require less maintenance and have a longer lifespan than traditional materials. However, there are some challenges associated with using recycled plastic in construction. For example, the quality and consistency of recycled plastic can vary depending on the source and the recycling process. This can pose challenges for builders and manufacturers who need to ensure that the plastic meets the required standards for strength and durability. Overall, the use of recycled plastic as a building material can provide environmental and economic benefits. By using recycled plastic, the construction industry can help to conserve natural resources, reduce waste, and reduce greenhouse gas emissions. However, careful consideration must be given to the quality and consistency of recycled plastic to ensure that it meets the required standards for strength and durability in construction applications. The incorporating recycled materials into green buildings is an important sustainable practice that can significantly reduce the environmental impact of the building industry. By using recycled steel, concrete, wood, glass, insulation, and plastic in building designs, architects and builders can create sustainable buildings that are environmentally responsible and cost-effective.

REFERENCES

1. Ancient Construction Techniques of India: A Regional Study. [accessed 2023 May 30]. https://worldarchitecture.org/architecture-news/eepff/sheppard-robson-creates-europe-s-largest-green-wall-for-this-mixeduse-building-in-london.html
2. Ergün Hatir M, İnce İ. 2021. Lithology mapping of stone heritage via state-of-the-art computer vision. Journal of Building Engineering. 34:101921. doi:10.1016/J.JOBE.2020.101921
3. Miller AZ, Sanmartín P, Pereira-Pardo L, Dionísio A, Saiz-Jimenez C, Macedo MF, Prieto B. 2012. Bioreceptivity of building stones: A review. Science of The Total Environment. 426:1–12. doi:10.1016/J.SCITOTENV.2012.03.026
4. Johnson B, Friedman M, Hopkins TW, Bauer SJ. 1987. Strength and microfracturing of westerly granite extended wet and dry at temperatures to 800°C and pressures to 200 Mpa [Internet]. [accessed 2023 May 12]. /ARMAUSRMS/proceedings-abstract/ARMA87/All-ARMA87/129843
5. Zhiming Y, Jinxu L, Xinya F, Shukui L, Yuxin X, Jie R. 2017. Investigation on mechanical properties and failure mechanisms of basalt fiber reinforced aluminum matrix composites under different loading conditions. Journal of Composite Materials. 52(14):1907–1914. [accessed 2023 May 12]. doi:10.1177/0021998317733807
6. Duggal SK. 1998. Building materials [Internet]. A.A. Balkema; [accessed 2023 May 29]. www.routledge.com/Building-Materials/Duggal/p/book/9789054107644
7. Rodney Fernando P. 2018. Synthesis and characterization of sustainable man-made low cost clay bricks with bamboo leaf ash. Engineering Physics. 2(1):15. doi:10.11648/J.EP.20180201.14
8. Aigbomian EP, Fan M. 2013. Development of wood-crete building materials from sawdust and waste paper. Constr Build Materials. 40:361–366. doi:10.1016/J.CONBUILDMAT.2012.11.018
9. Choo, B.S., & MacGinley, T.J. (1990). Reinforced Concrete: Design theory and examples (2nd ed.). CRC Press. https://doi.org/10.1201/9781315274829
10. Ahmad J, Zhou Z. 2022. Mechanical properties of natural as well as synthetic fiber reinforced concrete: A review. Constructions of Building Materials. 333:127353. doi:10.1016/J.CONBUILDMAT.2022.127353
11. Laning A. Primarily used to reduce shrinkage cracking, polypropylene, nylon, and polyester fibers offer other benefits as well Synthetic fibers.
12. Bird NJ, Robert Michell A, Michael Peters A, Annamaneni KK, Pedarla K. 2023. Compressive and flexural behavior of glass fiber-reinforced concrete. Journal of Physics. 12025. doi:10.1088/1742-6596/2423/1/012025
13. Narayanan P, Sekhar tirumala S, Sravana P, Rao SP, Sekhar ST. 2010. Strength properties of glass fiber concrete strength properties of glass fibre concrete [Internet]. 5(4). [accessed 2023 May 13]. www.researchgate.net/publication/288971117
14. Hilles MM, Ziara MM. 2019. Mechanical behavior of high strength concrete reinforced with glass fiber. Engineering Science and Technology, an International Journal. 22(3):920–928. doi:10.1016/J.JESTCH.2019.01.003
15. Chen PW, Chung DDL. 1993. Carbon fiber reinforced concrete for smart structures capable of non-destructive flaw detection. Smart Mater Struct [Internet]. 2(1):22. [accessed 2023 May 30]. doi:10.1088/0964-1726/2/1/004
16. Wang Z, Ma G, Ma Z, Zhang Y. 2021. Flexural behavior of carbon fiber-reinforced concrete beams under impact loading. Cement and Concrete Composites. 118:103910. doi:10.1016/J.CEMCONCOMP.2020.103910
17. Li YF, Li JY, Ramanathan GK, Chang SM, Shen MY, Tsai YK, Huang CH. 2021. An experimental study on mechanical behaviors of carbon fiber and microwave-assisted pyrolysis recycled carbon fiber-reinforced concrete. Sustainability. 13:6829 [Internet]. [accessed 2023 May 30]. doi:10.3390/SU13126829

18. Islam T, Rashed HMMA. 2019. Classification and application of plain carbon steels. Reference Module in Materials Science and Materials Engineering. doi:10.1016/B978-0-12-803581-8.10268-1
19. Zeng S, Xu D, Chang J, Venkatesan V, Ganesan R. 1964. A general study of light gauge steel building-case study you may also like development of hot rolled weathering angle steel for transmission towers a general study of light gauge steel building-case study. Journal of Physics Conference Series. 72004. doi:10.1088/1742-6596/1964/7/072004
20. Zhang Z, Wong YC, Arulrajah A, Horpibulsuk S. 2018. A review of studies on bricks using alternative materials and approaches. Construction of Buildings Materials. 188:1101–1118. doi:10.1016/J.CONBUILDMAT.2018.08.152
21. Ramachandran G. 2020. Experimental study on fly ash bricks by using granite and marble powder. International Research Journal of Engineering and Technology [Internet]. [accessed 2023 May 30]. www.irjet.net
22. Arbab M, Finley JJ. 2010. Glass in architecture. International Journal of Applied Glass Science [Internet]. 1(1):118–129. [accessed 2023 May 29]. doi:10.1111/J.2041-1294.2010.00004.X
23. Henderson J. 1985. The raw materials of early glass production. Oxford Journal of Archaeology [Internet]. [accessed 2023 May 29] 4(3):267–291. doi:10.1111/J.1468-0092.1985.TB00248.X
24. Mayatskaya IA, Yazyeva SB, Zakieva NI, Lapina AP. 2018. Modern glass constructions and comfortable urban environment. Materials Science Forum [Internet]. 931:754–758. [accessed 2023 May 29]. doi:10.4028/WWW.SCIENTIFIC.NET/MSF.931.754
25. Neville A, Aïtcin PC. 1998. High performance concrete—An overview. Materials and Structures/Materiaux et Constructions [Internet]. 31(2):111–117. [accessed 2023 May 30]. doi:10.1007/BF02486473/METRICS
26. Neville A, Tcin P-CA. 1998. High performance concrete-An overview.
27. Graybeal B. 2011. Ultra-high performance concrete.
28. Xue J, Briseghella B, Huang F, Nuti C, Tabatabai H, Chen B. 2020. Review of ultra-high performance concrete and its application in bridge engineering. Construction of Buildings Materials. 260:119844. doi:10.1016/J.CONBUILDMAT.2020.119844
29. Hager MD, Zwaag S van der (Sybrand), Schubert U (Ulrich). 2010. Self-Healing Materials. Adv. Mater., 22: 5424–5430. https://doi.org/10.1002/adma.201003036
30. De Belie N, Gruyaert E, Al-Tabbaa A, Antonaci P, Baera C, Bajare D, Darquennes A, Davies R, Ferrara L, Jefferson T, et al. 2018. A review of self-healing concrete for damage management of structures. Advance Mater Interfaces [Internet]. 5(17):1800074. [accessed 2023 Mar 23]. doi:10.1002/ADMI.201800074
31. Seifan M, Samani AK, Berenjian A. 2016. Bioconcrete: Next generation of self-healing concrete. Applying Microbiology Biotechnology [Internet]. 100(6):2591–2602. [accessed 2023 Mar 23]. doi:10.1007/S00253-016-7316-Z/TABLES/2
32. Self-Healing Concrete—Materials—Materials Library—Institute of Making. [accessed 2023 Mar 21]. www.instituteofmaking.org.uk/materials-library/material/self-healing-concrete
33. Garg, R., Garg, R., & Eddy, N. O. (2023). Microbial induced calcite precipitation for self-healing of concrete: a review. Journal of Sustainable Cement-Based Materials, 12(3), 317–330. https://doi.org/10.1080/21650373.2022.2054477
34. Gurung A. 2017. Experimental study on flexural behavior of bendable concrete. International Journal of Scientific Engineering and Applied Science (IJSEAS) [Internet]. [accessed 2023 May 30] (3). www.ijseas.com
35. Solanki DS, Kangda MZ, Sarur S, Noroozinejad Farsangi E. 2022. A state-of-the-art review on the experimental investigations of bendable concrete. Journal of Building Pathology and Rehabilitation [Internet]. 7(1):1–15. [accessed 2023 May 30]. doi:10.1007/S41024-021-00158-7/METRICS

36. Ekrami N, Garat A, Fung AS. 2015. Thermal analysis of Insulated Concrete Form (ICF) Walls. Energy Procedia. 75:2150–2156. doi:10.1016/J.EGYPRO.2015.07.353
37. Shahedan NF, Abdullah MMAB, Mahmed N, Kusbiantoro A, Binhussain M, Zailan SN. 2017. Review on thermal insulation performance in various type of concrete. AIP Conference Proceedings [Internet]. 1835(1):020046. [accessed 2023 Mar 23]. doi:10.1063/1.4981868
38. Amer-Yahia C, Majidzadeh T. 2012. Inspection of insulated concrete form walls with ground penetrating radar. Constructions of Building Materials. 26(1):448–458. doi:10.1016/J.CONBUILDMAT.2011.06.044
39. Project Management News: Pros & Cons of Using Insulated Concrete Forms. [accessed 2023 Mar 21]. https://esub.com/blog/the-pros-and-cons-of-using-insulated-concrete-forms/
40. Wangler T, Roussel N, Bos FP, Salet TAM, Flatt RJ. 2019. Digital concrete: A review. Cement and Concrete Composites. 123:105780. doi:10.1016/J.CEMCONRES.2019.105780
41. Han B, Zhang L, Ou J. 2017. Light-emitting concrete. Smart and Multifunctional Concrete Toward Sustainable Infrastructures [Internet]. 285–297. [accessed 2023 Mar 23]. doi:10.1007/978-981-10-4349-9_16
42. Purnama DD, Iduwin T, Hidayawanti R. 2020. Fluorescent concrete with strontium phosphorus powder and plastic waste. IOP Conference Series Mater Science Engineering [Internet]. 771(1):012061. [accessed 2023 Mar 23]. doi:10.1088/1757-899X/771/1/012061
43. Exploratory Study on Photo Luminescence Induced Concrete. [accessed 2023 Mar 23]. www.iaeme.com/IJCIET/index.asp622www.iaeme.com/ijmet/issues.asp?JType=IJCIET&VType=10&IType=3www.iaeme.com/IJCIET/index.asp623www.iaeme.com/IJCIET/issues.asp?JType=IJCIET&VType=10&IType=3
44. Han B, Zhang L, Ou J. 2017. Light-emitting concrete. In: Smart and multifunctional concrete toward sustainable infrastructures. pp. 285–297. doi:10.1007/978-981-10-4349-9_16
45. Iqbal DM, Wong LS, Kong SY. 2021. Bio-cementation in construction materials: A review. Materials [Internet]. 14(9). [accessed 2023 May 30]. doi:10.3390/MA14092175

2 Advances in Thermal Energy Storage in Buildings

Abhishek Gautam, C. Suresh, R. P. Saini, and Yongseok Jeon

2.1 INTRODUCTION

Over the coming decades, the generation of electricity in emerging markets and developing economies such as India, Indonesia, South Africa, Brazil, and Nigeria will have a significant impact on the environment and global greenhouse gas emissions. At the present, the majority of people spend 90% of their time indoors and rely on mechanical heating and cooling, rendering buildings the largest energy consumers in the world. Furthermore, 30% of overall energy consumption now accounts for energy utilization in building applications.[1] Building energy utilization exceeds 40% of total energy usage in several nations, including the United States of America and the United Kingdom.

In 2020, the construction and operation of buildings accounted for 36% (149 EJ) of worldwide energy consumption, with 127 EJ predicted to be used for building operations and 22 EJ for manufacturing building materials, as shown in Figure 2.1. The building sector emits 37% of global emissions connected to building energy as compared to the other sectors.[2] Most of the energy in buildings is utilized for cooling and heating purposes to achieve the desired comfort level. Thus, innovative methods for energy conservation in buildings must be developed. Therefore, sustainable buildings have emerged as a major concern for many industrialized and developing nations in the twenty-first century. Building sustainability is a concept with many different facets, including economic, ecological, social, and technological considerations.

In order to mitigate the negative impacts of buildings on the environment, society, and economy, the integration of thermal energy storage (TES) has been identified as an efficient pillar for sustainable buildings. TES systems can store heat or cold energy, which can be further retrieved for various applications like air heating or cold storage. The integration of TES in buildings helps in reducing air conditioning energy consumption, shifting the peak loads, and minimizing the mismatch between energy supply and demand. The details about the various heat transfer modes involved in building envelopes and other aspects are introduced under the following paragraphs.

DOI: 10.1201/9781003415695-2

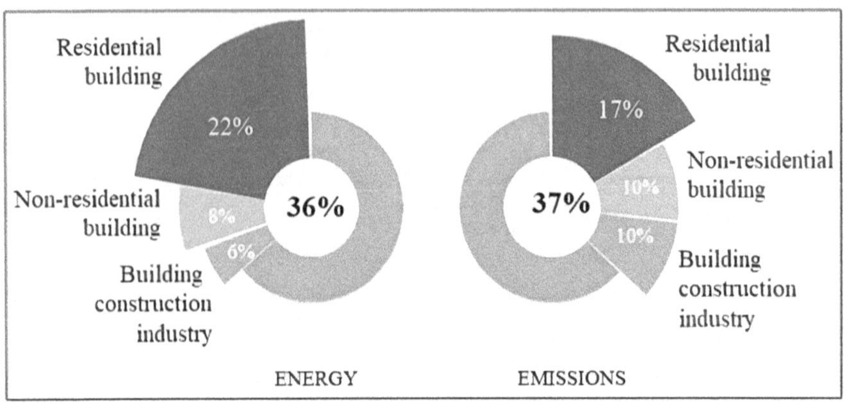

FIGURE 2.1 Energy consumption in building and construction materials and CO_2 emissions related to that energy utilized.[2]

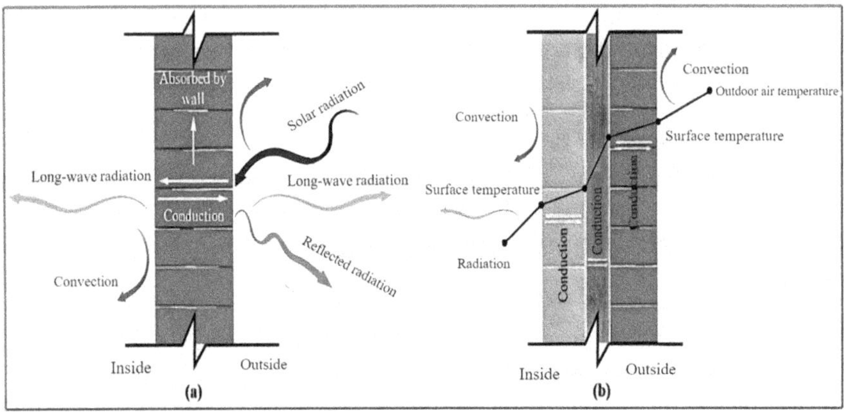

FIGURE 2.2 Schematic showing different heat transfer modes in (a) solid wall (b) composite wall.[3]

2.1.1 Heat Transfer in Building Envelopes

The heat transfer process through building envelopes is a complex mechanism as well as a dynamic that occurs by conduction, convection, and radiation. For example, consider the wall of a building, and the solar radiation or heat flux hits the surface of the wall, as shown in Figure 2.2.

From Figure 2.2, it is observed that heat flux from solar radiation is transferred to the surface of the wall through convection and radiation, and some of the flux is reflected back to the environment. Further, the heat is conducted through the wall and exchanged to the indoor environment through convection and radiation again. These heat transfer modes control the temperature of the indoor air, which affects the level of thermal comfort. Numerous factors such as solar heat gain, interior and outdoor temperature, material thermophysical properties, and exposed surface area affect the heat exchange rate and direction through the building envelopes.[4] Thermal conductivity density, thermal resistance and transmittance, heat capacity, and surface characteristics are the thermos-physical properties of materials that influence the heat rate of heat transfer in the building envelopes. The material thickness impacts the wall's ability to store and transfer heat in addition to the thermophysical qualities.[5] Moreover, the orientation of the wall can also affect heat gain and loss through it, therefore this should be taken into account while designing an energy-efficient building envelope.[6]

Buildings can be categorized into two categories based on how they acquire heat: those in hot regions and those in cold climates. In hot climates, the majority of the heat is drawn from the outside through building envelopes such as roofs, windows, walls, and ceilings. The inclusion of energy storage materials or storage systems in the interior or exterior of building envelopes must inevitably result in lower heat transfer to the building's indoor or outdoor climate depending on climate conditions (hot or cold), which in turn results in less energy used for heating and cooling to obtain the desired comfort levels.[7] The ability of a storage material to absorb and transmit heat is one of its most important energy-saving features. Different construction materials respond differently to the environment depending on their inherent properties. Therefore, it was confirmed that the introduction of TES can significantly enhance thermal comfort by reducing energy consumption and achieving energy savings.

2.1.2 Classification of TES Systems

TES systems are classified into three categories, that is, sensible, latent, and thermochemical energy storage, as shown in Figure 2.3. Further, the integration of TES systems into the buildings is classified by active and passive systems, as well as the duration of storage and ground integration.[8]

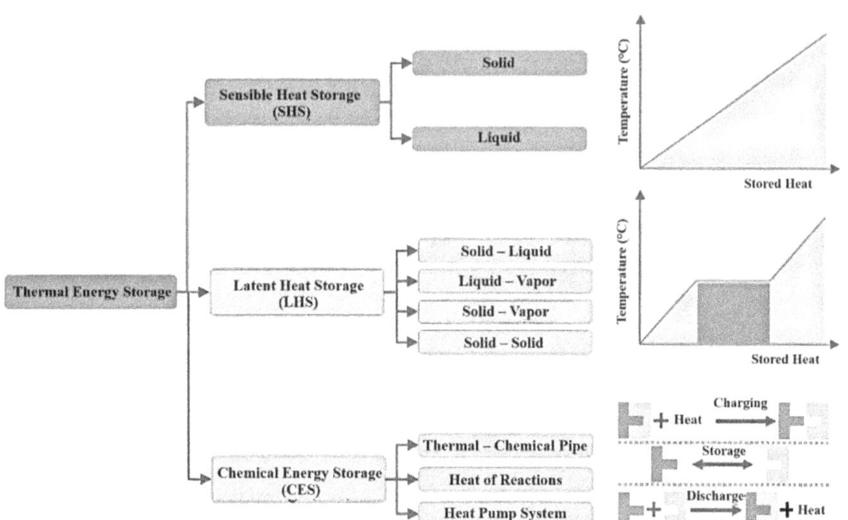

FIGURE 2.3 Classification of different TES systems.[9]

Sensible heat storage: Sensible heat storage stores or releases the heat energy by changing the temperature or internal energy of the material without undergoing a phase change. The amount of energy accumulated or retrieved is mainly dependent on the temperature difference of the storage building material, specific heat capacity, and mass.[10]

Latent heat storage: Latent heat storage stocks or retrieves the heat energy by undergoing a phase change of a material. The materials used in latent heat storage are called phase change materials (PCMs) and will undergo a phase transition by absorbing a large amount of energy. The quantity of energy stored is mainly dependent on the latent heat of fusion, temperature, and mass of the PCM. Further, there are four types of PCMs according to their phase transition process, that is, solid–solid, solid–liquid, solid–gas, and liquid–gas. Among those, solid–liquid is considered to be the most suitable one for building applications due to less volume change during phase transition and high energy density.[11]

Thermochemical energy storage: Thermochemical energy storage stores and recovers the energy during exothermic and endothermic reversible reactions. However, thermochemical energy storage is not yet viable for large-scale installation because of its complex reaction mechanism, safety concerns, high initial investment costs, and low reliability.[12]

Active TES systems in buildings: In active systems, the heat transfer fluid (HTF) (air or liquid) is pumped over the storage material or components to store or recover the energy for obtaining the desired comfort level. In active systems, mechanical devices are used for driving the HTF in terms of forced convective heat transfer, as shown in Figure 2.4.

Passive TES systems in buildings: In a passive system, natural convection and conduction play major roles in energy transfer, and there is a need for external devices that consume the energy in a passive system. The storage materials (sensible or latent) can be integrated into the building construction materials or components for storing or recovering the energy and to obtain the desired comfort level, as shown in Figure 2.4.[13]

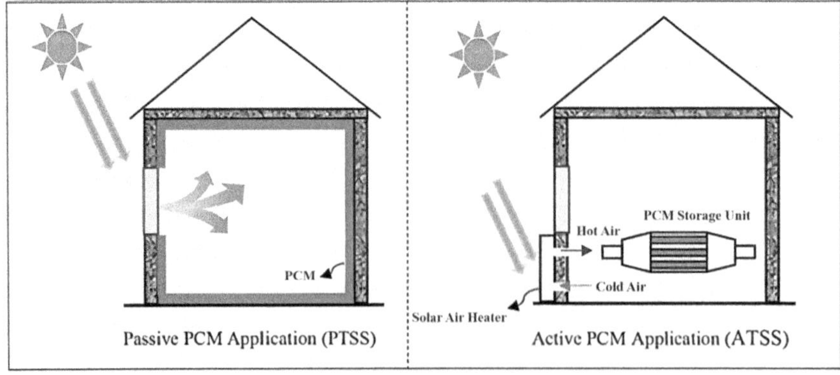

FIGURE 2.4 Passive and active TES system for building heating applications.[13]

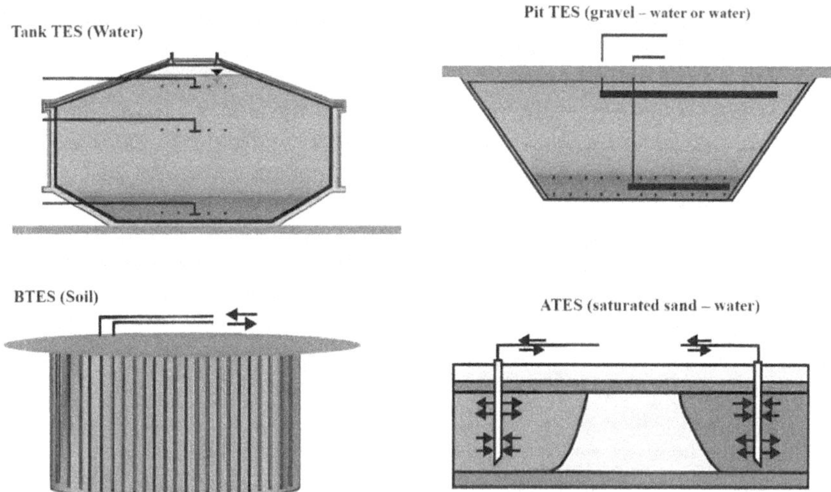

FIGURE 2.5 Classification of different TESs based on the location of the storage tank.[15]

Duration of storage: Based on the storage duration, TES systems are classified into shorter or longer periods. In a shorter duration, the energy obtained from renewable energy sources is stored in terms of heat or cold. And that stored energy can be reused for meeting the daily or weekly energy demands during peak energy periods or on cloudy days for heating or cooling applications. In long-term TES, the energy is conserved for months or even a whole season. Long-term energy storage involves storing energy that is available in the summer and recovering it in the winter and vice-versa for end-use applications such as building heating or cooling purposes.[14]

Ground-integrated TES: Ground-integrated TES systems are generally long-term and large-scale TES systems and are classified into different categories according to the location of energy storage, that is, above-ground TES (TES tank on the surface of the earth) and underground storage (borehole storage, pit TES, and aquifer storage), as shown in Figure 2.5. These are mainly used for district heating and cooling of large societies.[15]

Based on the aforementioned discussion, it is found that TES has a great potential for energy savings and hence a wide scope of research. Its integration into the buildings has favorable outcomes along with the scope of increased renewable energy share to meet the thermal energy demand. There are a number of emerging as well as commercial TES technologies available, and a significant amount of literature has been published on the advancement of the same. The integration of TES with buildings is a crucial decision, and significant preliminary knowledge is required. In view of the same, the present chapter discusses the various crucial aspects of TESs in the context of their integration into the building. The chapter includes a discussion about the various types of TESs, that is, sensible TES, latent TES, and thermochemical energy storage, which is followed by the various developed and developing mechanisms of TES in buildings. Further, the advantages, drawbacks, and challenges associated with TES in buildings are presented. Lastly, the conclusions drawn from the whole discussion are presented, along with several recommendations. The present chapter may be useful for researchers, academics, manufacturers, and other stakeholders in the TES systems working in the building sector.

2.2 TYPES OF THERMAL ENERGY STORAGE

2.2.1 Sensible Heat Storage in Buildings

Building envelopes act as a barrier between the outdoor and indoor environmental conditions. It protects the building's interior environment from extremes of the exterior environment, which include wind, temperature, light, noise, and water. In order to mitigate the heat wave penetration easily from outside to the inside of the building, the building envelope should offer good thermal resistance. As a result, it aids in obtaining a suitable thermal comfort condition in the building's indoor environment. The building envelopes are actually the roof, floor, walls, ceilings, doors, windows, and glazing of a building. These building components come together to form a building's thermal mass. The thermal mass of the building elements controls the indoor environmental conditions of the building by absorbing, storing, and releasing the thermal energy. Further, thermal mass with a high TES capacity will have high thermal mass effectiveness that will lead to a reduction in indoor temperature, resulting in better comfort.[16] Traditionally, the building's structural components store heat energy as sensible thermal energy.

In sensible TES, building components stock the excess heat when the temperature rises and dissipate the same stored heat to the surroundings as the temperature drops. Thus, these kinds of energy-efficient building envelopes can enhance the structural internal thermal behavior by serving as thermal insulation against outside heat waves and solar gains. The deployment of energy-efficient building materials such as energy storage materials in the construction of building envelopes reduces the heat gained from the outside air and solar radiation, as shown in Figure 2.6.

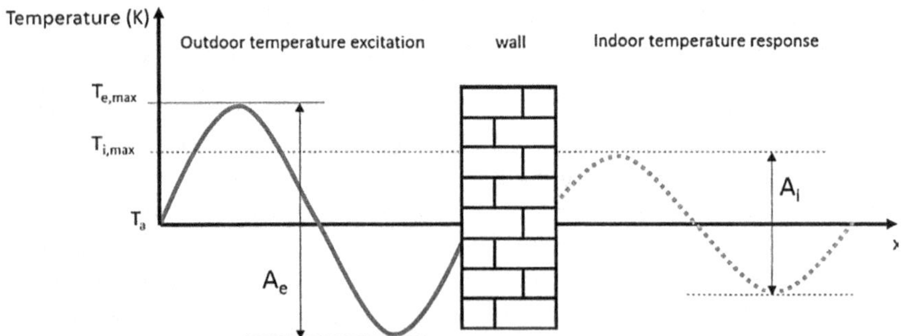

FIGURE 2.6 Schematic showing the propagation of a heat wave through the energy-efficient building wall.[17]

Sensible TES systems have a significant impact on addressing the problems associated with peak energy demand, energy conservation, and management. Sensible TES technologies have been utilized for building applications in the past centuries and have become more popular recently. The real examples of sensible TES are the archaeological structures or buildings that can be found all over the world. Each one has a unique historical relevance as well as a scientific foundation for its existence. These constructions are primarily made of stone, which serves as the fundamental source of TES.

The main objective of the massive sturdy constructions using rocks or stone as a building material is to keep the indoor temperature low as compared to the outdoor conditions. This is the reason why someone might feel colder inside a large magnificent building than in the outside environment. Actually, the reason behind the concept is absorbing the sensible heat by the massive building elements in the form of heat, and that results in maintaining the indoor temperature below the outdoor conditions. It's fascinating to note that our ancient ancestors had already absorbed the philosophy behind sensible TES technologies. Current sensible TES systems are being developed as a mimic of those systems, but the use of advanced materials and technologies can make them more effective and long-lasting.[18]

Sensible TES systems are classified into two types based on the storage medium, that is, i) solid-medium sensible TES and ii) liquid-medium sensible TES.

2.2.1.1 Solid Storage Media

Solid materials include concrete, clay, mortar, rocks, sand, etc. for the construction of buildings and thermal applications. Solid-medium TES is mostly preferred for fulfilling the TES requirements in building space cooling or heating applications.

The following are some of the objectives in developing solid storage materials[19]:

i) To avoid leakage at elevated temperatures.
ii) To operate at higher temperatures.

However, solid storage materials exhibit some limitations, which are as follows:

i) Solid materials are found with low specific heat capacities that result in lower heat absorption capacity.
ii) Lower energy storage density as compared to the liquid storage medium.
iii) Higher heat loss during long-term TES.

Solid storage materials are inexpensive and easily available on the market. But due to the lower energy density, the system occupies more space and requires a large storage tank installation for supplying the higher energy demand of heating and cooling to the building. This will lead to the increase in installation and operational costs of the system.

2.2.1.2 Liquid Storage Media

Water is the most often used liquid storage material in real life because of its high specific heat capacity, availability, and affordability. For instance, it can be determined that water has an energy storage density of about 290 MJ/m^3 as the temperature rises

TABLE 2.1
Summary of Different Sensible Storage Materials Used in the Construction of Buildings and for Thermal Applications[20]

Materials	Density (kg/m³)	Thermal conductivity (W/m.K)	Specific heat (KJ/kg.K)	Volumetric specific heat (J/K.m³)
Concrete	2000	1.35	1	2000
Rock	28000–1500	0.85–3.5	1	2150
Sand and gravel	1700–2200	2	0.91–1.18	2072
Cement mortar	1800	1	1	1800
Clay	1200–1800	1.5	1.67–2.5	3252
Limestone	1600–2600	0.85–2.3	1	2100
Lime mortar	1600	0.8	1	1600
Reinforced concrete	2400	2.5	1	2400
Ceramic tile	2000	1	0.8	1600
Ceramic brick	1800	0.73	0.92	1656
Plywood	1000	0.24	1.6	1600
Asphalt sheet	2300	1.2	1.7	3910
Gypsum coating	1000	0.4	1	1000
Gypsum board	900	0.25	1	900
Wood	700	0.18	1.6	1120
Oil	888	0.14	1.88	1669
Water	990	0.63	4.19	4148

to 70°C. The majority of medium-temperature applications for solar TES rely on liquid material (water) to achieve the requisite sensible heat storage. The properties of sensible heat storage materials mainly utilized for the construction of buildings and for building applications are summarized in Table 2.1.

The amount of energy accumulated in the sensible TES systems is determined by using Equation 2.1 Hence, higher density and specific heat should be required to attain a large volumetric storage capacity (MJ/m³).

$$Q = \int_{T_i}^{T_f} mC_p dT \tag{2.1}$$

Where m represents the mass of the storage media, C_p is the specific heat, and T_i and T_f are the initial and final temperatures of the storage media. Figure 2.7 demonstrates the best sensible storage materials that are currently available on market based on the average specific heat and diffusivity. The use of the ground as an insulating medium can also be advantageous, as demonstrated by large-scale underground TES systems. However, there are some drawbacks that limit the usage of solids, such as their lower energy storage density compared to that of water, comparatively expensive operating and maintenance costs, and potential long-term self-discharge concerns.

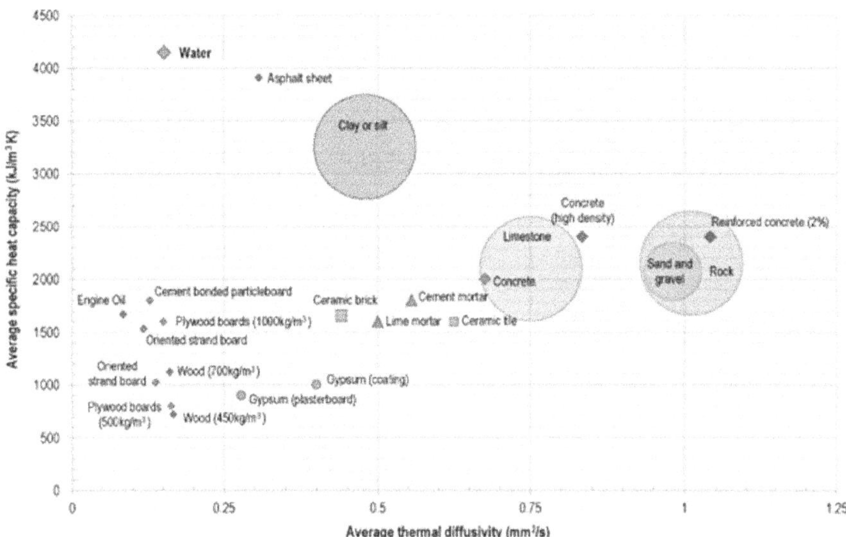

FIGURE 2.7 Different sensible storage materials employed for building heating and cooling applications.[21]

2.2.2 Latent Heat Storage in Buildings

Nowadays, latent TES systems receive greater attention due to the advantages such as higher energy density and lower temperature swings as compared to sensible TES systems. In addition to their large heat storage capacity, PCMs provide a number of advantages over other TES technologies and materials, including a broad availability range, high latent fusion, non-toxicity, chemical stability, low price, and environmental friendliness. The PCMs allow for storing heat in the latent form up to 5–14 times more efficiently than sensible heat TES materials. Because of these advantages, the application of latent TES in buildings has emerged as the best technique in recent days for regulating indoor temperature and as well reducing energy consumption.[22] Figure 2.8 illustrates how a PCM-embedded wall regulates a building's interior temperature throughout the day. In the daytime, heat waves induced by solar insolation penetrate the PCM-embedded wall, and that leads to the melting of PCM by absorbing the heat. This melting of the PCM results in delaying and minimizing heat transfer to the building's indoor environment and causes the maintenance of desired comfort temperatures as well reduces the energy demand for cooling. During the nighttime, when lower temperatures are reported, the heat energy stored in the PCM is released to both indoor and outdoor areas, resulting in better thermal comfort by minimizing the heating load of a building. The main distinction between PCMs and conventional thermal mass is that a masonry wall absorbs and releases the heat energy slowly, while PCMs exhibits faster heat

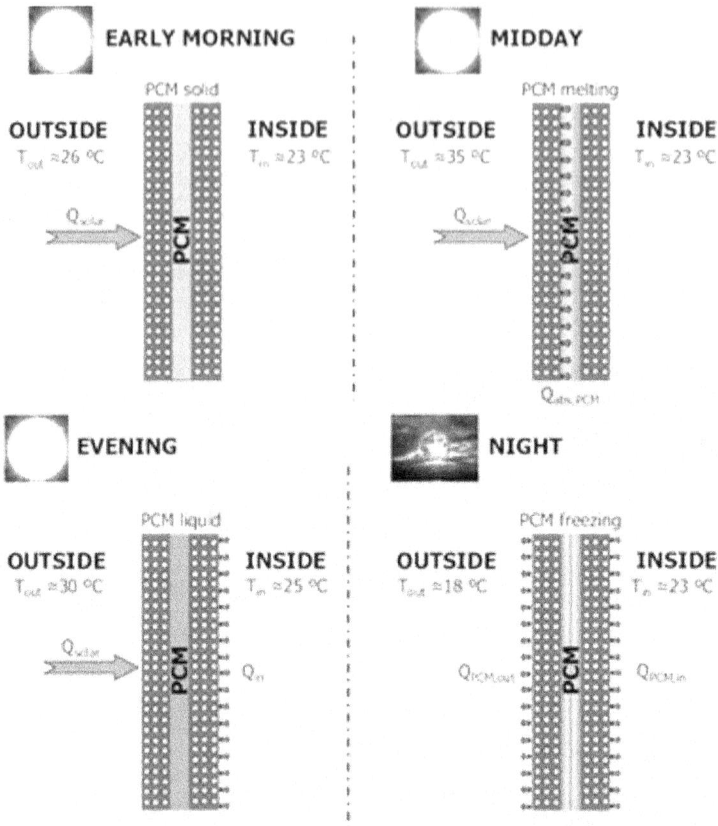

FIGURE 2.8 Schematic showing the working principle of a PCM-embedded wall.[23]

exchange. The strengths and weaknesses of employing PCMs in building envelopes are as follows:

Strength

- Enhances thermal capacity and thermal inertia of building envelops such as Trombe walls and lightweight walls.
- Decreases the building peak loads, which helps in shifting the peak load demands and also maintains constant power supply, especially when the major contribution of energy is from renewable energy sources.
- Decreases the building heating and cooling load during winter and summer seasons. This is due to the PCM's ability to absorb solar insolation and transform it into heat. Further, this stored heat in PCM helps in maintaining the required thermal comfort level during winter and summer with minimum energy consumption.
- PCMs are available with a wide range of phase transition temperatures. Therefore, they have a significant capacity for regulating heat that enables the building's adaptation to various external climatic conditions.

- At the end of their useful lifespan, PCM components can be easily disassembled and are environmentally friendly also.
- Buildings with PCMs installed can have stronger and more durable construction. This is owing to PCMs' ability to control the building envelope's temperature fluctuations, which in turn lowers the rate of material deterioration due to chemical interactions such as reinforcement corrosion, sulfate attack, and alkali–silicon reactions.[24,25]

Weakness

- The most common organic paraffin PCMs employed in building components suffer from some major issues like flammability, leakage during melting, and thermal expansion.
- PCMs encounter hysteresis and super-cooling phenomena during the phase transition, which affect their capacity to store and release the heat.
- PCMs' poor thermal conductivity prevents the accumulated heat from being released in a timely manner. Buildings using PCMs need to have better long-term reliability and durability.
- The temperature of the PCM will rise quickly once it has completely melted. Thus, the PCM-embedded envelope is vulnerable to overheating on hot days due to the heat from the discharge process being transported into the room.
- It is impossible to control the energy storage trigger mechanism in PCMs. Thus, it might be challenging to find PCMs that perform well in both summer and winter.[26,27]

2.2.2.1 Classification of PCM

PCMs are classified into different categories based on their composition, that is, organic, inorganic, and eutectic, as shown in Figure 2.9. Further, these are again subdivided into various groups based on their chemical composition, as depicted in Figure 2.9.[28]

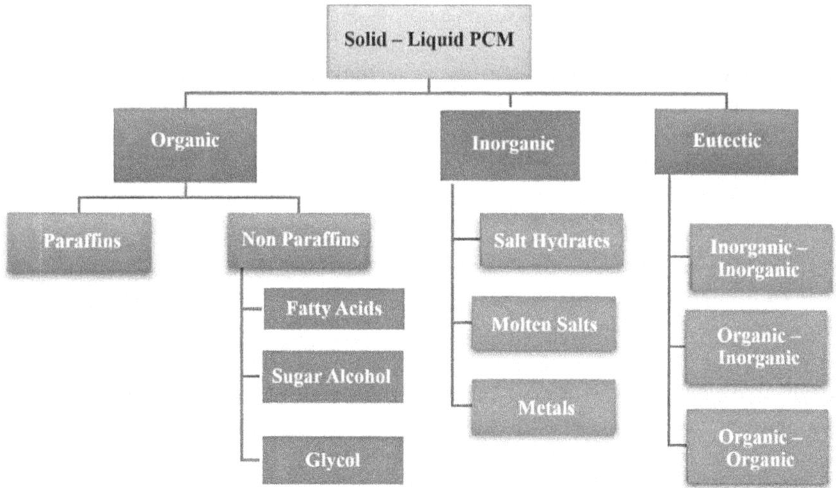

FIGURE 2.9 Classification of solid–liquid PCMs.[28]

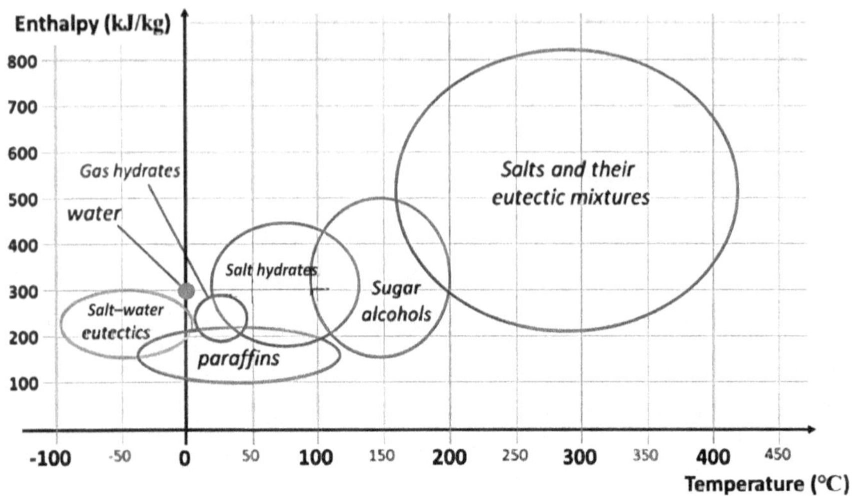

FIGURE 2.10 Schematic showing enthalpies of different commonly available PCMs.[29]

Identifying each PCM's operational range and considering its thermal capacity at the same time is crucial for meeting the required demand. Figure 2.10 demonstrates the enthalpies as a function of melting temperature for commonly used PCMs.

2.2.2.1.1 Organic PCMs

Generally, there are two categories of organic PCMs, that is, paraffin and non-paraffin. Paraffin-based PCMs emerge as the most widely utilized PCMs in thermal storage applications due to their phase transition temperature range of 0–100°C. They also possess additional qualities that make them a viable choice for TES applications. Their chemical properties are better stability, higher energy storage density during phase transitions, non-corrosive and non-toxic qualities, solidification without supercooling, and a high nucleation rate. They are limited to particular applications due to their low density, low thermal conductivity, and inability to operate above 150°C. Further, due to the high flammability, poor thermal conductivity, and variable levels of toxicity, fatty acids, esters, and alcohols that are under the group of non-paraffins, these PCMs are often not used in building cooling or heating applications.

2.2.2.1.2 Inorganic PCMs

Inorganic PCMs are generally salt hydrates and metals. Metallic PCMs are not practical to be employed in building heating or cooling applications due to their heavy weight and inadequate temperature range. Compared to organic PCMs, they have better energy density and efficiency and are also found with the advantages of a wide phase transition temperature range, greater latent heat, and thermal conductivity. In addition, they are less expensive, non-toxic, and non-flammable organic PCMs. However, the demerits of lower thermal stability, supercooling, corrosiveness, decomposition, and phase segregation dominate their merits and hinder their usage in building applications.

2.2.2.1.3 Eutectics

A eutectic is a mixture of two or more components that has a melting point lower than each component alone, that is, organic, inorganic, or both. During the crystallization state, a combination of components is produced that can behave as a single entity, and each constituent can freeze and melt independently. The components of the eutectic produce a very homogeneous crystal mixture as they freeze, and there is no distinction between the two components while both PCMs melt simultaneously. Eutectics have the major benefit of allowing the melting points of their PCMs to be changed by varying the weight ratios of their components. Their best qualities include not being concerned with phase segregation and super-cooling, which results in an increase in usage in building applications. Commercially available PCMs in the market with phase transition temperature and latent heat capacity are depicted in Figure 2.11.

2.2.2.2 PCM Selection Criteria for TES

A suitable PCM must be chosen in order to control the indoor thermal conditions of the buildings. Not all the available PCMs can be utilized as storage material, especially in building thermal comfort applications. The thermal characteristics of PCMs vary according to their chemical composition. Therefore, to attain the required thermal comfort levels in varied climatic conditions and seasons, it is essential to select the proper PCM with a suitable phase transition range and energy storage capacity. It can be challenging to choose a PCM with the required thermophysical characteristics for a specific TES application. Hence, the optimum characteristics required for building applications should be considered while selecting the PCM. The optimal PCMs for building energy storage should have the desired thermodynamic, kinetic, chemical, technological, and economic properties,[30] as shown

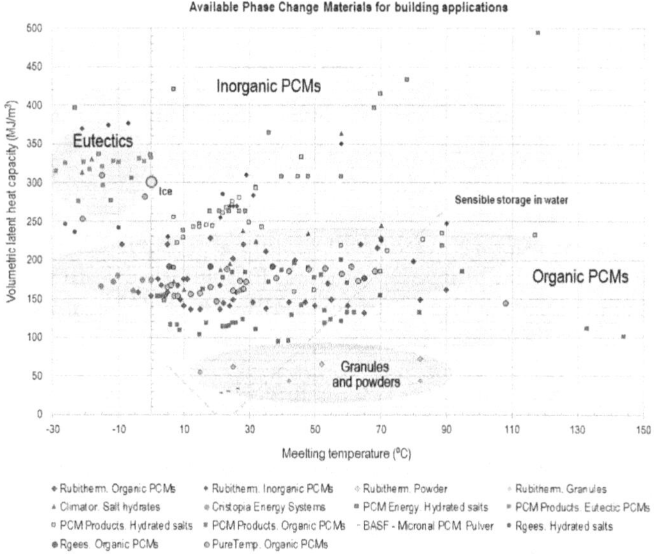

FIGURE 2.11 Schematic of different PCMs available in the market with phase transition temperatures and latent heat capacities for building applications.[21]

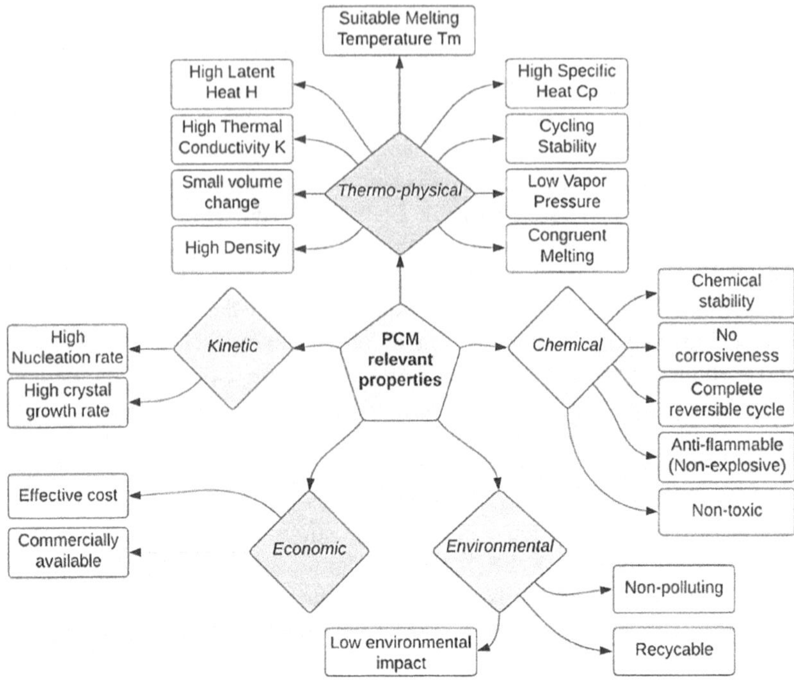

FIGURE 2.12 PCM selection requirements for building applications.

in Figure 2.12. Further, the procedure to be followed for choosing the specific PCM based on the assessments of the materials for building applications is shown in the flow chart as Figure 2.13.

When choosing PCMs for building thermal comfort applications, the operating temperature range of the PCMs is another crucial factor to be considered. The PCMs' operating temperature range has to be in conformity with the regional climate, where they are located in the buildings, or the sort of system they are employed in. Building thermal comfort applications allow the usage of PCMs with a phase transition nearer to the range of human comfort temperatures (18–30°C).

There are three different PCM temperature ranges that have been recommended for usage in buildings: up to 21°C for building cooling, 22–28°C for human thermal comfort application, and finally, 29–60°C for hot water requirements in buildings.

2.2.2.3 Benefits of PCM-Based TES for Buildings

The PCM-based TES systems offer many benefits for buildings. This technique is the most effective way of accumulating thermal energy in building components or elements. The following sections show some of the advantages.

2.2.2.3.1 Enhancing Energy Management and Utilization Efficacy

This technology can assist us in increasing energy efficiency by conserving energy derived from numerous resources that would otherwise be squandered, such as

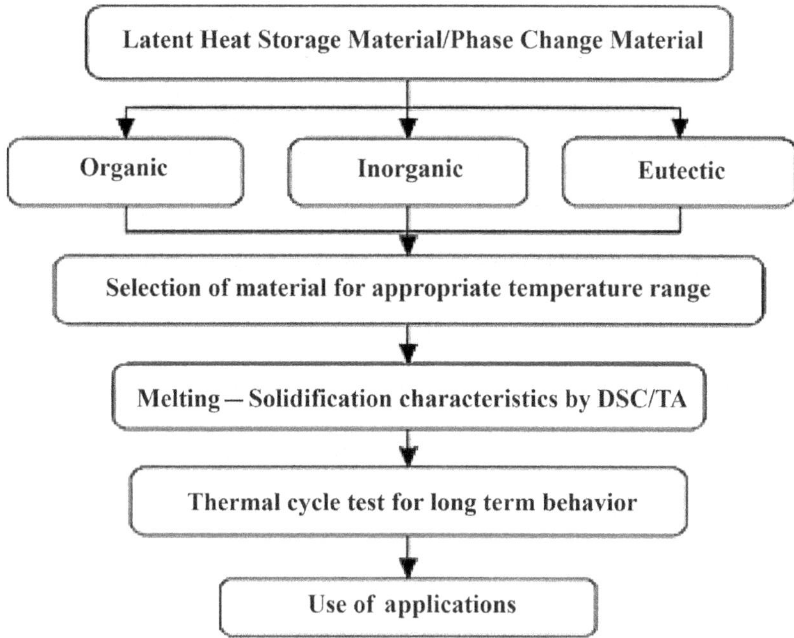

FIGURE 2.13 Flow chart showing the procedure for assessing PCMs in building applications.

heating and cooling systems utilized in a variety of applications, waste heat from equipment and processes, renewable energy, and so on.

2.2.2.3.2 Energy Stored for Later Use

Energy can be stored using this method for either short- or long-term consumption in the future. Energy that is accumulated throughout the summer and utilized during the winter is an example of long-term storage.

2.2.2.3.3 Shifting the Heating and Cooling Demand

With the help of this technology, it is also possible to adjust the heating and cooling load as needed, thereby balancing the needs for energy supply and demand. As a result, it may be possible to lower operating costs for energy generation, transmission, and distribution, which leads to enhancing system performance. Therefore, this technology may be advantageous to both the supplier and the customer.

2.2.2.3.4 Flexibility

This technology is flexible and can be utilized in buildings as either an active or passive technology. It enables buildings to perform an automatic thermal exchange, allowing heat to be captured during the day and released at night to maintain interior comfort levels, or it can act as an insulator to absorb excess heat during the day.

It enables more effective integration of solar systems and other renewable energy technologies into electrical systems. Different organic and inorganic PCMs used in building components are given in Tables 2.2 and 2.3, respectively.

TABLE 2.2
Different Organic PCMs for Building Comfort Applications[31]

Group		Compound	Melting temp. (°C)	Heat of fusion (kJ/Kg)	C_p (kJ/Kg K)	K (W/m K)	ϱ(kg/m³)
Paraffin	**Cooling**	Paraffin C14	4.5	165	–	–	–
		Paraffin C15-C16	8	153	2.2 (s.)	–	–
		Polyglycol E400	8	99.6	–	0.187 (liq. 38.6°C) 0.185 (liq. 69.9°C)	1125 (liq. 25°C) 1228 (s. 3°C)
		Dimethyl-sulfoxide (DMS)	16.5	85.7	–	–	1009 (s./liq.)
		Paraffin C17-C18	20–22	152	2.2 (s.)	–	–
		Polyglycol E600	22	127.2	–	0.189 (liq. 38.6°C) 0.187 (liq.67.0°C)	1126 (liq. 25°C) 1232 (s, 4°C)
		Paraffin C13-C24	22–24	189	2.1 (s.)	0.21 (s.)	760 (liq. 70°C) 900 (s. 20°C)
		1-Dodecanol	26	200	–	–	–
		Paraffin C18	28 27.5	244 243.5	2.2 (s.)	0.148 (liq. 67.0°C) 0.15 (s.)	774 (liq. 70°C) 814 (s. 20°C)
	Heating	Paraffin C20-C33	48–50	189	–	0.21 (s.)	769 (liq. 70°C) 912 (s. 20°C)
		Paraffin C22-C45	58–60	189	2.4 (s.)	0.21 (s.)	795 (liq. 70°C) 920 (s. 20°C)
		Paraffin wax	64	173.6 266	–	0.167 (liq. 63.5°C) 0.346 (s. 33.6°C) 0.339 (s. 45.7°C)	790 (liq. 65°C) 916 (s. 24°C)
		Polyglycol E6000	66	190	–	–	1085 (liq. 70°C) 1212 (s. 25°C)
		Paraffin C21-C50	66–68	189	–	0.21 (s.)	830 (liq. 70°C) 930 (s. 20°C)

(Continued)

TABLE 2.2 (Continued)
Different Organic PCMs for Building Comfort Applications[31]

Group		Compound	Melting temp. (°C)	Heat of fusion (kJ/Kg)	C_p (kJ/Kg K)	K (W/m K)	ϱ(kg/m³)
		1-Tetradecanol	38	205	–	–	–
		Paraffin C16-C28	48–50	189	–	0.21 (s.)	765 (liq. 70°C) 910 (s. 20°C)
		Biphenyl	71	119.2	–	–	991 (liq. 73°C) 1166 (s. 24°C)
		Propionamide	79	168.2	–	–	–
		Naphthalene	80	147.7	2.8 (s.)	0.132 (liq. 83.8°C) 0.341 (s. 49.9°C) 0.310 (s. 66.6°C)	976 (liq. 84°C) 1145 (s. 20°C)
Fatty Acids	Cooling	Propyl palmitate	10 16–19	186	–	–	–
		Caprylic acid	16	148.5	–	0.149 (liq. 38.6°C) 0.145(liq. 67.7°C)	901 (liq. 30°C) 862 (liq. 80°C)
			16.3	149		0.148 (liq. 20°C)	981 (s. 13°C) 1033 (s. 10°C)
		Isopropyl palmitate	11	95–100	–	–	–
		Capric-lauric acid + Pentadecane (90%–10%)	13.3	142.2	–	–	–
		Isopropyl stearate	14–18	140.142	–	–	–
		Capric-lauric acid (65%–35%)	18 17–21	148 143	–	–	–
		Butyl stearate	19	140	–	–	–
		Capric-lauric acid (45%–55%)	21	143	–	–	–
		Dimethyl sebacate	21	120–135	–	–	–

(Continued)

TABLE 2.2 (Continued)
Different Organic PCMs for Building Comfort Applications[31]

Group	Compound	Melting temp. (°C)	Heat of fusion (kJ/Kg)	C_p (kJ/Kg K)	K (W/m K)	ϱ(kg/m³)
	Octadecyl 3-mencaptopropylate	21	143	–	–	–
	Mystriric acid-capric acid (34%–66%)	24	147.7	–	0.164 (liq. 39.1°C)	888 (liq. 25°C)
					0.154 (liq. 61.2°C)	1018 (s.)
	Octadecylthioglycate	26	90	–	–	–
Heating	Vinyl stearate	27–29	122	–	–	–
	Lauric acid	42–44	178	2.3 (liq.)	0.147 (liq. 50°C)	862 (liq. 60°C)
		44	177.4	1.7 (s.)		870 (liq. 50°C)
	Myristic acid	49–51	204.5	2.4 (liq.)	–	861 (liq. 55°C)
		54	187			844 (liq. 80°C)
		58	186.6	1.7 (s.)		990 (s. 24°C)
	Palmitic acid	64	185.4	2.8 (liq.)	0.162(liq. 68.4°C)	850 (liq. 65°C)
		61	203.4	1.9 (s.)	0.159(liq. 80.1°C)	847 (liq. 80°C)
		63	187		0.165 (liq. 80°C)	989 (s. 24°C)
	Stearic acid	69	202.5	2.2 (liq.)	0.172 (liq. 70°C)	848 (liq. 70°C)
		60–61	186.5	1.6 (s.)		
		70	203			965 (s. 24°C)

TABLE 2.3

Different Inorganic PCMs for Building Heating and Cooling Applications[31]

Application	Compound	Type of melting*	Melting temperature (°C)	Heat of fusion (kJ/Kg)	C_p (kJ/Kg K)**	K (W/m K)**	ρ (g/m³)
Cooling	$LiClO_3$–$3H_2O$	C	8.1	253	1.35 (s.)	–	1720
	$ZnCl_2$–$3H_2O$	–	10	–	–	–	–
	K_2HPO_4–$6H_2O$	–	13	–	–	–	–
	NaOH–7/2CO_2	–	15	–	–	–	–
	Na_2CrO_4–$10H_2O$	–	18	–	1.31 (s.)	–	–
	KF–$4H_2O$	C	18.5	231	1.84 (s.)	–	1447 (liq.) 1445 (s.)
	$Mn(NO_3)_2$–$6H_2O$	–	25.8	125.9	–	2.34(s.)	1800 (liq.)
	$CaCl_2$–$6H_2O$	i	29	188.34	1.43 (s.) 2.31 (liq.)	0.54 (liq.) 1.09 (s.)	1562 (liq.) 1802 (s.)
Heating	$LiNO_3$–$3H_2O$	–	30	296	–	0.58 (liq.) 1.37 (s.)	1780 (liq.) 2140 (s.)
	Na_2CO_3–$10H_2O$	i	33	247	1.79 (s.)	–	1442 (s.)
	Na_2SO_4–$10H_2O$	i	32.4	251	1.44 (s.)	0.5–0.7	1420 (s.)
	$CaBr_2$–$6H_2O$	i	34	115.5	–	–	1956 (liq.) 2194 (s.)
	$K(CH_3COO)$ 3/2H_2O	–	42	–	–	–	–
	K_3PO_4 $7H_2O$	–	45	–	–	–	–
	$Zn(NO_3)_2$–$6H_2O$	C	36.4	147	1.34 (s.) 2.26 (liq.)	–	2065 (s.)
	$Ca(NO_3)_2$–$4H_2O$	i	42.7	–	–	–	–
	Na_2HPO_4–$7H_2O$	i	48	281	1.70 (s.) 1.95 (liq.)	0.514 (s.) 0.476 (liq.)	1520 (s.) 1442 (liq.)

(Continued)

TABLE 2.3 (Continued)
Different Inorganic PCMs for Building Heating and Cooling Applications[31]

Application	Compound	Type of melting*	Melting temperature (°C)	Heat of fusion (kJ/Kg)	C_p (kJ/Kg K)**	K (W/m K)**	ρ (g/m³)
	$Na_2S_2O_3$–$5H_2O$	i	48–49	209.9	3.83 (liq.)	–	1666
			48	201–206			
	$Zn(NO_3)_2$–$2H_2O$	C	54	–	–	–	–
	NaOH–H_2O	C	58	–	2.18 (s.)	–	–
	$Na(CH_3COO)$–$3H_2O$	–	58	226	–	–	1450
				267	2.79–4.57	0.63 (s.)	1280
	$Cd(NO_3)_2$–$4H_2O$	C	59.5	–	–	–	–
	$Fe(NO_3)_2$–$6H_2O$	–	60	–	–	–	–
	$Na_2B_4O_7$–$10H_2O$	i	68.1	–	–	–	–
	Na_3PO_4–$12H_2O$	i	69	–	–		–
	$Ba(OH)_2$–$8H_2O$	i	78	265.7	–	0.653 (liq.)	1937 (liq.)
						0.678 (s.)	2070 (s.)
	$KAl(SO_4)_2$–$12H_2O$	i	80	–	–	–	–
	$Al_2(SO_4)_3$–$18H_2O$	–	85.8	–	–	–	–
	$Al(NO_3)_3$–$8H_2O$	–	88	–	–	–	–

* c = congruent, i = incongruent

s = Solid, l = liquid

2.2.2.4 Current Problems in Embedding PCMs in Building Envelopes

PCMs are expected to provide more comfortable temperatures and energy savings. However, the current problems such as low thermal conductivity, phase segregation, supercooling, thermal stability, flammability, leakage, and seasonal adaptability[32] as shown in Figure 2.14 have a significant impact on the utilization and popularity of PCM in buildings.

2.2.2.4.1 Low Thermal Conductivity

Most of the organic PCMs have poor thermal conductivities (usually between 0.2 and 0.3 W/m K), which affects the rates of heat storage and release. In order to improve the thermal conductivity of PCMs, additives with high conductivity can be added to the PCMs to form a composite PCM, and another example is the encapsulation of PCM to enhance heat transfer. The additives include graphite powder, expanded graphite, metal particles, carbon fibers, and metal foams. The composite PCM's thermal conductivity and thermal stability are both improved by these additives.

2.2.2.4.2 Phase Segregation

The phase segregation process generally occurs in salt hydrates and industrially produces organic PCMs that are not pure, or eutectic PCMs. The storage capacity and heat transfer efficiency of salt hydrates will decrease as they are irreversibly split into two or more layers. This issue can be resolved by adding thickening or gelling material to the PCMs.

2.2.2.4.3 Supercooling

In inorganic salt hydrates and pure PCMs, keeping the PCM in a liquid form at a temperature lower than the solidification temperature is known as supercooling. The

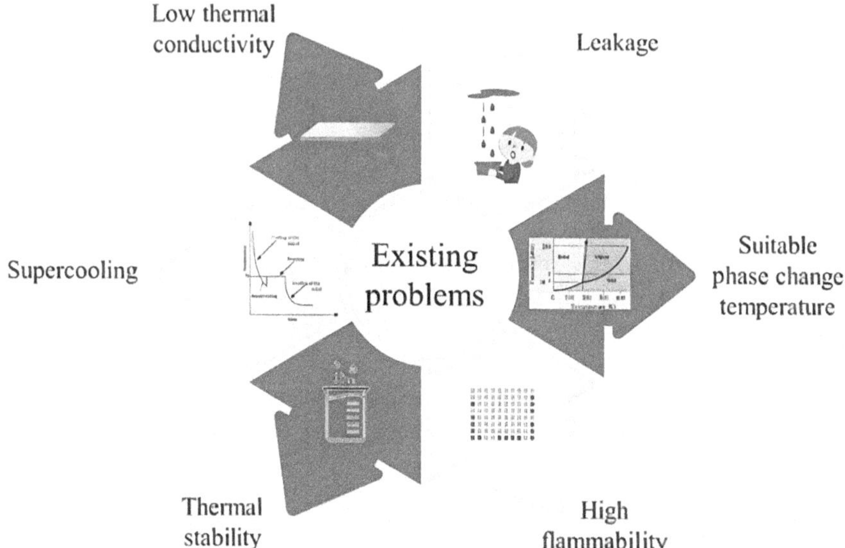

FIGURE 2.14 Schematic representing current issues in PCMs for adoption in buildings.[32]

latent heat release will be prevented by a significant temperature differential of the PCM. The supercooling effect can be reduced by adjusting the cooling rate, modifying the container's properties, or adding nucleate additives.

2.2.2.4.4 *Adaptability*

For PCMs, the issue of "season adaptation" possess a significant challenge especially when buildings are situated in countries with complex climate conditions, such as extreme summer and winter regions. In order to create a comfortable interior and maximize energy savings, the phase transition temperature of the PCM depends on the climate, thermal comfort level, and location of the PCM layer in the building envelope. The issue of seasonal adaptation has been addressed by the use of multiple PCMs (with different phase change temperatures) in building envelopes.

2.2.2.4.5 *Thermal Stability*

One important criterion for evaluating PCMs is thermal stability. In order to assess thermal stability, differential scanning calorimetry (DSC) experiments and accelerated melting/solidification cycles are used. Moreover, tests using Fourier transform infrared spectrometry (FT-IR) and thermal gravimetric analysis (TGA) are commonly utilized to verify chemical stability in addition to thermophysical properties. The thermal stability of PCMs can be improved through the shape stabilization process.

2.2.2.4.6 *Flammability*

Organic PCMs are widely utilized to control the inside temperature of buildings. However, the greatest obstacle to their widespread implementation is their flammability. Chemical modifications of the material can improve the flame-retardant characteristics of organic PCMs by adding synergistic dioxins.

2.2.2.4.7 *Leakage of PCMs*

The incorporation of PCMs in building applications is significantly hindered by PCM leakage during phase change. Leakage of the PCM will reduce storage capacity in addition to corroding the building's exterior and envelopes. Composite PCMs with shape stabilization could effectively address the leakage issue during the phase change process. However, because of the incompatibility between the PCM and the building materials as well as the structural stress, the leakage problem remains extremely significant.

2.2.2.5 PCM Embedding Techniques in Buildings

In order to avoid leakage or mixing with foreign particles and surrounding environment, to enhance heat transfer, and to hold the PCM during melting, a proper incorporation technique is required. A suitable incorporation technique is extremely essential for PCMs' thermal performance. There are five incorporation techniques that have been used in building applications, as shown in Figure 2.15.[33,34]

2.2.2.5.1 *Direct Mixing*

In direct mixing, the PCM is directly mixed with the building construction materials to form a composite construction, and this is only limited to passive building applications. It is considered the simplest and most affordable approach to incorporating

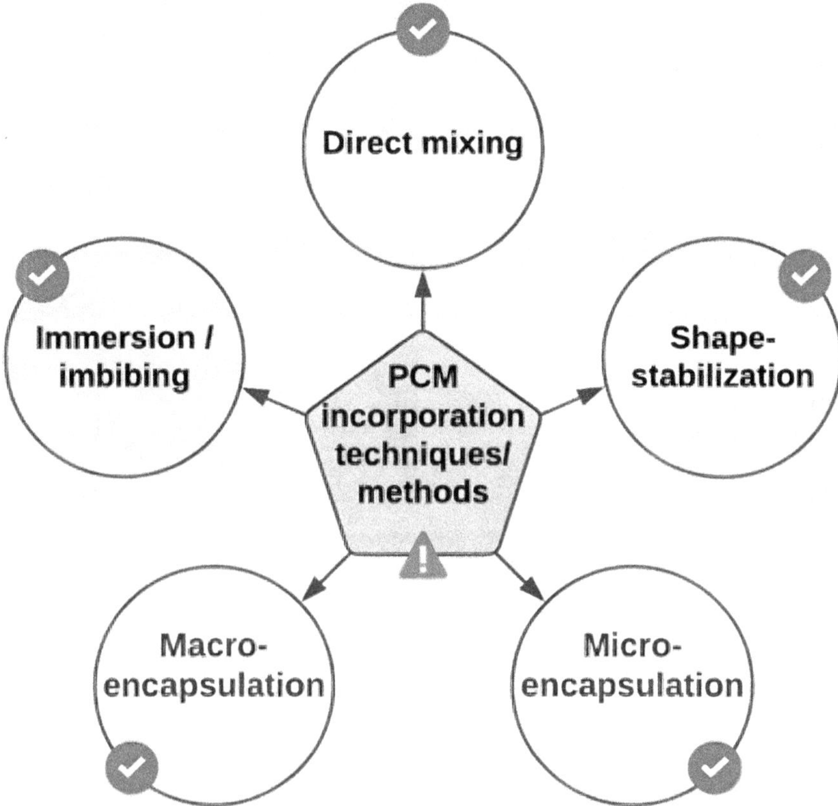

FIGURE 2.15 Different PCM incorporation techniques in building envelopes.

PCMs. However, there are certain drawbacks, that is, leakage, reduction in the strength of materials, and exposure to chemical reactions in the surroundings.

2.2.2.5.2 Immersion

The method utilized to increase the thermal inertia of building materials that have already been manufactured is immersion. In the immersion method, the building material is dipped into the liquid PCM, and that liquid PCM is absorbed by the materials through capillary action to form a composite PCM. The interaction between the PCM and the hydration products of the building element has an adverse impact on both durability and mechanical qualities, which is regrettably one major drawback. In general, the impregnation method is appropriate for incorporating PCMs into lightweight building materials.

2.2.2.5.3 Macro-Encapsulation

Encapsulation is a technique that produces an exterior shell or coating on the surface of the PCM. The process of encasing a PCM in a shell material having a dimension of greater than 5 mm is known as macro-encapsulation. In macro-encapsulation,

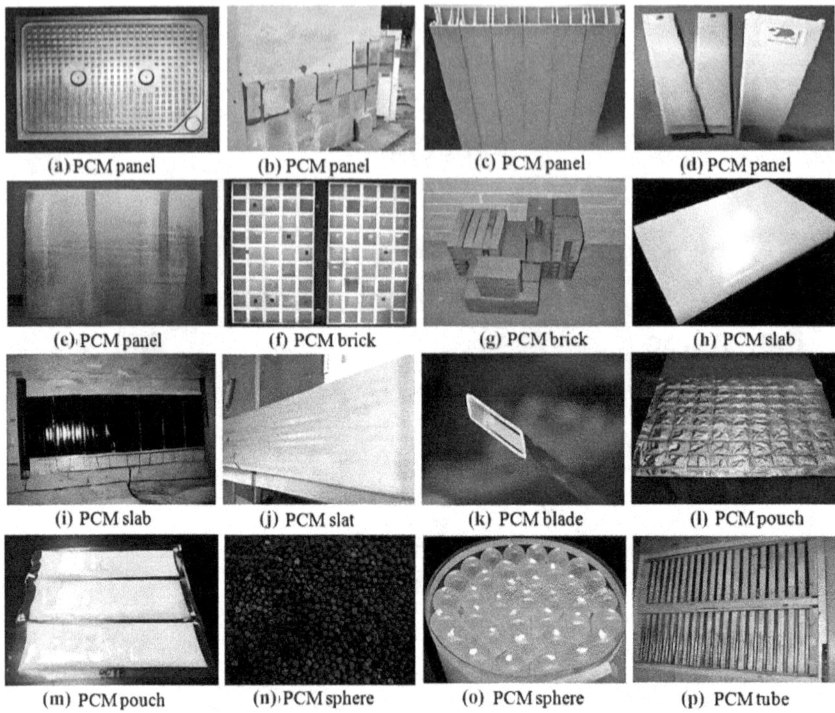

FIGURE 2.16 Photograph of different PCM encapsulation methods used for PCM-embedded buildings.[35]

the PCM is the core of the capsule, which prevents outdoor reactions and hazardous fires. Encapsulating the PCM also enhances the thermal conductivity and surface area based on the geometry and materials utilized. The geometry configuration of a macro-capsule might vary depending on the application, and different PCM encapsulation geometries for building applications are depicted in Figure 2.16. Further, salt hydrates are challenging to encapsulate because of their corrosive nature toward metals and water solubility. In contrast, organic PCMs exhibit insolubility in water and are non-corrosive with metals. Enveloping or encasing the organic PCMs with an inorganic material will lead to a reduction in the risk of flammability.[35]

2.2.2.5.4 Micro-Encapsulation

Small capsules made of synthetic or natural polymers with a diameter ranging from 1 μm to 1000 μm are used in this process to encapsulate PCMs, known as micro-encapsulation. A micro-encapsulated PCM consists of the PCM as the core and a synthetic or natural polymer shell as an envelope. Micro-encapsulation offers numerous benefits that include the augmentation of heat transfer due to an increase in the heat exchange surface area, no leakage concerns, and a reduction in reactivity with surroundings. Micro-capsules can also be manufactured in different shapes such as regular (spherical, tube, and cylinder) and irregular shapes.

2.2.2.5.5 Shape Stabilization

Instead of the PCM being coated with the shell, the PCM is blended or infused directly into the supporting metal. There are two different kinds of supporting material: one is porous, which avoids PCM leakage, while the other uses nano-materials, which enhance PCM thermal characteristics. Pores in porous materials are typically micro- or nanoscale in size. Higher porosity and smaller pores in a porous material show stronger surface tension and capillary action for absorbing PCM, which will lead to the prevention of leakage. The most popular shape-stabilizing processing methods are vacuum impregnation and melting impregnation. The key benefits associated with these approaches are their high specific heat, strong thermal conductivity, and potential to maintain PCM stability during the phase change process without the requirement of a container. Based on the discussions and comparison of all the encapsulation methods, macro- and micro-encapsulation techniques are the most widely employed in building applications.

2.2.3 Thermochemical Storage in Buildings

Thermochemical energy storage (TCES) systems rely on highly reversible chemical processes in which a heat source supplies heat to convert an AB chemical substance into products (A and B). Thereafter, the A and B products are separated and stored at the surrounding temperatures. As shown in Figure 2.17, the separated products (A and B) are mixed to produce the previous product (AB), which results in a reversible chemical reaction that releases a considerable amount of heat. A rearrangement of the chemical bonds of the molecules involved in a reaction (dissociation and recombination) characterizes a chemical reaction. With the help of a catalyst, endothermic reactions can store energy, and reverse exothermic reactions can release energy. As compared to both sensible and latent TES technologies, TCES offers the distinguishing characteristics of combined higher energy storage and low heat losses and is currently deemed to be the most promising alternative.[36] Another appealing feature of TCES is the ability to store heat energy at room temperature as long as required, without suffering from heat losses resulting from reversible chemical interactions. Under the same volume of storage, TCES has an energy storage density that is roughly 8–10 times higher than that of the sensible storage system and twice as high as that of latent storage materials. TCES has gained considerable interest as a technology for seasonal energy storage because of those benefits.[37]

The phenomenon of holding or capturing a gas or a vapor by an adsorptive material in a condensed state (solid or liquid) through less intense physical reactions is known as sorption storage. Further, TCES materials are divided into different categories based on the process involved, that is, physical sorption (zeolite and silica gel) and chemical reaction. For building applications, the research and development in TCES systems are still in the early stage. The main obstacles to deployment are the high cost of materials, inadequate heat and mass transport capabilities, improper working temperature, ineffective discharge power for building applications, delayed kinetic reactions, and low/moderate efficacy of storage material densities. The primary focus of material research for building applications is long-term solar energy storage.

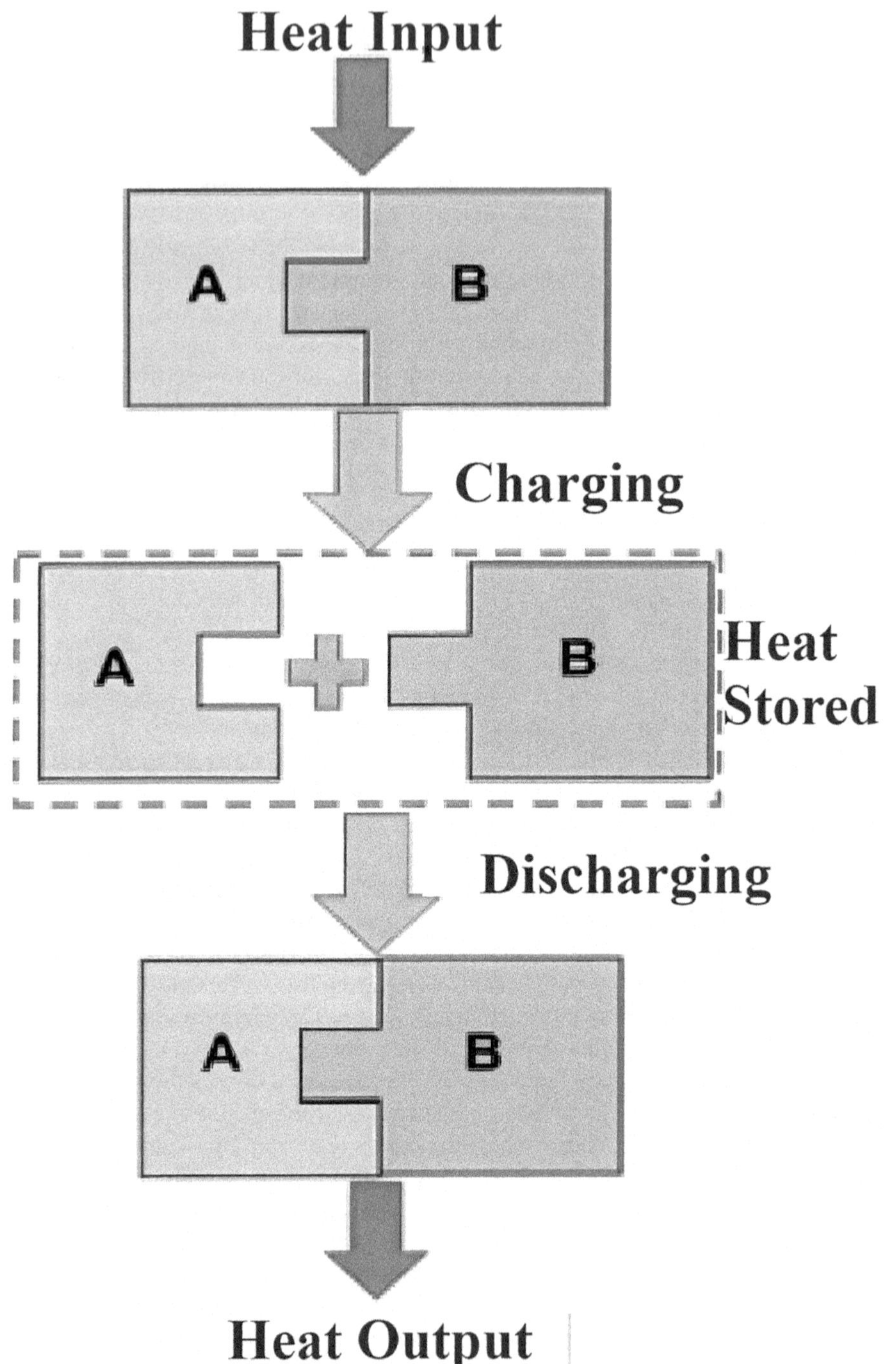

FIGURE 2.17 Schematic of the heat storage and release process in TCES[38]

2.3 TES METHODS AND THEIR APPLICATIONS IN BUILDINGS

The various types of TES materials discussed in the previous section can be implemented in buildings using various methods, as presented in Figure 2.18. The selection of the TES method in a building application depends upon various factors of building, location, climate, energy source, material availability, and economic viability. However, this section is limited to the discussion of various TES methods in building applications. As shown in Figure 2.18, the TES methods for building applications are broadly divided into passive and active methods. In passive methods of TES, the temperature difference between the storage material and surrounding establishes the driving force during the charging and discharging process. On the other hand, pumps or fans are used in the case of active methods. The combination of active and passive methods has also been frequently used for TES in buildings, wherein charging is carried out using an active method with discharging using a passive method and vice-versa. The utility of the active method is found more frequently as compared to the passive methods. The various methods of TES in buildings and their advancement are discussed in detail in the present section.

2.3.1 Passive Storage Techniques

The purpose of a building envelope is not only to provide structural support but also to provide protection from outdoor weather conditions. Passive methods of TES include the smart utilization of storage material under such building envelopes. In these methods, the indoor temperature must vary with time to allow heat transfer during storage and its release. Therefore, a thorough study of the indoor temperature variation characteristics is essential for its effective design and efficient functioning. The implementation of adequate passive storage techniques in buildings may reduce thermal energy consumption by 7% to 22%.[40,41] The thermal energy can be stored with the passive method in the form of latent heat using PCMs as well as in the form of sensible heat using high thermal mass. The thermal performance of a PCM is majorly dependent on the specific latent heat and specific heat capacity. The discussion about material properties is presented under the previous section, and details of storing techniques are discussed in the following subsections.

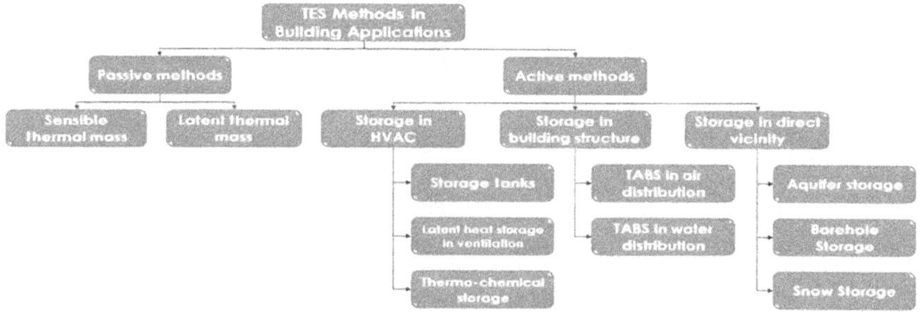

FIGURE 2.18 Classification of TES methods in building applications.[39]

2.3.1.1 Sensible Thermal Mass

The term 'thermal mass' refers to the construction material of the building itself having the ability to absorb and release heat for heating/cooling applications as per the availability and requirement. Such materials respond as a heat sink during the warm period and a heat source during the cool period. The TES behavior of these materials is influenced by the specific heat capacity, thermal conductivity, thermal diffusivity, and density of the material. The sensible thermal mass with the concept of a nearly zero-energy building (NZEB) is found suitable to accomplish heating and cooling in climatic conditions of a desert. However, comparatively heavy construction is recommended in such conditions to reduce the peak load with a constant value of the total heating and cooling load.[42] Commercial buildings of large occupancy such as schools and offices have a higher potential for energy savings by the use of heavy construction materials. It is also found to be an effective method to reduce temperature swings and increase thermal comfort.

The sensible thermal mass is basically categorized as high-thermal-mass materials and solar walls. High-thermal-mass materials are used to store a larger amount of heat, which provides a higher value of thermal inertia to the building components. Stones, earth materials, concrete, and alveolar bricks are considered high-thermal-mass materials. A detailed discussion about high-thermal-mass material is reported by Rempell and Rempell.[43] Solar walls are specifically designed building walls that provide a feasible approach to effectively utilizing the directional flow of heat within the building. Trombe walls and solar water walls are two different types of solar walls. A Trombe wall is made up of high-thermal-mass material covered by a glass cover, and an air channel is provided between them. The concept involved in a Trombe wall can be understood with the help of various configurations presented in Figure 2.19, and a detailed review of the same is reported by Saadatian et al.[44]

2.3.1.2 Latent Thermal Mass

The heat associated with a material during its phase change is referred to as latent heat. Mostly, the solid–liquid phase is used for storage due to ease of control, constant temperature process, and large heat storage capacity. The materials used to store the latent heat are known as PCMs, which store and release heat during the melting and solidification processes, respectively. Shape stabilization, stable composite form,

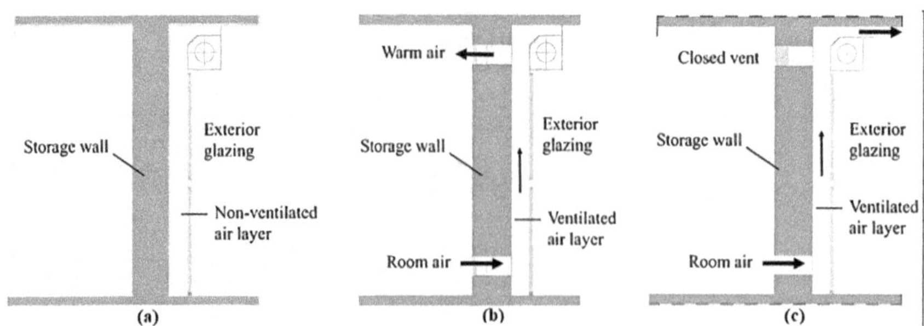

FIGURE 2.19 Various configurations of Trombe walls (a) without ventilation, (b) with air circulation for winter mode, (c) with cross ventilation for summer mode.[45]

direct incorporation, immersion, vacuum impregnation, and encapsulation are the different ways to incorporate PCMs in the building structure.

In direct incorporation, the PCM in liquid or powder form is directly added into a building material such as concrete, gypsum, or mortar. In the immersion method, the porous building structure material is subjected to immersion in a container filled with liquid PCM for its absorption through capillary action, and then, the obtained material is applied to the building structure. In contrast, the air is evacuated initially from porous aggregates using a vacuum pump and subjected to socking with a liquid PCM under a vacuum in vacuum impregnation. The obtained aggregate is further used for the building structure. The mixing of high-density polyethylene with a PCM at high temperatures to increase its thermal conductivity and use it after solidification for TES in buildings is called shape stabilization. Moreover, once a mass fraction of high-density polyethylene is optimized based on the requirement, the method is referred to as a stable composite.

The use of a micro-encapsulated PCM in the concrete wall of cubicles is also found favorable in terms of avoiding excess temperature and temperature swings.[46] However, the addition of a PCM with concrete can deliver a positive outcome up to a certain limit, as concrete itself has a large thermal mass. Moreover, its addition of lightweight material may result in comparatively better performance.[47] Figure 2.20 shows the variation in indoor air temperature (T_i^{free}) when sinusoidal heat input is given to maintain the room temperature at 22°C for a variable ambient temperature (T_e) for light as well as heavy construction material.

A significant amount of energy consumption for cooling can be reduced by the use of PCM ceiling panels in buildings.[48] Alternatively, the use of PCM under building walls can effectively reduce the heating load by 6% to 15% and 7% to 20% with a maximum comfort approach and minimum energy approach, respectively. However, geographical location and climatic conditions will be the influencing factors in such decisions.[49] In order to assess the effectiveness of PCM integration under wallboards, Evola et al.[50] suggested several indicators such as thermal discomfort intensity, storage efficiency, thermal discomfort frequency, and frequency of activation.

Conclusively, it can be said that the increase in sensible or latent thermal mass results in decreased heating/cooling load along with higher stability in indoor temperature.

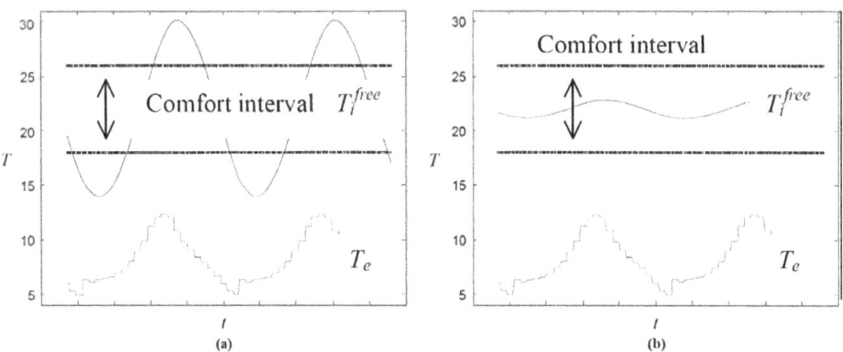

FIGURE 2.20 Variation in indoor air temperature (T_i^{free}) with a sinusoidal heat gain in the case of (a) light construction material or (b) heavy construction material.[39]

However, the common outcomes obtained in most of the studies revealed that the magnitude of energy saving using passive TES is lesser compared to that for other TES techniques. The outcomes of latent thermal mass are significantly dependent on the properties of the selected PCM under the given operating conditions.

2.3.2 Active Storage Techniques

In active storage techniques, some external source of energy is deliberately added by means of pumps or fans for the effective working of TES systems. As shown in Figure 2.18, the thermal energy can be stored using the active method in three different ways: storage in the heating, ventilation, and air conditioning (HVAC) system; building structure; and direct vicinity of the building. With HVAC systems, the TES unit is integrated with the HVAC system and it is used for heating as well as cooling purposes. In the case of storage in a building structure, the thermal energy is stored in walls, floors, or ceilings of the building, and the existing air/water distribution system is used to transfer it for required applications. Developing boreholes, aquifers, snow storage, and underground tanks for TES are categorized as storage in the direct vicinity of the building. Various positions/ways to integrate different storage methods are shown in Figure 2.21. A detailed discussion of these active methods of TES is presented in the following subsections.

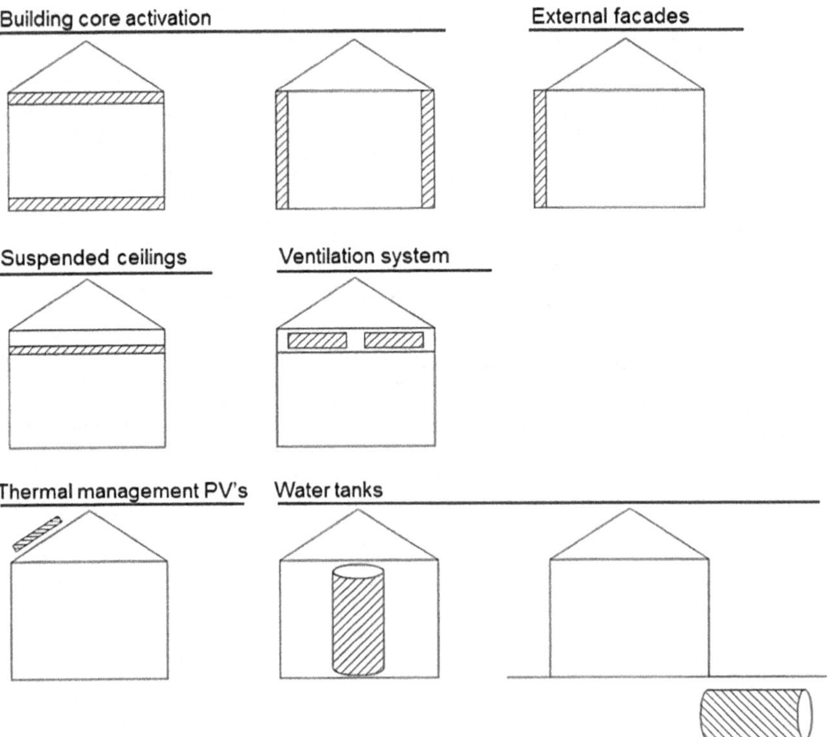

FIGURE 2.21 Different ways to integrate TES in buildings.[51]

2.3.2.1 Storage in Heating, Ventilation, and Air Conditioning Systems

The purpose of integrating HVAC systems in buildings is to maintain indoor thermal comfort, and its overall energy consumption can be reduced by the integration of TES. Most common technologies such as domestic water heaters and solar collectors with storage tanks are considered under this category of TES. In order to provide more clarity, these types of TES have been further categorized into storage tanks, latent storage in ventilation, and thermochemical storage.

2.3.2.1.1 Storage Tanks

A storage tank is the most common and widely used technique of TES. Depending upon the type of storage material, the thermal energy can be stored in the form of latent as well as sensible heat under the storage. Water is one of the most common media of TES due to its favorable properties, easy availability, and almost negligible cost.[52] Due to the requirement of large amounts of energy in a short span of time, water storage tanks are usually used in domestic water heaters with solar collectors, gas boilers, and conventional electric heaters. The installation of water storage tanks helps to reduce peak saving and heating power. The thermal performance of such storage tanks majorly depends on their design, and the use of macro-encapsulated PCM has also been reported in favorable outcomes in recent studies.[53–56]

The concentration and placement of PCM modules in hybrid water/PCM storage tanks play a crucial role in their performance. Studies suggest that the positioning of a PCM on the top of the storage tank delivers positive outcomes like longer stability and comparatively reduced size.[57] Moreover, the average energy density also increases due to the presence of PCMs.[58] However, several factors are crucial in the case of solar water heaters having PCMs in storage tanks. Heat losses during the night may increase due to higher temperature differences between the tank and the atmosphere. Therefore, proper insulation measures are a must to prevent such heat losses. Moreover, solar collectors work at reduced efficiency in the morning period due to the comparatively high temperature of the PCM; therefore, the overall performance of a system is negatively impacted.[59] Therefore, a decision regarding the use of PCMs with water in storage tanks depends on several other system-related factors apart from the properties of the PCM.

Various developments have also been reported about the performance of compact storage tanks.[60–62] The major drawback associated with compact storage tanks is a poor heat transfer rate. Therefore, a smooth tube coil is recommended to replace the finned tube coils in such systems. The utility of external heat exchangers may increase the efficiency of the storage tanks.[62] A study based on the simulations carried out using TRNSYS revealed that the replacement of the internal heat exchanger with an external one for a single-family house may reduce the energy saving by 300 to 400 kWh/year.[63] However, the importance of a control strategy for effective functioning as well as energy savings is also pointed out by the authors.

For cooling applications, paraffin waxes, chilled water, ice harvesters, ice storage, and eutectic salt-based storage systems are commonly used. The economic aspect plays a crucial role in the selection of these systems, as the chiller and other auxiliary equipment are its additional requirements. Various researchers have reported the potential and benefits of ice storage for cooling purposes. In general, ice can be stored in three different ways in storage tanks: pure ice, ice slurry, and encapsulated ice. Direct heat transfer with the working fluid is found to be more useful in the case of ice storage.

The favorable outcomes and associated challenges of TES in storage tanks for seasonal storage of solar heat have also been explored by various researchers. This method of TES is popular in European nations. Moreover, the exergy efficiency of the storage tank is also affected by stratification, which is the formation of a thermocline with a high thermal gradient. The higher the stratification (thinner layer of thermocline), the better the performance of the storage tank will be.[64]

2.3.2.1.2 Latent Heat Storage in Ventilation

Thermal energy can also be stored along with ventilation and AC systems of buildings, and PCMs are most used in this method due to their lower specific volume. When integrating the PCM storage tank with the air conditioner system for space heating, the major issue is due to the poor heat transfer rate of the PCM. Hamada and Fukai[65] recommended using a carbon fiber brush in the tank to overcome this issue with additional benefits of space saving and cost reduction.

Several researchers also reported the use of PCMs directly in the ventilation system.[66–69] Integrating the packed bed of a PCM into the ventilation system is a promising approach as is associated with the additional benefit of free cooling. Moreover, PCMs with a higher range of melting points are favorable to sustain a longer cycle. However, such PCMs are not suitable for regions with significant variations in ambient temperature in day and night periods. This is due to the early solidification of PCM and hence the reduced ventilation effects during the night period. In general, PCM cold storage is very sensitive to climatic conditions, building design, cooling load, and the specifications of HVAC systems. The provision of a bypass is necessary with PCM storage systems to handle the situations of lower ambient air temperature than air under PCM storage.[70] Nevertheless, the cold ambient air during the night period may be used to reduce the temperature of PCM cold storage. It will reduce conventional energy consumption and save energy consumption.[71,72]

The application of ice slurries for cold storage and its various factors are reported by Bellas and Tassou.[73] The authors compared the cooling performance of the absorption system with the ice slurry system for a building complex and found that cooling of a building can be carried out with ice slurry at a 50% lower running cost. However, the energy requirement for ice slurry production is found to be more than that for normal ice production. Finally, the authors found the cooling performance of the ice slurry system better for lesser energy consumption by the air distribution system, scope of peak load reduction, and flexibility to shift the ice production process from day to night.

In several applications, absorption cooling using ice is not found effective due to the operating conditions. Such issues arise mostly in applications where cooling is required below 0°C. In such a situation, an aqueous solution of an ammonium salt named clathrate hydrate slurry (CHS) is reported as a feasible option, which can also be used with absorption cooling. CHS can be used as a storage medium and heat transfer fluid. When comparing the performance of the CHS storage system with no storage, water storage, and ice storage for an office building, Hayashi et al. found the lowest annual electricity consumption for the same due to a reduction in pumping power requirement.[74] The outcomes of using micro-encapsulated PCM slurry are also found favorable through initial research and development. This may be due to the high surface area-to-volume ratio and its homogenous nature. However, more research and development is required to offer the same for commercial technologies.

2.3.2.1.3 Thermochemical Storage

In thermochemical energy storage systems, the involvement of heat during reversible chemical reactions (exothermic and endothermic reactions) is utilized to store the energy. As discussed in section 2.2.3, heat is absorbed by a compound during its split into two components and stored for a long duration of time without losses. These two components are brought together to form the initial compound and thermal energy is released due to the exothermic nature of this reaction. The thermochemical storage systems are basically of two types: closed systems and open systems. Both types of systems can store the energy with adsorption as well as absorption principles.

In closed systems, mass exchange is restricted to the environment, and the reactions are carried out under a vacuum environment. The evaporation of working fluid during discharge is essential in closed systems and is usually achieved through some low-grade heat source. The heat source should be sufficient to evaporate water at 5°C under vacuum conditions, as water is the commonly used working fluid in these systems. In contrast, mass exchange is allowed with the environment in the open systems. Therefore, the use of water becomes mandatory in such systems. Moreover, the local climatic conditions play a crucial role in the performance of these systems due to the use of ambient air during discharging.

2.3.2.2 Storage in Building Structures

The addition of storage facilities to building elements such as floors, ceilings, and walls can also be used to store thermal energy. The use of a water pipe network or electric heating cables in the floor for heating and active ceiling panels for cooling purposes are common examples of this storage method. These storage systems are also referred to as thermally activated building systems (TABSs). Based on the charging/discharging method, TABSs can be classified as air distribution systems and water distribution systems, which are discussed in this section.

In TABS with air distribution systems, the thermal mass integrated into a building structure for TES is charged/discharged using the network of the air distribution system. This type of storage method is found to be better compared to the methods discussed in the previous section (storage in HVAC) in terms of indoor thermal comfort and energy efficiency.[75] However, energy efficiency is significantly influenced by the coordination between thermal mass temperature and air mass flow rate. The performance of these systems is equally influenced by the properties of the storage material and configuration of the storage system.[76] The major observations reported in the literature are based on numerical studies; therefore, more experimental studies are required for this type of storage method.

The TABSs with water distribution systems have major applications in building cooling using activated ceiling panels with PCMs. The combination of water as a heat transfer fluid and PCM as a storage material increases the allowable temperature range of the cooling system compared to the air distribution system.[77] A concept of an active PCM hollow slab was reported by Pomianowski and Jensen. It was found that such a slab gives better results compared to a conventional concrete slab for cooling.[78] Moreover, Mazo et al. investigated the use of PCM in a radiant floor heating system coupled with a heat pump.[79] The authors developed a model of the same and obtained the performance of single-zone building through simulations. Based on the obtained results, it reported that the use of PCM may shift the peak to off-peak hours, and hence, associated energy costs may be reduced by up to 18%.

2.3.2.3 Storage in the Direct Vicinity

Various storage mechanisms discussed under sections 2.3.2.1 and 2.3.2.2 directly or indirectly become integral parts of the building. Contrary, the TESs that fall under this category are located outside the building structure. Such types of TESs are suitable for larger buildings or groups of buildings. Aquifer storage, borehole storage, and snow storage are different types of TESs in the direct vicinity of buildings, and these types are discussed in detail in this section.

2.3.2.3.1 Aquifer Storage

Aquifers are geologically formed bodies of saturated rocks that contain groundwater and have a great potential to store thermal energy for a long duration. In such aquifers, one or more wells are drilled to inject and extract the groundwater as a heat carrier. The charging/discharging processes are carried out either through alternative flow in production and injection wells or continuous flow in one direction. The charging/discharging processes with a continuous flow are adopted for cooling applications. The working principle of typical low-temperature aquifer storage is shown in Figure 2.22. The existence of aquifers in large volumes (hundreds to thousands of

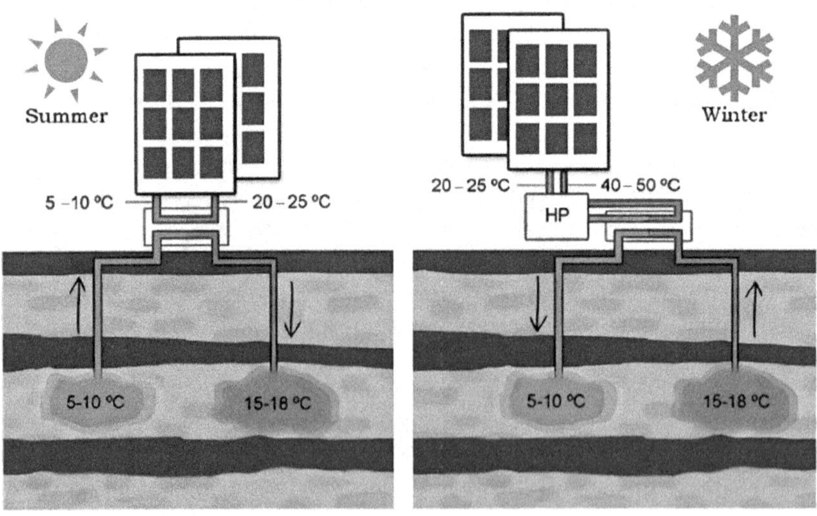

FIGURE 2.22 Working principle of a typical low-temperature aquifer storage system.[81]

cubic meters) facilitates them to store huge amounts of thermal energy. Therefore, their application is found suitable for seasonal storage, that is, the heat stored in summer can be utilized for heating applications during winter. For the same reason, aquifer storage is found appropriate for building applications.[80]

Economic feasibility, high energy storage capacity, and higher charging/discharging rates are several additional benefits associated with aquifer storage systems. On the other hand, conflicts associated with groundwater use and water chemistry are the major concerns allied with it. Novo et al.[82] claimed that aquifer storage's average heat storage capacity is between 30 and 40 kWh/m^3. However, the rate of heat loss to the environment decreases with an increase in the size of the aquifer. Moreover, the effect of natural convection and thermal diffusion can prevent this by increasing its size beyond 20000 m^3.

2.3.2.3.2 Borehole Storage

In borehole storage systems, a bed of rocks is used for heat storage, and a borehole having a pipe is used to circulate the heat carrier, as shown in Figure 2.23. Borehole storage and aquifer storage are the underground TES concepts and are considered beneficial for seasonal TES. As stated earlier, aquifer storage is usually used for a large time periods (seasonal TES), whereas borehole storage can be used for small-scale, large-scale, or seasonal storage systems. Small-scale borehole storage can be

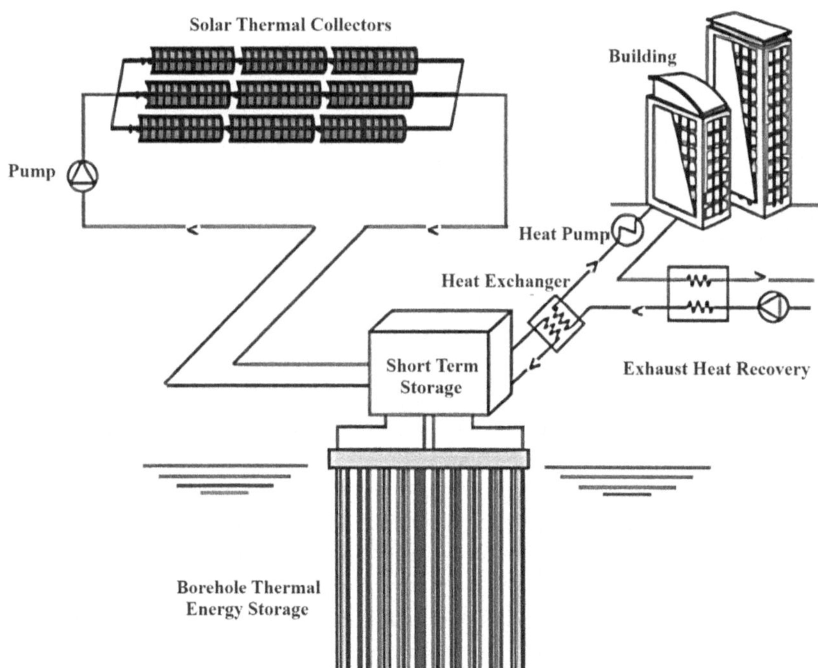

FIGURE 2.23 Schematic of a typical solar-based borehole storage system.[83]

a single borehole for cooling, a single borehole for heating with a heat pump, or a single borehole for heating with a heat pump as well as direct cooling. Large-scale systems have more than one borehole for heat extraction with a heat pump. The size of borehole systems for seasonal storage is comparatively much larger, and energy is stored for later extraction.

2.3.2.3.3 Snow Storage

Snow storage is an energy-efficient method of cooling with the help of snow collected and stored during the winter season. It is an ancient method of cold storage, which includes harvesting ice/snow from rivers/lakes and storing it in barns. In the modern era, artificial snow also comes into the picture for space cooling and food storage applications. Along with the building sector, the seasonal snow storage method has a strong potential in the industrial and agricultural sectors. Similar to the borehole and aquifer storage concepts, energy extraction takes place by the circulation of energy carriers through stored snow.

2.4 ADVANTAGES AND CHALLENGES OF TES

In this section, the advantages and challenges associated with TES are discussed in brief. The integration of TES systems in buildings delivers many advantages for different groups such as building owners, the environment and society, and energy providers. The reduction in costs associated with heating/cooling, improved quality of the indoor environment, and lesser expense of electricity to fulfill the energy needs of HVAC equipment are benefits for the building owners or consumers. Moreover, TES integration in buildings is beneficial from a societal and environmental point of view, as it can help to increase the utility of renewable energy sources to fulfill the energy needs of the buildings. A lower requirement of energy from power plants helps in the reduction of conventional fuel consumption and pollutant emissions. Moreover, TES integration in buildings is also beneficial for the energy provider, as it may reduce the peak load electricity demand, help to increase the efficiency of the energy production plants, and enhance the utility load factor.

Apart from the benefits, there are several challenges also associated with TES technologies and their implementation in buildings. The selection of a TES technology is the major challenge associated with its integration in buildings as it depends on a number of factors such as storage period, operating conditions, compatibility of storage material with building materials, and many more. There is no 'one-size-fits-all' theory available in the literature to define the selection criteria of the TES technology in buildings. Thermal load profiles, excess heat availability, electricity costs, provision of auxiliary, building type, and occupancy are several specific parameters that should be taken into consideration during the selection of TES technology. Economic feasibility is another major challenge in the adaptability of TES in buildings. Due to the higher initial costs of such technologies compared to that of their conventional counterparts, this option is usually not considered by the common people during the development of the building. However, the payback period may be nearly five years or less for the same, whereas operating and maintenance costs are almost negligible in the case of TES systems for thermal comfort. The retrofitting of TES technologies to existing buildings is also a major issue, and stakeholders are working on this problem to take advantage of the benefits discussed in the previous paragraph.

The various benefits revealed in the first paragraph of this section may be achieved by overcoming the various challenges discussed in the previous paragraph. Social awareness, the adoption of concepts like NZEB, guidelines for building construction, and policy making are several steps that require efforts regarding TES integration.

2.5 CONCLUSIONS

In the present chapter, a detailed discussion of the various aspects of TES integrated into buildings is presented. It includes the importance of integrating the TES in buildings in terms of energy conservation and environmental benefits along with various types of TES systems. The chapter also discusses the various existing and underdeveloped mechanisms of TES in buildings. Various advantages, drawbacks, and challenges associated with TES in buildings are also incorporated in the chapter. It can be concluded from the discussion that the utilization of TES in buildings is beneficial to increase energy efficiency in a sustainable manner. It also provides a platform to increase the share of renewable energy resources to meet the total energy demand of the buildings. Apart from this, the emission of harmful pollutants is reduced, the efficiency of HVAC components is increased, peak load magnitude is reduced, and temperature swing in indoor thermal comfort is decreased with the use of TES. A wide range of TES technologies is compatible with the buildings and commercially available. Moreover, the results of several emerging technologies are also found favorable, and their commercialization may be a milestone in the adaptability of TES in buildings. However, the economic feasibility, compatibility of TES materials with building materials, public awareness, and involvement of related policies in essential guidelines for building development are several challenges associated with TES in buildings, and stakeholders are working on them. Moreover, the selection of a TES material and mechanism is a crucial decision that requires a substantial amount of preliminary research about the existing options. The TES technologies for buildings need more research and development to make them competitive with their conventional counterparts. The present chapter may be useful for the stakeholders working in the field of TES and energy efficiency in buildings.

REFERENCES

1. Sharshir SW, Joseph A, Elsharkawy M, Hamada MA, Kandeal AW, Elkadeem MR, Kumar Thakur A, Ma Y, Eid Moustapha M, Rashad M, Arıcı M. 2023. Thermal energy storage using phase change materials in building applications: A review of the recent development. Energy Build. 285. doi:10.1016/j.enbuild.2023.112908
2. Li Y, Li S, Xia S, Li B, Zhang X, Wang B, Ye T, Zheng W. 2023. A review on the policy, technology and evaluation method of low-carbon buildings and communities. Energies (Basel). 16(4). doi:10.3390/en16041773
3. Jannat N, Hussien A, Abdullah B, Cotgrave A. 2020. A comparative simulation study of the thermal performances of the building envelope wall materials in the tropics. Sustainability (Switzerland). 12(12). doi:10.3390/SU12124892

4. Asan H. 2006. Numerical computation of time lags and decrement factors for different building materials. Build Environment. 41(5):615–620. doi:10.1016/j.buildenv.2005.02.020
5. Schiavoni S, D'Alessandro F, Bianchi F, Asdrubali F. 2016. Insulation materials for the building sector: A review and comparative analysis. Renewable and Sustainable Energy Reviews. 62:988–1011. doi:10.1016/j.rser.2016.05.045
6. Chi F, Wang Y, Wang R, Li G, Peng C. 2020. An investigation of optimal window-to-wall ratio based on changes in building orientations for traditional dwellings. Solar Energy. 195:64–81. doi:10.1016/j.solener.2019.11.033
7. Akeiber H, Nejat P, Majid MZA, Wahid MA, Jomehzadeh F, Zeynali Famileh I, Calautit JK, Hughes BR, Zaki SA. 2016. A review on phase change material (PCM) for sustainable passive cooling in building envelopes. Renewable and Sustainable Energy Reviews. 60:1470–1497. doi:10.1016/j.rser.2016.03.036
8. Lizana J, Chacartegui R, Barrios-Padura A, Valverde JM. 2017. Advances in thermal energy storage materials and their applications towards zero energy buildings: A critical review. Appling Energy. 203:219–239. doi:10.1016/j.apenergy.2017.06.008
9. Nair AM, Wilson C, Huang MJ, Griffiths P, Hewitt N. 2022. Phase change materials in building integrated space heating and domestic hot water applications: A review. Journal of Energy Storage. 54. doi:10.1016/j.est.2022.105227
10. Suresh C, Saini RP. 2020. Review on solar thermal energy storage technologies and their geometrical configurations. International Journal of Energy Research. 44(6):4163–4195. doi:10.1002/er.5143
11. Bre F, Lamberts R, Flores-Larsen S, Koenders EAB. 2023. Multi-objective optimization of latent energy storage in buildings by using phase change materials with different melting temperatures Energy-efficient building Latent energy storage Phase change material Multiple melting temperatures Multi-objective optimization Climate-representative locations [Internet]. doi:10.5281/zenod
12. Amghar N, Sánchez Jiménez PE, Pérez Maqueda LA, Perejón A. 2023. Thermochemical energy storage using calcium magnesium acetates under low CO_2 pressure conditions. Journal of Energy Storage. 63:106958. doi:10.1016/j.est.2023.106958
13. Gholamibozanjani G, Farid M. 2020. A comparison between passive and active PCM systems applied to buildings. Renew Energy. 162:112–123. doi:10.1016/j.renene.2020.08.007
14. Novo AV., Bayon JR, Castro-Fresno D, Rodriguez-Hernandez J. 2010. Review of seasonal heat storage in large basins: Water tanks and gravel-water pits. Applying Energy. 87(2):390–397. doi:10.1016/j.apenergy.2009.06.033
15. Dahash A, Ochs F, Janetti MB, Streicher W. 2019. Advances in seasonal thermal energy storage for solar district heating applications: A critical review on large-scale hot-water tank and pit thermal energy storage systems. Applying Energy. 239:296–315. doi:10.1016/j.apenergy.2019.01.189
16. Pinel P, Cruickshank CA, Beausoleil-Morrison I, Wills A. 2011. A review of available methods for seasonal storage of solar thermal energy in residential applications. Renewable and Sustainable Energy Reviews. 15(7):3341–3359. doi:10.1016/j.rser.2011.04.013
17. Verbeke S, Audenaert A. 2018. Thermal inertia in buildings: A review of impacts across climate and building use. Renewable and Sustainable Energy Reviews. 82:2300–2318. doi:10.1016/j.rser.2017.08.083
18. Navarro L, de Gracia A, Niall D, Castell A, Browne M, McCormack SJ, Griffiths P, Cabeza LF. 2016. Thermal energy storage in building integrated thermal systems: A review. Part 2. Integration as passive system. Renew Energy. 85:1334–1356. doi:10.1016/j.renene.2015.06.064

19. Tatsidjodoung P, Le Pierrès N, Luo L. 2013. A review of potential materials for thermal energy storage in building applications. Renewable and Sustainable Energy Reviews. 18:327–349. doi:10.1016/j.rser.2012.10.025
20. Asan H, Santar YS. 1998. Effects of wall's thermophysical properties on time lag and decrement factor.
21. Lizana J, Chacartegui R, Barrios-Padura A, Valverde JM. 2017. Advances in thermal energy storage materials and their applications towards zero energy buildings: A critical review. Appl Energy. 203:219–239. doi:10.1016/j.apenergy.2017.06.008
22. Pinel P, Cruickshank CA, Beausoleil-Morrison I, Wills A. 2011. A review of available methods for seasonal storage of solar thermal energy in residential applications. Renewable and Sustainable Energy Reviews. 15(7):3341–3359. doi:10.1016/j.rser.2011.04.013
23. Rathore PKS, Gupta NK, Yadav D, Shukla SK, Kaul S. 2022. Thermal performance of the building envelope integrated with phase change material for thermal energy storage: An updated review. Sustainable Cities and Society. 79. doi:10.1016/j.scs.2022.103690
24. Devaux P, Farid MM. 2017. Benefits of PCM underfloor heating with PCM wallboards for space heating in winter. Apply Energy. 191:593–602. doi:10.1016/j.apenergy.2017.01.060
25. Soares N, Costa JJ, Gaspar AR, Santos P. 2013. Review of passive PCM latent heat thermal energy storage systems towards buildings' energy efficiency. Energy Building. 59:82–103. doi:10.1016/j.enbuild.2012.12.042
26. Wang X, Li W, Luo Z, Wang K, Shah SP. 2022. A critical review on phase change materials (PCM) for sustainable and energy efficient building: Design, characteristic, performance and application. Energy Building. 260. doi:10.1016/j.enbuild.2022.111923
27. Wang P, Liu Z, Zhang X, Hu M, Zhang L, Fan J. 2023. Adaptive dynamic building envelope integrated with phase change material to enhance the heat storage and release efficiency: A state-of-the-art review. Energy Building. 286. doi:10.1016/j.enbuild.2023.112928
28. Faraj K, Khaled M, Faraj J, Hachem F, Castelain C. 2021. A review on phase change materials for thermal energy storage in buildings: Heating and hybrid applications. Journal of Energy Storage. 33. doi:10.1016/j.est.2020.101913
29. Baetens R, Jelle BP, Gustavsen A. 2010. Phase change materials for building applications: A state-of-the-art review. Energy Building. 42(9):1361–1368. doi:10.1016/j.enbuild.2010.03.026
30. Khadiran T, Hussein MZ, Zainal Z, Rusli R. 2016. Advanced energy storage materials for building applications and their thermal performance characterization: A review. Renewable and Sustainable Energy Reviews. 57:916–928. doi:10.1016/j.rser.2015.12.081
31. Tatsidjodoung P, Le Pierrès N, Luo L. 2013. A review of potential materials for thermal energy storage in building applications. Renewable and Sustainable Energy Reviews. 18:327–349. doi:10.1016/j.rser.2012.10.025
32. Li C, Wen X, Cai W, Yu H, Liu D. 2023. Phase change material for passive cooling in building envelopes: A comprehensive review. Journal of Building Engineering. 65. doi:10.1016/j.jobe.2022.105763
33. Konuklu Y, Ostry M, Paksoy HO, Charvat P. 2015. Review on using microencapsulated phase change materials (PCM) in building applications. Energy Building. 106:134–155. doi:10.1016/j.enbuild.2015.07.019
34. Kuznik F, David D, Johannes K, Roux JJ. 2011. A review on phase change materials integrated in building walls. Renewable and Sustainable Energy Reviews. 15(1):379–391. doi:10.1016/j.rser.2010.08.019
35. Liu Z, Yu Z (Jerry), Yang T, Qin D, Li S, Zhang G, Haghighat F, Joybari MM. 2018. A review on macro-encapsulated phase change material for building envelope applications. Build Environment. 144:281–294. doi:10.1016/j.buildenv.2018.08.030

36. Aneke M, Wang M. 2016. Energy storage technologies and real life applications – A state of the art review. Apply Energy. 179:350–377. doi:10.1016/j.apenergy.2016.06.097
37. Zhang H, Baeyens J, Cáceres G, Degrève J, Lv Y. 2016. Thermal energy storage: Recent developments and practical aspects. Program Energy Combustion Science. 53:1–40. doi:10.1016/j.pecs.2015.10.003
38. Desai F, Sunku Prasad J, Muthukumar P, Rahman MM. 2021. Thermochemical energy storage system for cooling and process heating applications: A review. Energy Conversation Manage. 229. doi:10.1016/j.enconman.2020.113617
39. Heier J, Bales C, Martin V. 2015. Combining thermal energy storage with buildings—A review. Renewable and Sustainable Energy Reviews. 42:1305–1325. doi:10.1016/j.rser.2014.11.031
40. Ghoreishi AH, Ali MM. 2013. Parametric study of thermal mass property of concrete buildings in US climate zones. Architure Science Reviews. 56(2):103–117. doi:10.1080/00038628.2012.729310
41. Ghoreishi AH, Ali MM. 2011. Contribution of thermal mass to energy performance of buildings: A comparative analysis. International Journal of Sustainable Building Technology and Urban Development. 2(3):245–252. doi:10.5390/SUSB.2011.2.3.245
42. Zhu L, Hurt R, Correia D, Boehm R. 2009. Detailed energy saving performance analyses on thermal mass walls demonstrated in a zero energy house. Energy Building. 41(3):303–310. doi:10.1016/j.enbuild.2008.10.003
43. Rempel AR, Rempel AW. 2013. Rocks, clays, water, and salts: Highly durable, infinitely rechargeable, eminently controllable thermal batteries for buildings. Geosciences (Switzerland). 3(1):63–101. doi:10.3390/geosciences3010063
44. Saadatian O, Sopian K, Lim CH, Asim N, Sulaiman MY. 2012. Trombe walls: A review of opportunities and challenges in research and development. Renewable and Sustainable Energy Reviews. 16(8):6340–6351. doi:10.1016/j.rser.2012.06.032
45. Stazi F, Mastrucci A, Di Perna C. 2012. The behaviour of solar walls in residential buildings with different insulation levels: An experimental and numerical study. Energy Building. 47:217–229. doi:10.1016/j.enbuild.2011.11.039
46. Konuklu Y, Ostry M, Paksoy HO, Charvat P. 2015. Review on using microencapsulated phase change materials (PCM) in building applications. Energy Building. 106:134–155. doi:10.1016/j.enbuild.2015.07.019
47. Baetens R, Jelle BP, Gustavsen A. 2010. Phase change materials for building applications: A state-of-the-art review. Energy Building. 42(9):1361–1368. doi:10.1016/j.enbuild.2010.03.026
48. Yahaya NA, Ahmad H. 2011. Numerical investigation of indoor air temperature with the application of PCM Gypsum board as ceiling panels in buildings. Procedia Engineering. 20:238–248. doi:10.1016/j.proeng.2011.11.161
49. Peippo K, Kauranen P, Lund PD. 1991. A multicomponent PCM wall optimized for passive solar heating.
50. Evola G, Marletta L, Sicurella F. 2013. A methodology for investigating the effectiveness of PCM wallboards for summer thermal comfort in buildings. Build Environment. 59:517–527. doi:10.1016/j.buildenv.2012.09.021
51. Navarro L, de Gracia A, Colclough S, Browne M, McCormack SJ, Griffiths P, Cabeza LF. 2016. Thermal energy storage in building integrated thermal systems: A review. Part 1. Active storage systems. Renew Energy. 88:526–547. doi:10.1016/j.renene.2015.11.040
52. Wu P, Wang Z, Li X, Xu Z, Yang Y, Yang Q. 2020. Energy-saving analysis of air source heat pump integrated with a water storage tank for heating applications. Build Environment. 180. doi:10.1016/j.buildenv.2020.107029
53. Pop OG, Balan MC. 2021. A numerical analysis on the performance of DHW storage tanks with immersed PCM cylinders. Appling Thermal Engineering. 197. doi:10.1016/j.applthermaleng.2021.117386

54. Carmona M, Rincón A, Gulfo L. 2020. Energy and exergy model with parametric study of a hot water storage tank with PCM for domestic applications and experimental validation for multiple operational scenarios. Energy Conversation Management. 222. doi:10.1016/j.enconman.2020.113189
55. Pelella F, Zsembinszki G, Viscito L, William Mauro A, Cabeza LF. 2023. Thermo-economic optimization of a multi-source (air/sun/ground) residential heat pump with a water/PCM thermal storage. Apply Energy. 331. doi:10.1016/j.apenergy.2022.120398
56. Koželj R, Mlakar U, Zavrl E, Stritih U, Stropnik R. 2021. An experimental and numerical analysis of an improved thermal storage tank with encapsulated PCM for use in retrofitted buildings for heating. Energy Building. 248. doi:10.1016/j.enbuild.2021.111196
57. Cabeza LF, Ibáñez M, Solé C, Roca J, Nogués M. 2006. Experimentation with a water tank including a PCM module. Solar Energy Materials and Solar Cells. 90(9):1273–1282. doi:10.1016/j.solmat.2005.08.002
58. Mehling H, Cabeza LF, Hippeli S, Hiebler S. 2003. PCM-module to improve hot water heat stores with stratification. www.elsevier.com/locate/renene
59. Talmatsky E, Kribus A. 2008. PCM storage for solar DHW: An unfulfilled promise? Solar Energy. 82(10):861–869. doi:10.1016/j.solener.2008.04.003
60. Koukou MK, Dogkas G, Vrachopoulos MG, Konstantaras J, Pagkalos C, Stathopoulos VN, Pandis PK, Lymperis K, Coelho L, Rebola A. 2020. Experimental assessment of a full scale prototype thermal energy storage tank using paraffin for space heating application. International Journal of Thermofluids. 1–2. doi:10.1016/j.ijft.2019.100003
61. Osterman E, Stritih U. 2021. Review on compression heat pump systems with thermal energy storage for heating and cooling of buildings. Journal of Energy Storage. 39. doi:10.1016/j.est.2021.102569
62. Martin V, Setterwall F. 2009. Compact heat storage for solar heating systems. Journal of Solar Energy Engineering, Transactions of the ASME. 131(4):0410111–0410116. doi:10.1115/1.3197841
63. Persson T. 2022. Modellering och simulering av tappvattenautomater i solvärmesystem.
64. Gautam A, Saini RP. 2020. A review on technical, applications and economic aspect of packed bed solar thermal energy storage system. Journal of Energy Storage. 27. doi:10.1016/j.est.2019.101046
65. Hamada Y, Fukai J. 2005. Latent heat thermal energy storage tanks for space heating of buildings: Comparison between calculations and experiments. Energy Conversation Management. 46(20):3221–3235. doi:10.1016/j.enconman.2005.03.009
66. Medved S, Arkar C. 2008. Correlation between the local climate and the free-cooling potential of latent heat storage. Energy Building. 40(4):429–437. doi:10.1016/j.enbuild.2007.03.011
67. Arkar C, Medved S. 2007. Free cooling of a building using PCM heat storage integrated into the ventilation system. Solar Energy. 81(9):1078–1087. doi:10.1016/j.solener.2007.01.010
68. Mosaffa AH, Infante Ferreira CA, Talati F, Rosen MA. 2013. Thermal performance of a multiple PCM thermal storage unit for free cooling. Energy Conversation Management. 67:1–7. doi:10.1016/j.enconman.2012.10.018
69. Takeda S, Nagano K, Mochida T, Shimakura K. 2004. Development of a ventilation system utilizing thermal energy storage for granules containing phase change material. Solar Energy. 77(3):329–338. doi:10.1016/j.solener.2004.04.014
70. Persson J, Westermark M. 2012. Phase change material cool storage for a Swedish Passive House. Energy Building. 54:490–495. doi:10.1016/j.enbuild.2012.05.012
71. Turnpenny JR, Etheridge DW, Reay DA. 2020. Novel ventilation cooling system for reducing air conditioning in buildings. Part I: testing and theoretical modelling. Apply Thermal Engineering [Internet]. 20:1019–1037. www.elsevier.com/locate/apthermeng

72. Turnpenny JR, Etheridge DW, Reay DA. 2021. Novel ventilation system for reducing air conditioning in buildings. Part II: Testing of prototype. Apply Thermal Engineering [Internet]. 21:1203–1217. www.elsevier.com/locate/apthermeng
73. Bellas I, Tassou SA. 2005. Present and future applications of ice slurries. International Journal of Refrigeration. 28(1):115–121. doi:10.1016/j.ijrefrig.2004.07.009
74. Hayashi K, Takao S, Ogoshi H, Matsumoto S. 2000. Research and development of high-density cold latent heat medium transportation technology. IEA Annex-10-PCMs and Chemical Reactions for Thermal Energy Storage 4th Workshop, Tsu, Japan.
75. Henze GP, Felsmann C, Kalz DE, Herkel S. 2008. Primary energy and comfort performance of ventilation assisted thermo-active building systems in continental climates. Energy Building. 40(2):99–111. doi:10.1016/j.enbuild.2007.01.014
76. Chae YT, Strand RK. 2013. Modeling ventilated slab systems using a hollow core slab: Implementation in a whole building energy simulation program. Energy Building. 57:165–175. doi:10.1016/j.enbuild.2012.10.036
77. Lehmann B, Dorer V, Koschenz M. 2007. Application range of thermally activated building systems tabs. Energy Building. 39(5):593–598. doi:10.1016/j.enbuild.2006.09.009
78. Pomianowski M, Heiselberg P, Jensen RL. 2012. Dynamic heat storage and cooling capacity of a concrete deck with PCM and thermally activated building system. Energy and Buildings. 53: 96–107. doi.org/10.1016/j.enbuild.2012.07.007.
79. Mazo J, Delgado M, Marin JM, Zalba B. 2012. Modeling a radiant floor system with phase change material (PCM) integrated into a building simulation tool: analysis of a case study of a floor heating system coupled to a heat pump. Energy and Buildings. 47:458–66. doi.org/10.1016/j.enbuild.2011.12.022.
80. Hasnain SM. 1998. Review on sustainable thermal energy storage technologies, part I: heat storage materials and techniques. Energy Conversation Management. 39(10):1127–1138.
81. Beernink S, Bloemendal M, Kleinlugtenbelt R, Hartog N. 2022. Maximizing the use of aquifer thermal energy storage systems in urban areas: Effects on individual system primary energy use and overall GHG emissions. Apply Energy. 311. doi:10.1016/j.apenergy.2022.118587
82. Novo AV., Bayon JR, Castro-Fresno D, Rodriguez-Hernandez J. 2010. Review of seasonal heat storage in large basins: Water tanks and gravel-water pits. Apply Energy. 87(2):390–397. doi:10.1016/j.apenergy.2009.06.033
83. Pokhrel S, Amiri L, Zueter A, Poncet S, Hassani FP, Sasmito AP, Ghoreishi-Madiseh SA. 2021. Thermal performance evaluation of integrated solar-geothermal system; a semi-conjugate reduced order numerical model. Apply Energy. 303. doi:10.1016/j.apenergy.2021.117676

3 Progress in Ventilated Walls and Double-Skin Facades for Sustainability

Abhishek Mishra, Shubheksh Mishra, Sunil Kumar Tripathi, Biswajit Roy, Rajgourab Ghosh, Chanakya Upadhyay, Sonika Pandey, Prabhanshu Kumar, Amit Kaushik, Akhilesh K. Chaube, Rajeev Singh, P. P. Singh, and Anshuman Mishra

3.1 INTRODUCTION

In the building industry, it is critical to make use of sustainable materials and recycled ones. This can include wood that was harvested in an ethical manner, steel that has been recycled, flooring made of bamboo, and paints and finishes that are non-toxic and have low emissions. According to the findings of the International Energy Agency's study of global carbon dioxide (CO_2) emissions, the amount of CO_2 released into the atmosphere as a result of the combustion of fossil fuels rose by 60% between the years 1990 and 2017 (Yoon et al., 2023). The need to reduce greenhouse gas emissions is urgent because carbon dioxide stays in the atmosphere for over 10 years. All energy-consuming sectors have developed energy-efficient solutions, including smart building design strategies in the building sector, to reduce climate change's environmental impact. It is also important that houses be designed to produce the least amount of greenhouse gas waste during the building process and to incorporate strategies for the recycling and correct disposal of materials. Ventilated double-skin facades, also known as VDFs, make use of the thermal interaction that occurs between the interior and exterior of a building in order to lower the amount of heating or cooling a building requires and the amount of greenhouse gas emissions it produces. Depending on the parameters, they also offer naturally occurring ventilation and daylight, as well as thermal and acoustic insulation. This chapter reviews previous research on VDFs' thermal behavior, discussing classification criteria as well as the parameters that affect it.

- ***Ventilated double-skin facades (VDFs)***

A building's facade protects its interior. It's crucial to the building's functionality and occupants' safety. It helps conserve energy and improve the building's appearance. Despite their aerodynamic unknowns, facade systems are being improved and

DOI: 10.1201/9781003415695-3

used more due to their importance. Double-skin facade (DSF) systems clad buildings. The inner and outer skins are separated by a gap. Tall buildings with mostly glass and steel facades use these. A few centimeters to 2 m can separate the facades. DSF systems save energy because the gap air insulates the building's interior and exterior. An improperly designed DSF system may perform worse thermodynamically than single-skin building facades.

Double-skin facades (DSFs) are a promising passive building technology that improves energy efficiency and thermal comfort through measured and simulated performance (Ghaffarianhoseini et al., 2016). Double-skin facade refers to decoupled layers of transparent, translucent, or opaque constructions. Double-skin facades follow curtain wall construction principles. The gap width between the facades is crucial because a DSF system can reduce effective heat transfer by 27% and 24% in summer and winter, respectively (Zhu et al., 2019). The control of greenhouse gas emissions, the improvement of the building's thermal and acoustic insulation, as well as the provision of natural ventilation and exposure to daylight are all essential components of a sustainable house. Studies have shown that making full use of the electric energy generated by solar PV panels that are integrated with a DSF system can reduce the total amount of energy needed for heating, cooling, and lighting by up to 44.1%. This can be accomplished by integrating the panels into the DSF system (Yoon et al., 2023).

- ***Natural ventilated systems***

Natural ventilation systems need weather-sensitive planning. Mechanical solutions are more reliable but require control and installation. Mechanically driven openings, fans, and air conditioning units can control temperature and humidity. Each floor offers more flexibility, making room configurations easier. Its cost and time to clean are drawbacks. Air conditioning, natural lighting, and light control in the facade increase design complexity and effort. This chapter discussed comprehensive analysis of the design, technical aspects, and construction characteristics for energy efficiency and thermal performance.

- ***Thermodynamics and greenhouse effect***

The building has two transparent facades, which enable a significant amount of solar radiation to enter the interior. Due to the fact that the sunlight is reflected, absorbed, and refracted by both facades, it is possible that they will cause a greenhouse effect in the gap (Škvorc et al., 2023). The outer facade reflects some solar radiation from the inner facade. The outer facade partially reflects solar radiation onto the inner facade. In cooler climates, this may be beneficial, but in warmer climates, it may overheat. Shape the building's exterior or interior facades to control (lower and higher) the temperature. Controlling solar radiation on building facades adjusts heating and cooling.

- ***House materials and climate***

The process of reducing emissions caused by the built environment includes using materials that have a low carbon footprint, which is an important component of that process. Because eco-friendly materials are readily available, the future of environmentally friendly building materials looks bright (Modus Staff, 2021). As environmental awareness grows, so do sustainable building materials. Sustainable building materials help architects, builders, and homeowners reduce their environmental footprint.

- ***House construction materials***

Recycled wood and clay can build energy-efficient, resource-saving buildings. Ten eco-friendly construction materials that are made from sustainable building materials can provide better way for the future (The Constructor, 2023). Renewable, eco-friendly, and low-impact plant materials for house construction are available. Bamboo, cork, hempcrete, timber, coconut timber, etc. are sustainable plant-based construction materials (The Constructor, 2023; Nowotna et al., 2019). Fast-growing bamboo can be harvested sustainably in a few years. It's tough and versatile. Bamboo can be used for flooring, walls, roofing, and structures. Cork comes from cork oak bark. The bark is harvested without cutting down the tree, making it renewable. Cork is fireproof, acoustic, and insulating. Flooring, wall tiles, and insulation use it. Grain crops produce straw bales. They can be stacked and used as load-bearing or infill walls, providing excellent insulation. Hempcrete is a mixture of hemp fibers and lime binder. It is lightweight, breathable, and has excellent thermal insulation properties. Hempcrete is used as an alternative to traditional concrete for walls and insulation, reducing carbon emissions associated with concrete production. Sustainable timber is used for framing, flooring, siding, and interior finishes. Linoleum flooring is made from linseed oil, wood flour, and jute fibers. It is biodegradable, non-toxic, and durable. Mature coconut palm trees without fruit produce coconut timber. Coconut trees grow faster than hardwood, making it more sustainable. Coconut wood is used for flooring, furniture, and structures. Eco-friendly materials address the environmental, social, and economic sustainability in building construction (Manandhar et al., 2019; Nowotna et al., 2019).

- ***Renewable plant materials***

Green composites—biocomposites with a bio-based polymer lattice reinforced by common strands—are a growing polymer research field. Concrete is a composite material made of coarse granular material (filler aggregate) encased in a hard matrix (cement or binder) that bonds the aggregate particles together. Energy-saving materials like sheep's wool, cellulose, plywood, fibrous materials with external gypsum, wooden panels, and clay plaster with straw are healthy, cheap, and often locally available (Manandhar et al., 2019). Green architecture reduces the amount of resources used in a building's construction, use, and operation, as well as its emissions, pollution, and waste (Ragheb et al., 2016). These renewable plant materials offer a variety of benefits, including a lower impact on the environment, the use of renewable sources, increased energy efficiency, and an improvement in the quality of the air inside the building. Figure 3.1 presents a graphical representation of a sustainable house with low-energy-performance characteristics. It is essential, when using plant-based materials, to ensure that they have been responsibly sourced and that they meet relevant quality standards for the structural integrity and safety of the building.

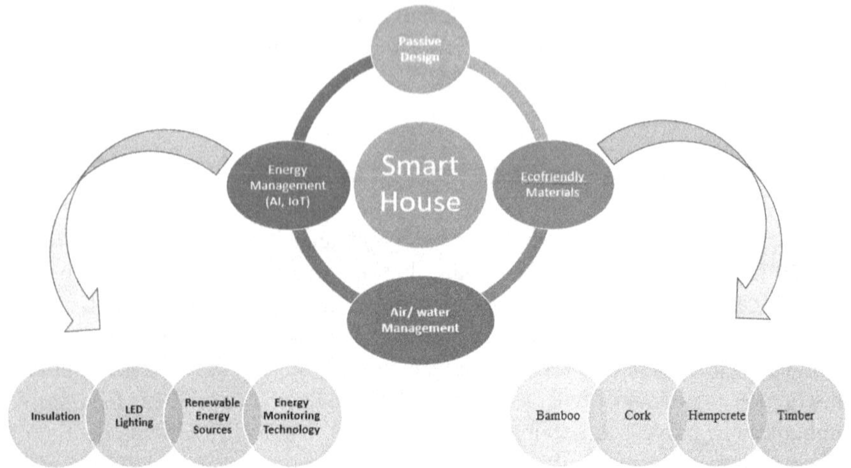

FIGURE 3.1 Graphical representation of the sustainable house characteristics.

3.2 CHALLENGES AND MITIGATION

Remember, a sustainable construction approach also includes minimizing waste, optimizing energy efficiency through proper design and insulation, and considering the life cycle of the materials used. Concrete is the second-most-used material after water and the second-largest CO_2 emitter. Due to global heat waves, storms, and flooding, climate-change-challenged construction is rising. Additional research is required to complete a comprehensive analysis of the significant benefits that using double-skin facades (DSFs) can provide (Ghaffarianhoseini et al., 2016).

According to a literature review, DSF systems on buildings are poorly understood, and tall buildings complicate the issue (Škvorc et al., 2023). Because tall buildings are so common in cities, interference effects between multiple buildings must be studied experimentally or numerically. These effects can severely disrupt signal transmission. How do we build a sustainable house? The cement industry would be the third-largest carbon dioxide emitter if it were a country. Concrete supports business and society worldwide. Its sustainability, exploitation, and development are essential to a healthy global economy and environment, although they are challenges to a sustainable environment. Concerns, consequences, and potential solutions regarding the emission of carbon dioxide from the building sector and bringing the greenhouse gas emissions produced by the cement industry down to net zero use mitigation strategies that focus on the value chain (Ali et al., 2020; Miller et al., 2021). A lack of appropriate data input, limited experience with the tools, difficulty using the tools, calculations, and integration into design and construction processes, requiring details for various elements and components for completing the whole installation. Kazemian recommends life-cycle analysis and carbon footprint calculations of any kind (Kazemian et al., 2022). As future performance-based design approaches for durability are expected, monitoring durability performance in tropical exposure environments is urgently needed to calibrate them (Tam Chat Tim, 2014). Green concrete encourages sustainable and creative use of waste and unconventional concrete materials (Liew et al., 2017).

Standards, demonstration projects, training, public awareness, cross-disciplinary collaborations, and more research and development are needed to promote green concrete in large-scale infrastructure projects worldwide. Many carbon dioxide regulating and environmentally friendly alternatives exist. Beyond fixing concrete issues, they will make buildings greener. Skill shortage, unscheduled maintenance/replacement, policies, standards, and regulations are possible requirements for sustainable construction. Sustainable construction requires training in new eco-friendly materials. These barriers could improve new policy actions. Removing these barriers would help the sector achieve waste policy goals and adopt circular economy thinking (European Environment Agency, 2023). An European Environment Agency (2023) briefing examines how circular economy-inspired actions can help achieve waste policy goals such as waste prevention and increasing recycling quantity and quality while reducing hazardous materials. By incorporating sustainable materials, we can create an environmentally friendly and energy-efficient home.

In order to ensure that relevant quality standards are met for houses and buildings and to make a sustainable future, this chapter discusses energy-efficient houses, performance-based design, adoptable technology approaches, healthy environments, and plant-based materials.

3.3 ENERGY-EFFICIENT HOUSES

These houses are designed to minimize energy consumption and maximize energy efficiency. This includes proper insulation, energy-efficient windows, LED lighting, and the use of energy-efficient appliances and HVAC (heating, ventilation, and air conditioning) systems. Renewable energy sources such as solar panels or wind turbines can be installed to generate clean energy on-site. Characteristics of water heating and solar systems represent efficient house structures.

- ***Heating systems***

Water heating systems aim to efficiently convert energy into heat to raise the temperature of water. Water heating can be achieved using various energy sources, including electricity, natural gas, oil, or solar energy. Water heating systems can either have a storage tank to store heated water or be tankless, instantly heating water as it flows through the system. Water heating systems can experience heat loss, especially in systems with storage tanks. Proper insulation and efficient design help minimize heat loss. A low-energy-performance house, also known as a passive house or energy-efficient house, is designed and constructed to minimize energy consumption and maximize energy efficiency. Such houses are built using various design principles, materials, and technologies to reduce the need for heating, cooling, and overall energy usage. Here are some characteristics and details commonly associated with low-energy-performance houses mentioned in Table 3.1. Hamburg et al. (2021) addressed the issue of heat loss from domestic hot water pipes, exploring factors contributing to heat loss and proposing solutions to minimize it. Marszal-Pomianowska et al. (2021) conducted a field study in two Danish detached houses, examining the comfort aspects of domestic water in terms of flow, temperature, and energy at various draw-off points. P.A. Hohne et al. (2019) provided a review of water heating technologies, focusing on their application in the South African context and assessing their energy efficiency and suitability for the region.

- ***Renewable energy sources***

Solar energy is a renewable energy source derived from the sun's radiation. It is continuously available and does not deplete natural resources. The sun provides an abundant source of energy capable of meeting the world's energy needs many times over. Solar energy production does not release greenhouse gases or pollutants, contributing to a cleaner environment and reduced carbon footprint. Solar panels require minimal maintenance, primarily consisting of periodic cleaning to ensure optimal performance. While the upfront costs of installing solar energy systems can be significant, they can result in long-term cost savings by reducing reliance on traditional energy sources and lowering electricity bills. Low-energy-performance houses are mentioned in Table 3.1.

Solar low-energy smart houses explore the concept of energy efficiency and sustainability through the integration of solar technologies and smart systems (Chwieduk et al., 2020). In a building with solar PV/T panels and a heat storage unit, Behzadi A. et al. (2020) analyzed the energy savings and overall performance of such a system. Azhar K. Mohammed et al. (2021) compared electrical and solar water systems, particularly solar water heaters, to assess the energy savings and cost-effectiveness of adopting solar water heating technologies. Another study examined the effects of storing solar energy in residential and commercial buildings, evaluating different energy storage technologies and their economic and environmental implications for maximizing the utilization of solar energy (Fares et al., 2017). In general, the

TABLE 3.1
Characteristics and Details Commonly Associated with Low-Energy-Performance Houses Mentioned in Categories

Categories	Details	Reference
Water heating	Heat loss due to domestic hot water pipes	Hamburg et al., 2021
	Domestic water in Danish detached houses	Marszal-Pomianowska et al., 2021
	Application to the South African context	Hohne et al., 2019
Solar energy	Solar low-energy smart house	Chwieduk et al., 2020
	Building with solar PV/T panels and a heat storage unit	Behzadi et al., 2020
	Energy savings for electrical and solar water systems	Mohammed et al., 2021
	The impacts of storing solar energy	Fares et al., 2017

findings of these studies highlight how important it is for homes in various parts of the world to have a low energy performance. Ongoing research and development efforts continue to improve the efficiency and cost-effectiveness of solar energy systems, making them increasingly viable for widespread adoption.

3.4 DEVELOPING ENERGY-EFFICIENT HOUSES AND BUILDINGS WITH WALLS

The significance of residential and commercial construction is determined by a number of factors, including wall design, architectural modeling, and air flow, among others.

- ***Characteristics of house walls***

The wall can enhance the visual appeal, architectural design, structural support; insulate the interior; maintain comfortable temperatures; and protect against external elements such as weather and noise. In Table 3.2, characteristics of house walls were discussed in detail. A study by Leang E et al. (2021) examines the effect of a

TABLE 3.2
Characteristics and Details Commonly Associated with House Walls, House Design, Modeling, and Air Flow Systems Are Discussed in Detail

Categories	Details	Reference
House wall	Composite wall with phase change materials	Leang et al., 2021
	Wall-brick-timber	Zheng et al., 2021
	Construction of a passive exterior wall	Lan et al., 2022
	Thermal performance analysis of exterior wall materials	Zheng et al., 2019
	Wall with insulation systems	Carlier et al., 2020
Design of house	Design of buildings in the Polish region	Godlewski et al., 2021
	Design strategy for cold	Pan et al., 2020
	Building envelope design in the Lebanese climate	Sassine et al., 2022
	Climate-adaptive container building design	Shen et al., 2020
Modeling	Investment and energy calculations	Niskanen et al., 2020
	Building design for energy performance	Mahmoud et al., 2020
	Deep learning for smart apartments	Komala et al., 2022
	Eco-friendly speed control algorithm development	Kim et al., 2021
	Building energy performance simulations	Abbas et al., 2022
Air flow	Windows with portable air purifiers	Wang et al., 2021
	Indoor air quality	Kanama et al., 2021
	Sustainable air heat pump system	Jeong et al., 2020

composite Trombe wall with phase change materials on the thermal behavior of a low-energy-consumption house. It investigates how the wall design influences the energy efficiency and thermal performance of the individual house. Further, the effect of the envelope structure on the indoor thermal environment of a low-energy residential building in a humid subtropical climate is explored to assess how the design of the building envelope affects energy consumption and thermal comfort (Zheng et al., 2021). Another study investigates the thermal performance measurement and construction of a passive exterior wall with low energy (Lan et al., 2022). It explores different strategies and materials used in the construction of the exterior wall to achieve energy efficiency and thermal comfort in buildings. A thermal performance study analyzes the exterior wall materials used in residential buildings in Huizhou considering the local climate conditions. It examines how the choice of wall materials impacts energy efficiency and thermal comfort in buildings (Zheng et al., 2019). Another study focuses how different wall designs and materials affect the energy efficiency of buildings using switchable insulation technology (Carlier et al., 2020). The findings of these studies, taken as a whole, shed light on how important it is for residential walls to be energy-efficient in many different parts of the world.

- ***Design of houses***

The importance of a house's design is directly proportional to the degree to which it optimizes the functionality and efficiency of living spaces, safety features, etc., which in turn enable the house to adapt to changing requirements and ways of life. In Table 3.2, the design of houses was discussed in detail in several studies, focusing on important characteristics. In the Polish region, one study explores the design considerations for buildings, taking into account factors such as climate, culture, and local architectural traditions. It discusses strategies and principles for designing energy-efficient buildings (Godlewski et al., 2021). Pan W et al. (2020) discusses the techniques and approaches for designing buildings that can effectively withstand cold temperatures and minimize heat loss. Further, Sassine E et al. (2022) investigates strategies for designing envelope systems that can effectively regulate temperature, optimize energy performance, and ensure occupant comfort in the specific climatic conditions of Lebanon. Container buildings are designed to respond and adapt to changing climate conditions based on innovative design approaches and technologies for creating flexible structures that can adapt to different climates (Shen et al., 2020). In general, these studies emphasize the significance of the design and architecture of the home for the development of energy-efficient buildings.

- ***Importance of modeling***

The modeling process will result in the production of visual representations and simulations, which will improve both comprehension and investigation. It supports an iterative process of refinement and improvement, enables accurate representation of the house's systems, and facilitates scalability for analysis at various levels of detail. In the aspect of modeling, several studies have discussed sustainability and related topics in Table 3.2. Here is a summary of those studies that were discussed.

A study by Niskanen J et al. (2020) explores the investment calculations, energy modeling, and collaboration strategies employed by Swedish housing companies for passive houses, which are highly energy-efficient buildings. It focuses on the sustainability aspects of passive house design and construction. A building design study in the UK discusses the opportunities and limitations of building energy performance simulation tools during the early stages of building design (Mahmoud et al., 2020). The use of deep-learning techniques to develop an innovative photo energy model for estimating energy distribution in smart apartments is important for sustainability in residential buildings (Komala et al., 2022). The development of an eco-friendly speed control algorithm for autonomous vessel route planning helps to minimize energy consumption and reduce environmental impact (Kim et al., 2021). Abbas S. et al. (2022) investigates energy-efficient building structures using building energy performance simulations. This study examines how simulation tools can be used to assess and optimize the energy performance of buildings, contributing to sustainability goals. Collectively, these studies illuminate the universal importance of modeling for efficient household sustainability management.

Tools for building energy simulations provide the opportunity to integrate double-skin facade (DSF) technologies into whole-building simulations through the use of dedicated modules or other possible workarounds (Gennaro et al., 2023). Research compared simulation results with experimental data to assess their ability to predict thermophysical quantities. The results showed that no instrument performs better for any of the quantities analyzed. Due to a daytime peak underestimate, cavity air temperature is the least accurate parameter across all software. In the shade, the IES-VE model is most accurate for supply air and thermal buffer modes, while the EnergyPlus model is most accurate for outdoor air curtain mode. When it comes to surface temperatures and the amount of solar radiation that is transmitted, it seems that TRNSYS is the software that performs the best (Gennaro et al., 2023).

- ***Importance of air flow***

Air flow is very important in the process of developing an energy-efficient house because it encourages natural ventilation, lessens the need for artificial cooling systems, and improves indoor air quality by lessening the amount of pollutants and moisture. Additionally, it contributes to the comfort of the occupants as well as their overall well-being, which in turn helps to improve energy efficiency. Mechanical cooling and ventilation are reduced as a result. An important study investigates the factors influencing occupants' window operation in apartments equipped with portable air purifiers (Wang et al., 2021). The authors examine the impact of occupants' behavior on energy consumption and indoor air quality, contributing to sustainable building operation. Kanama N et al. (2021) discussed the indoor air quality campaign conducted in an occupied low-energy house. The authors explore strategies to improve indoor air quality and thermal comfort, considering the spatial and temporal aspects of building occupancy. Further, a study by Jeong M G et al. (2020) examines the effect of a sustainable air heat pump system on energy efficiency, housing environment, and productivity traits on a pig farm. It explores the application of sustainable heating systems in agricultural settings to enhance energy efficiency and environmental sustainability. Overall, these studies show how vital it is to improve indoor air quality in order to improve energy efficiency.

The porous double-skin facade (PDSF) is a DSF system that is widely used and is considered important. Its outer facade is porous and allows air to flow through the gap between the facades, in contrast to the solid inner facade, which is typically made of glass panels and steel frames (Škvorc et al., 2023). Porosity of the desired level can be accomplished by cutting openings in the exterior facade. Because of the higher dynamic pressure inside the building, the single glass panel that is located on the interior of the building's facade has the potential to cause significant damage if it is subjected to the elements of the environment such as precipitation, snow, or hail. Because of this, it is essential that the outer facade shield the inner facade from the various components that make up the environment.

3.5 ECO-FRIENDLY LIVING PRACTICES

It is important that important technologies for efficient house aspects focus on incorporating technologies that improve smartness and efficiency in residential buildings, and all together, residential management should require less money. The consumption of energy is growing at an alarming rate all over the world, and one of the major contributors to this trend is the demand for electricity to power lighting fixtures found inside of buildings. Research has identified five high-performance light-transmitting concrete samples including different amounts of optical fiber, and their performances in terms of daylight and electricity saving were analyzed based on simulations performed with the Diva for Rhino software (Navabi et al., 2023). The use of transparent facades is the most common method currently available to reduce the amount of electrical energy that is consumed in contemporary buildings such as museums; however, there are other methods available to deal with this issue. Nevertheless, these solutions come with a number of drawbacks, including insufficient protection for the space, heat gains during the summer, and glare; energy loss caused by light-transmitting seams and visual discomfort are the primary drawbacks associated with transparent facades. Thus, a new method is needed to pass natural light to improve visual comfort without damaging the building's exterior walls' thermal insulation. Light-transmitting concrete is one solution.

- ***Cost-effective households***

Research facilities that are both environmentally friendly and economical to operate are given ample attention in several studies. In Table 3.3, the following cost-effective research topics were discussed by various studies. Gürel AE et al. (2020) discussed research focused on cost-effectiveness in household refrigerators, exploring ways to improve energy efficiency while keeping costs reasonable. Another study discussed the use of sound-absorbing and insulating low-cost panels, aiming to provide cost-effective solutions for noise reduction and insulation in buildings (Neri et al., 2021). García Kerdan I et al. (2020) explored the use of an artificial neural network structure, which offers a cost-effective approach for various applications such as predicting energy consumption or optimizing system performance.

TABLE 3.3
Characteristics and Details Commonly Associated with House Cost-Effectiveness and Smartness Research Are Discussed in Detail

Characteristics	Details	Reference
Cost-effectiveness	Household refrigerators	Gürel et al., 2020
	Sound-absorbing and insulating low-cost panels	Neri et al., 2021
	Artificial neural network structure	García Kerdan et al., 2020
	CO_2 emissions and cost by floor types in apartments	Jang et al., 2016
	Utilization of cost-efficient waste-wood composite	Tsai et al., 2018
Smart houses	Cyber security framework for energy management	Alkatheiri et al., 2021
	Design and implementation of the E-Switch	García-Vázquez et al., 2022
	Architecture, technologies, and systems	Li et al., 2018
	Renewable energy management	Al-Ali et al., 2011
	Design and implementation of real-time management	Elkholy et al., 2022

The relationship between CO_2 emissions and cost based on different floor types in apartments was investigated by Jang H J et al. (2016). This research aimed to identify cost-effective flooring options that minimize environmental impact. Tsai M et al. (2018) study the utilization of cost-efficient waste-wood composite materials, highlighting their potential as a sustainable and affordable alternatives in various construction applications.

- ***Development of smart houses***

The development of smart houses has become an important task in recent times, and several studies have emphasized it through their technology and research (Table 3.3). Here are some key contributions from these studies. Alkatheiri MS et al 2021 discussed a cyber security framework for energy management in smart houses. This highlights the importance of protecting smart-home systems from potential cyber threats and ensuring the secure management of energy resources. Another study focused on the design and implementation of the E-Switch, which is a technology aimed at enhancing the control and automation capabilities of smart houses (García-Vázquez et al., 2022). This contributes to the development of more efficient and user-friendly smart-home systems. Min Li et al. (2018) researched the architecture, technologies, and systems involved in smart houses. This comprehensive overview provides insights into the different components and integration methods necessary for the successful implementation of smart-home technologies. The importance of renewable energy management in smart houses highlights the need for the efficient utilization of sources, such as solar or wind power, to optimize energy consumption and reduce reliance on traditional energy grids (Al-Ali et al., 2011). Elkholy MH et al. (2022) focused on the design and implementation of a real-time management system for smart houses. This contributes to the development of advanced monitoring and control systems that enable real-time data analysis and decision making to optimize energy usage and improve overall smart-home performance. Taken together, these reports stress the need for environmentally friendly smart houses in many parts of the world.

- ***Water conservation and waste management***

Low-flow toilets, faucets, and showerheads save water. Rainwater harvesting systems can store rainwater for non-potable uses like watering gardens and flushing toilets. Water-efficient fixtures like low-flow toilets, faucets, and showerheads can cut water use. Rainwater harvesting systems can be used to water gardens and flush toilets. According to the findings, water efficiency programs were more effective in lowering per capita consumption (pcc) in financially strained and smaller households (Manouseli et al., 2019). The findings indicate that the design of rainwater harvesting (RWH) systems should be based on archival data and account for many years of rainfall changes (Gwoździej-Mazur et al., 2022). This will improve performance and ensure benefits for system users. It has been suggested that the work that will be done in the future on rainwater harvesting will address these three priority challenges. First, more empirical data on system operation should be included; second, maintenance aspects for the quality of collected rainwater should be explored; and third, understanding of socio-political support to improve system efficacy is required (Campisano et al., 2017). Smart irrigation systems using weather data and soil moisture sensors optimize watering schedules and reduce water waste. These systems adjust watering based on the weather to avoid overwatering.

A sustainable water-harvesting house can reduce freshwater demand, municipal water system strain, and resource conservation by using these methods.

3.6 ENERGY-EFFICIENT STRUCTURES WITH GLOBAL COVERAGE

Research has used thermal mass and passive design to improve residential building energy efficiency and thermal comfort in specific climates. It is essential to have a good understanding of the significance of energy-efficient buildings all over the world in general, with particular reference to the climate of the various regions.

- ***Energy-efficient house worldwide***

The following is an account of the specifics of each study, with an emphasis on the significant information that may be applicable to any part of the world. In Table 3.4, the discussion revolves around energy-efficient structures in various regions of the globe. Fedorik F et al. (2021) studied the hygrothermal performance of buildings in the Nordic region. Given the unique climate conditions in this region, the research

TABLE 3.4
Discussions Revolving around Energy-Efficient Structures for Various Categories Are Mentioned with References

Characteristics	Details	Reference
Nordic region	Hygrothermal performance in the Nordic region	Fedorik et al., 2021
Brazilian region	Passive house for the Brazilian Bioclimatic Zone 8	Cruz et al., 2020
Mediterranean region	Thermal ceramic panels and passive systems in housing	Echarri et al., 2017
Egypt region	Modern building techniques	Farouk Mohamed et al., 2020
Creates sustainability	Walls using natural and waste materials	Georgescu et al., 2022
	Design of the passive solar house	Zhu et al., 2022
	Zero-energy buildings	Manzoor et al., 2022
	Promoting eco-friendly houses	Koengkan et al., 2023
Eco-friendly approach	Technology roadmap for eco-friendly building materials	Shim et al., 2019
	Decarbonization and energy management	Shimoda et al., 2021
Eco-friendly approach	Eco-friendly buildings and decision-making criteria	Šatrevičs et al., 2023
	Impact of energy policies on residential properties	Fuinhas et al., 2022
Creates sustainability	Demand control strategies of ventilation systems	Hu et al., 2021
	Design and implementation of energy management systems	Zhen et al., 2021

aimed to identify and analyze building techniques and materials that optimize energy efficiency while considering moisture management. Another study explored the concept of a passive house designed specifically for the Brazilian Bioclimatic Zone 8 (Cruz et al., 2020). By considering the climate conditions and specific energy needs of the region, the research aimed to propose energy-efficient design strategies and technologies suitable for residential buildings in Brazil. Echarri V. et al. (2017) researched the use of thermal ceramic panels and passive systems in housing within the Mediterranean region. By utilizing thermal mass and passive design principles, the research aimed to enhance energy efficiency and thermal comfort in residential buildings, taking into account the specific climate conditions. Through the use of experimental data, the simulation model demonstrates that there is potential for a reduction in the amount of energy used for cooling, heating, and lighting of up to 44.1% by fully utilizing the electric energy generated by solar PV panels that are integrated with the DSF system (Yoon et al., 2023).

Overall, these studies highlight the significance of energy-efficient structures in various regions of the world.

- ***Roadmap for eco-friendly houses***

The roadmap for energy-efficient structures through eco-friendly approaches is important in framing the sustainability agenda. Georgescu S. V. et al. in 2021 explored the use of natural and waste materials in constructing walls for energy-efficient structures. Zhu Y. et al discussed this topic in 2022 and discussed the design principles and strategies used to create passive solar houses, which harness solar energy for heating and cooling purposes. "Zero-Energy Buildings" is likely a bibliometric review that examines the existing literature on zero-energy buildings and energy efficiency, along with a case study to support sustainability goals (Manzoor et al. in 2022). Koengkan M. et al. (2023) discussed various initiatives and measures aimed at promoting the construction of eco-friendly houses. These references demonstrate the diverse research efforts and approaches within the field of energy-efficient structures and sustainability. The study conducted by Shim, H. et al. in 2019 discusses a technology roadmap for eco-friendly building materials. The roadmap focuses on adopting environmentally sustainable approaches in the construction industry. It aims to reduce carbon emissions and promote energy efficiency in the residential sector in Japan. The researchers propose evaluating decarbonization scenarios and energy-management requirements through bottom-up simulations of energy end-use demand. This approach will help identify the most effective strategies for achieving sustainability goals in the construction and housing sectors. By combining effective policies and sustainable management practices, the eco-friendly approach contributes to the creation of a sustainable future.

- ***Eco-friendly home management***

Policy and management for an eco-friendly approach require the integration of eco-friendly buildings and decision-making criteria. Researchers such as Šatrevičs V. et al. (2021) emphasize the importance of incorporating sustainable building practices and considering eco-friendly criteria in decision-making processes. In terms of policy impact, Fuinhas J.A. et al. (2022) studied the influence of energy policies on the energy efficiency performance of residential properties in Portugal. This highlights the significance of well-designed energy policies in promoting sustainability in the residential sector. To achieve sustainability, strategies and implementation are essential. Hu Y. et al. (2021) explored demand control strategies for a phase change material (PCM)-enhanced ventilation system in residential buildings, aiming to optimize energy usage. Additionally, Zhen Y. et al. (2021) discussed the design and implementation of smart-home energy-management systems using green energy, indicating the importance of innovative technologies in achieving eco-friendly goals.

The use of renewable energy sources can be quite challenging due to the limited space available on the exterior walls of buildings and the shared nature of the roof space available on buildings. As a result of the utilization of a central system, heating, ventilation, and air conditioning (HVAC) equipment can be challenging to replace. As a consequence of this, the majority of the methods for retrofitting apartments are limited to the replacement of windows. According to the evaluation of the environmentally friendly renovation project that was funded by the government of South

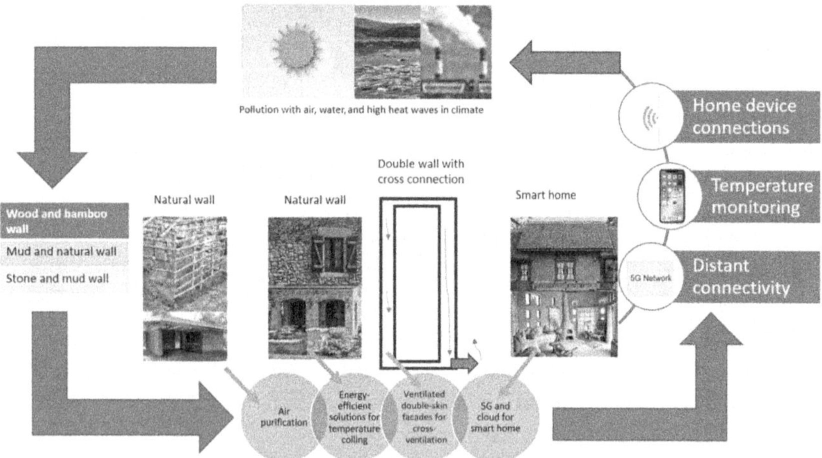

FIGURE 3.2 Graphical representation of the sustainable energy and climate-efficient house model.

Korea, approximately 10,000 cases involved the replacement of windows, while only 146 cases involved the addition of insulation or an HVAC system (Yoon et al., 2023). Figure 3.2 represents sustainable energy and a climate-efficient house model. By employing natural and circular economy building methods, we can lessen the impact that housing has on the surrounding environment. The elimination of waste, the maximization of resource efficiency, and the promotion of reuse and recycling are all results of designing buildings and infrastructure with a circular economy in mind. These studies, taken as a whole, shed light on how important eco-friendly buildings, effective policies, and sustainable management strategies are for the creation of a sustainable future. They help mitigate climate change, preserve biodiversity, and safeguard ecosystems, leading to long-term environmental and socio-economic benefits. The obvious knowledge gap concerning the effects of wind loads on high-rise structures that have porous double-skin facades is the primary focus of the current research, and the primary objective of this research is to close that gap. It is anticipated that the findings that were obtained will make a significant contribution to the existing body of knowledge on the subject, which will contribute to the advancement of research endeavors, the improvement of engineering applications, and the establishment of a reliable reference for international standards and codes. Here, we have presented energy-efficient solutions, including smart building design strategies in the house-building sector.

3.7 CONCLUSION AND FUTURE PROSPECTS

Eco-friendly management involves sustainable practices in construction. This may include efficient waste management, recycling, and eco-friendly technologies. Ventilated double-skin facades (VDFs) reduce a building's heating, cooling, and greenhouse gas emissions by using thermal interactions between the inside and outside. This includes creating and enforcing renewable energy, carbon emission, and natural resource protection policies. We can encourage the use of environmentally friendly materials in the construction of homes and get family members in the habit of water conservation by educating and raising their awareness about the issue. It protects ecosystems, mitigates climate change, and preserves biodiversity. We can reduce housing's environmental impact by using natural and circular economy construction methods. Designing buildings and infrastructure for a circular economy reduces waste, maximizes resource efficiency, and encourages reuse and recycling. Eco-friendly policy and management promote environmental conservation and reduce environmental impact to create sustainability. We can build homes and buildings that benefit ecosystems by incorporating nature into the built environment. Fostering behaviors that save energy and promote the responsible utilization of environmentally friendly materials will lead to sustainability.

REFERENCES

Abbas, S., Saleem, O., Rizvi, M. A., Kazmi, S. M. S., Munir, M. J., Ali, S. Investigating the energy-efficient structures using building energy performance simulations: A case study. Applied Sciences (Switzerland). 2022. 12(18). https://doi.org/10.3390/app12189386.

Al-Ali, A. R., El-Hag, A., Bahadiri, M., Harbaji, M., El Haj, Y. A. Smart home renewable energy management system. Energy Procedia. 2011. 12:120–126. https://doi.org/10.1016/j.egypro.2011.10.017.

Ali, K. A., Ahmad, M. I., Yusup, Y. Issues, impacts, and mitigations of carbon dioxide emissions in the building sector. Sustainable. 2020. 12:7427. https://doi.org/10.3390/SU12187427

Alkatheiri, M. S., Alqarni, M. A., Chauhdary, S. H. Cyber security framework for smart home energy management systems. Sustainable Energy Technologies and Assessments. 2021. 46:101232.

Amirmohammad, B., Ahmad, A. Feasibility study of a smart building energy system comprising solar PV/T panels and a heat storage unit. Energy. 2020. 210:118528. https://doi.org/10.1016/j.energy.2020.118528.

Building Materials. 10 Sustainable building materials for eco-friendly construction. 2023. https://theconstructor.org/sustainability/10-sustainable-building-materials-for-eco-friendly-construction/569864/

Campisano, A., David, B., Sarah, W., Burns, M. J., Eran, F., Kathy, D., Fisher-Jeffes, L. N., Enedir, G, Ataur, R, Hiroaki, F, Mooyoung, H. Urban rainwater harvesting systems: Research, implementation and future perspectives. Water Research. 2017. 115. 195–209. https://doi.org/10.1016/j.watres.2017.02.056.

Carlier, R., Dabbagh, M., Krarti, M. Impact of wall constructions on energy performance of switchable insulation systems. Energies. 2020. 13(22). https://doi.org/10.3390/en13226068.

Chwieduk, D., Chwieduk, M. Determination of the energy performance of a solar low energy house with regard to aspects of energy efficiency and smartness of the house. Energies. 2020. 13(12). https://doi.org/10.3390/en13123232

Cruz, A. S., de Carvalho, R. S., da Cunha, E. G. Passive house alternative proposal for the brazilian bioclimatic zone 8. International Journal of Sustainable Development and Planning. 2020. 15(6):827–833. https://doi.org/10.18280/ijsdp.150605

Echarri, V. Thermal ceramic panels and passive systems in mediterranean housing: Energy savings and environmental impacts. Sustainability (Switzerland). 2017. 9(9). https://doi.org/10.3390/su9091613.

Elkholy, M. H., Senjyu, T., Lotfy, M. E., Elgarhy, A., Ali, N. S., Gaafar, T. S. Design and implementation of a real-time smart home management system considering energy saving. Sustainability. 2022. 14(21):13840. https://doi.org/10.3390/su142113840.

European Environment Agency. Construction and demolition waste: Challenges and opportunities in a circular economy. www.eea.europa.eu/publications/construction-and-demolition-waste-challenges.

Fares, R., Webber, M. The impacts of storing solar energy in the home to reduce reliance on the utility. Nature Energy. 2017. 2:17001. https://doi.org/10.1038/nenergy.2017.1

Farouk Mohamed, A. Comparative study of traditional and modern building techniques in siwa oasis, egypt: Case study: Affordable residential building using appropriate building technique. Case Studies in Construction Materials. 2020. 12. https://doi.org/10.1016/j.cscm.2019.e00311.

Fedorik, F., Alitalo, S., Savolainen, J., Räinä, I., Illikainen, K. Analysis of hygrothermal performance of low-energy house in nordic climate. Journal of Building Physics. 2021. 45(3):344–367. https://doi.org/10.1177/1744259120984187

Fuinhas, J. A., Koengkan, M., Silva, N., Kazemzadeh, E., Auza, A., Santiago, R., . . . Osmani, F. The impact of energy policies on the energy efficiency performance of residential properties in portugal. Energies. 2022. 15(3). https://doi.org/10.3390/en15030802

García Kerdan, I., Morillón Gálvez, D. Artificial neural network structure optimisation for accurately prediction of exergy, comfort and life cycle cost performance of a low energy building. Applied Energy. 2020. 280. https://doi.org/10.1016/j.apenergy.2020.115862

García-Vázquez, F., Guerrero-Osuna, H. A., Ornelas-Vargas, G., Carrasco-Navarro, R., Luque-Vega, L. F., Lopez-Neri, E. Design and implementation of the E-Switch for a Smart Home. Sensors. 2021. 21(11):3811.

Gennaro, G., Catto Lucchino, E., Goia, F., Favoino, F. Modelling double skin façades (DSFs) in whole-building energy simulation tools: Validation and inter-software comparison of naturally ventilated single-story DSFs. Building and Environment. 2023. 231:110002. https://doi.org/10.1016/j.buildenv.2023.110002.

Georgescu, S., Şova, D., Campean, M., Coşereanu, C. A sustainable approach to build insulated external timber frame walls for passive houses using natural and waste materials. Forests. 2022. 13(4). https://doi.org/10.3390/f13040522

Ghaffarianhoseini, A., Ghaffarianhoseini, A., Berardi, U., Tookey, J., Li, D. H. W., Kariminia, S. Exploring the advantages and challenges of double-skin façades (DSFs). Renewable and Sustainable Energy Reviews. 2016. 60:1052–1065. https://doi.org/10.1016/j.rser.2016.01.130.

Godlewski, T., Mazur, Ł., Szlachetka, O., Witowski, M., Łukasik, S., Koda, E. Design of passive building foundations in the polish climatic conditions. Energies. 2021. 14(23). https://doi.org/10.3390/en14237855

Gürel, A. E., Ağbulut, Ü., Ergün, A., Ceylan, İ. Environmental and economic assessment of a low energy consumption household refrigerator. Engineering Science and Technology, An International Journal. 2020. 23(2):365–372.

Gwoździej-Mazur, J., Jadwiszczak, P., Kaźmierczak, B., et al. The impact of climate change on rainwater harvesting in households in Poland. Apply Water Science. 2022. 12:15. https://doi.org/10.1007/s13201-021-01491-5

Hamburg, A., Mikola, A., Parts, T., Kalamees, T. Heat loss due to domestic hot water pipes. Energies. 2021. 14(20). https://doi.org/10.3390/en14206446.

Hohne, P.A., Kusakana, K., Numbi, B.P. A review of water heating technologies: An application to the South African context. Energy Reports. 2019. 5:1–19. https://doi.org/10.1016/j.egyr.2018.10.013.

Hu, Y., Heiselberg, P. K., Larsen, T. S. Demand control strategies of a pcm enhanced ventilation system for residential buildings. Applied Sciences (Switzerland). 2020. 10(12). https://doi.org/10.3390/app10124336

Jang, H. J., Kim, T. H., Chae, C. U. CO_2 emissions and cost by floor types of public apartment houses in south korea. Sustainability (Switzerland). 2016. 8(5). https://doi.org/10.1016/j.jestch.2019.06.003

Jeong, M. G., Rathnayake, D., Mun, H. S., Dilawar, M. A., Park, K. W., Lee, S. R., Yang, C. J. Effect of a sustainable air heat pump system on energy efficiency, housing environment, and productivity traits in a pig farm. Sustainability (Switzerland). 2020. 12(22):1–13. https://doi.org/10.3390/su12229772.

Kanama, N., Ondarts, M., Guyot, G., Outin, J., Gonze, E. Indoor air quality campaign in an occupied low-energy house with a high level of spatial and temporal discretization. Applied Sciences (Switzerland). 2021. 11(24). https://doi.org/10.3390/app112411789.

Kazemian, A., Seylabi, E., Ekenel, M. Concrete 3D printing: Challenges and opportunities for the construction industry. In: Ghaffar, S.H., Mullett, P., Pei, E., Roberts, J. (eds) Innovation in construction. 2022. https://doi.org/10.1007/978-3-030-95798-8_12

Kim, S., Yun, S., You, Y. Eco-friendly speed control algorithm development for autonomous vessel route planning. Journal of Marine Science and Engineering. 2021. 9(6) https://doi.org/10.3390/jmse9060583.

Koengkan, M., Fuinhas, J. A., Radulescu, M., Kazemzadeh, E., Alavijeh, N. K., Santiago, R., Teixeira, M. Assessing the role of financial incentives in promoting eco-friendly houses in the lisbon metropolitan Area—Portugal. Energies. 2023. 16(4). https://doi.org/10.3390/en16041839.

Komala, C. R., Vimal, S., Ravindra, G., Hariramakrishnan, P., Razia, S., Geerthik, S., . . . Arappali, N. Deep learning for an innovative photo energy model to estimate the energy distribution in smart apartments. International Journal of Photoenergy. 2022. https://doi.org/10.1155/2022/1048378.

Lan, Y., Zou, R. Study on thermal performance measurement and construction of passive exterior wall with low energy: Teaching and laboratory building of shandong jianzhu university. Advances in Materials Science and Engineering. 2022. https://doi.org/10.1155/2022/3253085

Leang, E., Tittelein, P., Zalewski, L., Lassue, S. Impact of a composite trombe wall incorporating phase change materials on the thermal behavior of an individual house with low energy consumption. Energies. 2020. 13(18). https://doi.org/10.3390/en13184872.

Liew, K. M. Sojobi, A. O., Zhang, L. W. Green concrete: Prospects and challenges. Construction and Building Materials. 2017. 156:1063–1095. https://doi.org/10.1016/j.conbuildmat.2017.09.008.

Li, M., Gu, W., Chen, W., He, Y., Wu, Y., Zhang, Y. Smart home: Architecture, technologies and systems. Procedia Computer Science. 2018. 131:393–400. https://doi.org/10.1016/j.procs.2018.04.219.

Mahmoud, R., Kamara, J. M., Burford, N. Opportunities and limitations of building energy performance simulation tools in the early stages of building design in the UK. Sustainability (Switzerland). 2020. 12(22):1–28. https://doi.org/10.3390/su12229702

Manandhar, R., Kim, J.-H. Environmental, social and economic sustainability of bamboo and bamboo-based construction materials in buildings. Journal of Asian Architecture and Building Engineering. 2019. 18(2):49–59. https://doi.org/10.1080/13467581.2019.1595629

Manouseli, D., Kayaga, S. M., Kalawsky, R. Evaluating the effectiveness of residential water efficiency initiatives in England: Influencing factors and policy implications. Water Resources Management. 2019. 33:2219–2238. https://doi.org/10.1007/s11269-018-2176-1

Manzoor, B., Othman, I., Sadowska, B., Sarosiek, W. Zero-Energy buildings and energy efficiency towards sustainability: A bibliometric review and a case study. Applied Sciences (Switzerland). 2022. 12(4). https://doi.org/10.3390/app12042136

Marszal-Pomianowska, A., Jensen, R. L., Pomianowski, M., Larsen, O. K., Jørgensen, J. S., Knudsen, S. S. Comfort of domestic water in residential buildings: Flow, temperature and energy in draw-off points: Field study in two danish detached houses. Energies. 2021. 14(11). https://doi.org/10.3390/en14113314

Miller, S. A., Habert, G., Myers, R. J., Harvey, J. T. Achieving net zero greenhouse gas emissions in the cement industry via value chain mitigation strategies. One Earth. 2021. 4:1398–1411. https://doi.org/10.1016/j.oneear.2021.09.011

Modus Staff. 2021. https://ww3.rics.org/uk/en/modus/natural-environment/renewables/the-future-of-sustainable-building-materials.html.

Mohammed, A.K., Hamakhan, I. A. Analysis of energy savings for residential electrical and solar water heating systems. Case Studies in Thermal Engineering. 2021. 27:101347. https://doi.org/10.1016/j.csite.2021.101347.

Navabi, D., Amini, Z., Rahmati, A., Tahbaz, M., Butt, T. E., Sharifi, S., Mosavi, A. Developing light transmitting concrete for energy saving in buildings. Case Studies in Construction Materials. 2023. 18:e01969. https://doi.org/10.1016/j.cscm.2023.e01969.

Neri, M., Levi, E., Cuerva, E., Pardo-Bosch, F., Zabaleta, A. G., Pujadas, P. Sound absorbing and insulating low-cost panels from end-of-life household materials for the development of vulnerable contexts in circular economy perspective. Applied Sciences (Switzerland). 2021. 11(12). https://doi.org/10.3390/app11125372

Niskanen, J., Rohracher, H. Mainstreaming passive houses: More gradual reconfiguration than transition. Journal of Environmental Policy and Planning. 2022. 24(6):612–624. https://doi.org/10.1080/1523908X.2021.2019575

Nowotna, A., Pietruszka, B., Lisowski, P. IOP Conference series: Earth and environmental science. Central Europe Towards Sustainable Building (CESB19), 2–4 July 2019, Prague, Czech Republic. https://doi.org/10.1088/1755–1315/290/1/012024.

Pan, W., Mei, H. A design strategy for energy-efficient rural houses in severe cold regions. International Journal of Environmental Research and Public Health. 2020. 17(18):1–18. https://doi.org/10.3390/ijerph17186481

Ragheb A et al., 2016- P.A. Hohne, K. Kusakana, B.P. Numbi. A review of water heating technologies: An application to the South African context. Energy Reports. 2019. 1–19. https://doi.org/10.1016/j.egyr.2018.10.013.

Ragheb, A., El-Shimy, H., Ragheb, G. Green architecture: A concept of sustainability. Procedia—Social and Behavioral Sciences, 2016. 216:778–787. https://doi.org/10.1016/j.sbspro.2015.12.075.

Sassine, E., Dgheim, J., Cherif, Y., Antczak, E. Low-energy building envelope design in Lebanese climate context: The case study of traditional lebanese detached house. Energy Efficiency. 2022. 15(8). https://doi.org/10.1007/s12053-022-10065-6

Šatrevičs, V., Voronova, I., Bajare, D. Investigation of social opinion on green lifestyle and eco-friendly buildings. decision making criteria. Journal of Sustainable Architecture and Civil Engineering. 2021. 28(1):56–71. https://doi.org/10.5755/j01.sace.28.1.28092.

Shen, J., Copertaro, B., Zhang, X., Koke, J., Kaufmann, P., Krause, S. Exploring the potential of climate-adaptive container building design under future climates scenarios in three different climate zones. Sustainability (Switzerland). 2020. 12(1). https://doi.org/10.3390/SU12010108.

Shim, H., Kim, T., Choi, G. Technology roadmap for eco-friendly building materials industry. Energies. 2019. 12(5). https://doi.org/10.3390/en12050804.

Shimoda, Y., Sugiyama, M., Nishimoto, R., Momonoki, T. Evaluating decarbonization scenarios and energy management requirement for the residential sector in japan through bottom-up simulations of energy end-use demand in 2050. Applied Energy. 2021. 303. https://doi.org/10.1016/j.apenergy.2021.117510

Škvorc, P., Kozmar, H. The effect of wind characteristics on tall buildings with porous double-skin façades. Journal of Building Engineering. 2023. 69:106135. https://doi.org/10.1016/j.jobe.2023.106135.

Tim, T. C. Challenges and opportunities in tropical concreting. Procedia Engineering. 2014. 95:348–355. https://doi.org/10.1016/j.proeng.2014.12.193.

Tsai, M., Wonodihardjo, A. S. Achieving sustainability of traditional wooden houses in indonesia by utilization of cost-efficient waste-wood composite. Sustainability (Switzerland). 2018. 10(6). https://doi.org/10.3390/su10061718

Wang, Y., Tahmasebi, F., Cooper, E., Stamp, S., Chalabi, Z., Burman, E., Mumovic, D. An investigation of the influencing factors for occupants' operation of windows in apartments equipped with portable air purifiers. Building and Environment. 2021. 205. https://doi.org/10.1016/j.buildenv.2021.108260

Yoon, Y., Seo, B., Mun, J., Cho, S. Energy savings and life cycle cost analysis of advanced double skin facade system applied to old apartments in South Korea. Journal of Building Engineering. 2023. 71:106535. https://doi.org/10.1016/j.jobe.2023.106535.

Zhen, Y., Maragatham, T., Mahapatra, R. P. Design and implementation of smart home energy management systems using green energy. Arabic Journal of Geoscience. 2021. 14:1886. https://doi.org/10.1007/s12517-021-08206-9.

Zheng, B., Sui, J., Tan, Y., Zhang, L. Thermal performance analysis of exterior wall materials of huizhou residential buildings adapted to local climate. Annales De Chimie: Science Des Materiaux. 2019. 43(2):89–94. https://doi.org/10.18280/acsm.430204

Zheng, W., Wei, F., Su, S., Cai, J., Wei, J., Hu, R. Effect of the envelope structure on the indoor thermal environment of low-energy residential building in humid subtropical climate: In case of brick–timber vernacular dwelling in china. Environmental Technology and Innovation. 2022. 28. https://doi.org/10.1016/j.eti.2022.102884

Zhu, J., & He, G. (2019). Heat transfer coefficients of double skin facade windows. Science and Technology for the Built Environment, 25(9), 1143–1151. https://doi.org/10.1080/23744731.2019.1624447.

Zhu, Y., Liu, L., Qiu, Y., Ma, Z. Design of the passive solar house in qinba mountain area based on sustainable building technology in winter. Energy Reports. 2022. 8:1763–1777. https://doi.org/10.1016/j.egyr.2022.03.026.

4 Building-Integrated Greenery Systems

Bhartendu Mani Tripathi, Samakshi Verma, and Sonu Kumar

4.1 INTRODUCTION

A sustainable "building-integrated greenery system" solution involves incorporating live plants into exterior or interior architecture. Lower energy expenses, improved air quality, and improved aesthetics are benefits of building-integrated greenery systems. Vertical gardens, green walls, and green roofs are a few examples of available building-integrated greenery systems. A green roof covers a structure's roof with a layer of flora that may insulate a building and reduce heat absorption. On the other hand, green walls are vertical installations of plants on the interior or external walls of a structure. Although vertical gardens can be also constructed indoors or outdoors, they are freestanding, unlike green walls. It is crucial to consider several aspects when installing an extensive system, including the environment, the type of plants, and the upkeep needs. Certain plants do better in specific environments, while others may need more or less upkeep. Considering the building's structural integrity and any required irrigation and drainage systems is crucial. Using a building-integrated greenery system in the design of a building may be a terrific approach to promote sustainability and improve the usability and aesthetic appeal of the structure. Figure 4.1 shows the different greenery systems for integration in buildings.

4.2 WHY PLANTS?

Green indoor and outdoor spaces are essential in architecture because they develop a deep connection to nature. These green spaces soften the constructed environment, making it more welcoming and calmer while reducing the hazards connected with sick building syndrome and adding to people's psychological well-being. Plants not only enhance the environment physically, but they also improve psychological wellness. Roger Ulrich's research showed that exposure to living plants can considerably improve overall health. Plants provide several benefits to both buildings and residents. NASA research has shown that having at least one living plant per 100 square feet of floor space can successfully filter the air in an office. This purification technique is very effective against volatile organic compounds (VOCs), a significant contributor to sick building syndrome. Tropical plants excel at eliminating these toxic VOCs through natural processes such as photosynthesis, which also lowers CO_2

DOI: 10.1201/9781003415695-4

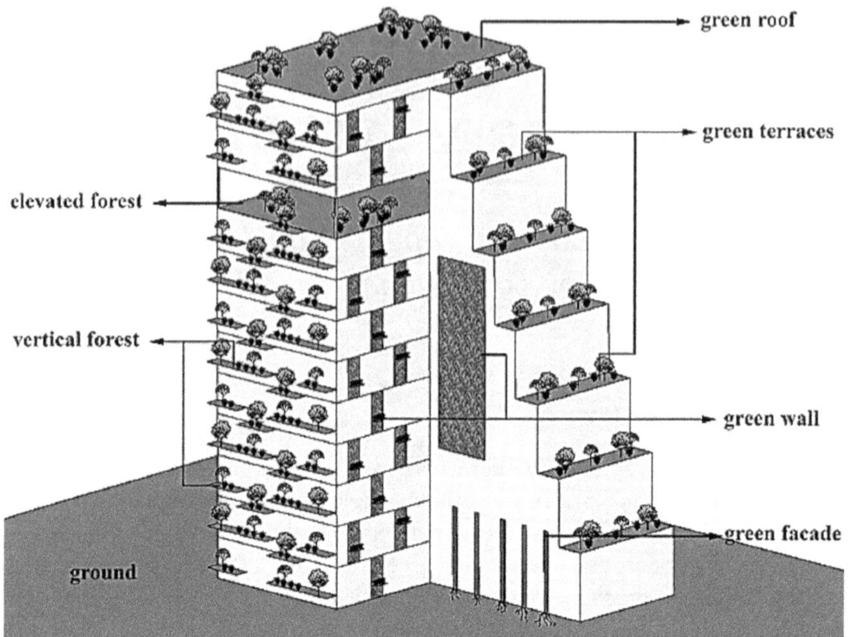

FIGURE 4.1 Different greenery systems for integration in buildings [1].

levels and increases oxygen release, enabling fresh and clean breathing for building occupants. In particular, trees contribute to environmental well-being by absorbing CO_2 during photosynthesis, providing shade for outdoor spaces, and functioning as noise barriers in metropolitan contexts. Plants also help to cool buildings by lowering air temperatures and lessening the demand for air-conditioning systems during hot summer days.

Furthermore, additional quantitative benefits connected with plants need to be better understood. These include stress reduction, as studies have shown that being around plants reduces stress levels, increases work productivity, and fewer complaints about symptoms linked with sick-building syndrome. In conclusion, integrated greenery systems offer an innovative approach to building design that integrates plants and vegetation into the built environment, providing various benefits to building occupants and the surrounding environment. By improving air quality and energy efficiency and creating green spaces in urban environments, IGSs (Integrated greenery systems) represent a sustainable solution to building design that is gaining popularity worldwide. Vegetation solves most indoor and outdoor problems. Optimal air quality plays a crucial role in architectural design, fostering a sense of well-being and nourishing the spirit. Plants, through photosynthesis, actively remove carbon dioxide while releasing oxygen, effectively cleansing the air of harmful chemicals. They contribute to enhanced air quality by replenishing oxygen levels, moderating humidity and temperature, capturing dust particles, absorbing pollutants, and generating ions. Studies led by Dr. Craig Knight and his team at the Universities of Exeter and Groningen have explored the impact of workspace environments on productivity and well-being. By comparing minimalist workspaces with enriched ones featuring elements like indoor plants, researchers observed a notable improvement of over 15% in well-being metrics such as workplace comfort, identity, engagement, and altruistic behavior. Interestingly, the research suggests that the benefits of incorporating plants into indoor and outdoor spaces extend beyond their direct environmental effects. Even modest greenery, such as grass and a few small trees, can significantly enhance individuals' ability to cope and adapt.

4.3 ADVANTAGES OF INTEGRATED GREENERY SYSTEMS

The enhancement of air quality is one of IGS's main advantages. Plants generate oxygen and absorb carbon dioxide and other air pollutants, purifying the air and fostering a healthier indoor atmosphere. In addition, plants can help to reduce noise pollution, making them ideal for urban environments where noise can be a significant problem.

IGS can also improve energy efficiency by providing natural insulation and reducing heat absorption. Green roofs, for example, can help reduce heat gain and loss through the roof, leading to lower energy costs and improved comfort for building occupants.

By offering shade and lowering heat absorption, green walls and living facades can also aid in controlling a building's temperature. The development of green spaces in urban settings is another advantage of IGS. IGS can restore nature to the city and make it more inviting and beautiful for tenants and the neighborhood by incorporating plants and vegetation into the building design. Other advantages of green environments include less stress, enhanced mental health, and increased productivity. The different advantages of utilizing integrated vegetation systems are as follows:

1. Better air quality: Plants release oxygen and absorb carbon dioxide, helping to improve the environment and reduce air pollution. Moreover, green roofs and living walls can filter airborne particles and contaminants.
2. Energy efficiency: Living walls and green roofs can increase insulation and temperature control, lowering the demand for heating and cooling systems and resulting in energy savings.
3. Stormwater management: By absorbing rainwater and lowering the amount of water entering stormwater systems, integrated greenery systems can help manage stormwater runoff and prevent flooding and the risk of water contamination.
4. Noise reduction: Green roofs and living walls are effective noise reduction strategies because plants can absorb and block out sounds.
5. Enhanced biodiversity: By offering a habitat for birds, insects, and other wildlife, living walls and green roofs can improve biodiversity in urban settings.
6. Better mental health: Research has shown that being around greenery can improve mental health by lowering stress and anxiety levels and elevating mood.
7. Aesthetics: Integrated plant systems can beautify and provide visual interest to structures and urban areas, increasing their allure and attractiveness to visitors.

Integrated greenery systems provide several advantages that might enhance human and environmental health and well-being.

4.4 DIFFERENT WAYS OF INTEGRATING GREENERY SYSTEMS WITHIN BUILDINGS

Greenery in buildings can be incorporated either in horizontal or vertical greenery systems. An example of a horizontal greenery system is the addition of flora and plants to horizontal surfaces like roofs, balconies, terraces, and even pavement. From conventional rooftop gardens to hanging baskets or planter boxes on balconies, horizontal greenery systems come in various shapes and sizes. The primary objective is to incorporate plants into the built environment since they may reduce heat gain and energy use, improve air quality, increase biodiversity, and enhance the area's visual appeal, among other advantages. When there is a lack of green space or room for traditional gardens, horizontal greenery systems are frequently employed in urban settings. They may also raise the standard of living for those who live in high-rise buildings by giving them access to a natural environment and a place to unwind and breathe fresh air. Ultimately, horizontal greenery systems provide several advantages to humans and the environment while offering a sustainable and eco-friendly solution to introduce greenery into urban spaces. Vertical greenery systems (VGSs) entail mounting vertical plant systems on building walls or facades, which has various advantages, including lowering the urban heat island effect, enhancing visual appeal, and improving air quality and biodiversity. Systems for vertical greenery can be as basic as trellises with climbing plants or as sophisticated as systems with soil, irrigation, and a wide range of plant species. They are adaptable to various climates and settings and can be constructed to fit practically any building or structure, from single-family houses to tall business towers. There are several advantages to vertical greenery systems. They can aid in decreasing building temperatures, which will minimize energy use and the urban heat island effect. In addition to providing a home for wildlife and fostering biodiversity in urban settings, they also contribute to improved air quality by filtering pollutants and absorbing carbon dioxide. Vertical greenery systems offer a sustainable and eco-friendly solution to add greenery to urban settings and several advantages for both people and the environment. Different types of IGSs, each with specific advantages and difficulties, can be included in buildings to improve their environmental performance and sustainability. For instance, green roofs need to be carefully planned and installed to ensure that they are structurally solid and do not leak. Green walls must be adequately irrigated and maintained to guarantee that the plants stay healthy and continue giving the desired advantages. The many strategies for incorporating a vegetation system in green buildings are depicted in Figure 4.2.

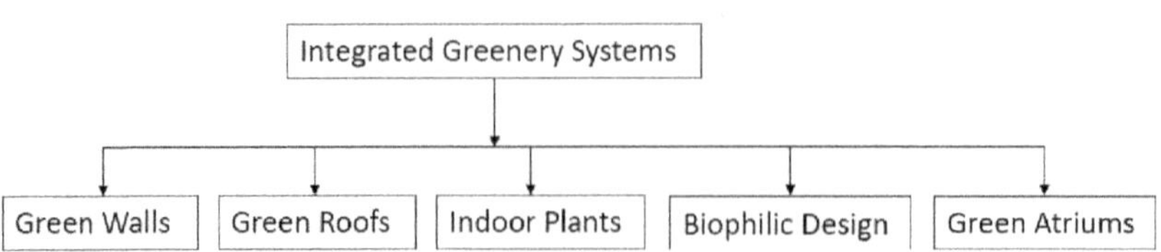

FIGURE 4.2 Different methods of integrating greenery systems in green buildings.

4.4.1 Green Roofs

Green or vegetative roofs are constructed atop a waterproofing membrane and covered in vegetation, such as plants and grasses, growing in a growing medium (soil, sand, and gravel). The different layers of the biodiverse roof are shown in Figure 4.3.

Green roofs provide a range of benefits to buildings and their surroundings, including:

1. Reducing the urban heat island effect: Green roofs can reduce the heat island effect. In this phenomenon, urban areas are significantly warmer than surrounding rural areas due to the absorption and retention of heat by buildings and other structures. Vegetation and soil on green roofs absorb heat and release moisture through evapotranspiration, which cools the air and reduces the temperature on the roof and surrounding areas.
2. Improving air quality: Plants on green roofs filter pollutants from the air, thereby improving the air quality in the surrounding area.
3. Reducing stormwater runoff: Green roofs absorb rainwater and release it slowly, reducing the amount of stormwater runoff and the strain on stormwater management systems.
4. Extending the life of the roof: The layer of vegetation on green roofs acts as a protective layer for the waterproofing membrane, thereby extending the roof reliability.

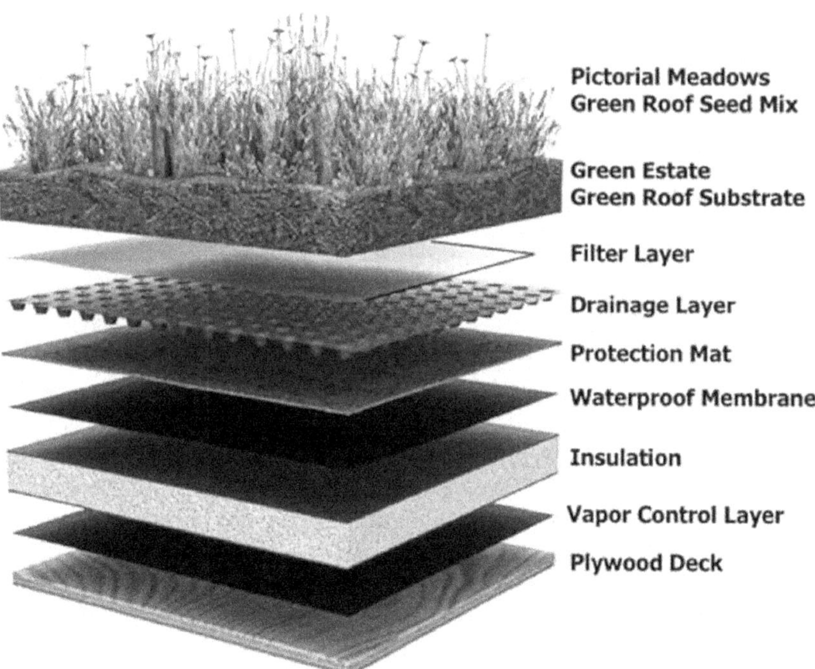

FIGURE 4.3 Different layers of vegetative roofing [2].

FIGURE 4.4 (a) Intensive green roofing. (b) Extensive green roofing [4], [5]

There are two types of green roofs, intensive and extensive green roofs. Both of them are vegetative roofs, which are roof systems that incorporate vegetation in some form. However, they differ in terms of their depth of soil, vegetation, and maintenance requirements.

A shallow soil covering (often 2 to 6 inches deep) and low-maintenance plants like sedum, grasses, and herbs define an extended green roof [3]. Large-scale green roofs are made to be lightweight, with little need for fertilizing or irrigation. They offer many advantages, such as enhanced air quality, energy savings, stormwater management, and aesthetic appeal. Unfortunately, only a small variety of plants may be planted due to the shallow soil layer. Extensive green roofs may offer several advantages over more intricate dense green roofs.

On the other hand, an extensive green roof is distinguished by a deeper soil layer (about 6 inches deep) and a wider variety of vegetation possibilities, including shrubs and small trees. Intensive green roofs require more upkeep than extensive green roofs because of the greater variety of plants and the deeper soil layer. However, they offer other advantages, including more extraordinary biodiversity and comprehensive stormwater management.

The structural integrity of the building, the desired degree of vegetation, and the care requirements all play a role in determining whether a green roof is intensive or extensive. Another category is semi-intensive green roofs, which falls between extensive and intensive green roof systems, characterized by small herbaceous plants, ground covers, grasses, and small shrubs, requiring moderate maintenance and occasional irrigation. A typical growing medium depth for a semi-intensive green roof is around 150 mm. This system can retain more stormwater than an extensive system and supports a larger range of species of vegetation. This green roof system provides the potential to make a formal garden, but it demands higher maintenance compared to an extensive system since the plants tend to need pruning, irrigation, and fertilization. The comparison between different green roofing system is given in Table 4.1.

TABLE 4.1
Comparison between Different Types of Green Roofing System [6].

Characterstics	Extensive	Semi-Intensive	Intensive
Depth of substrate	150 mm or less	About 150 mm	More than 150 mm
Accessibility	Often inaccessible	Partially accessible	Usually accessible
Fully saturated weight	70–170 Kg/m^2	170–290 Kg/m^2	290–970 Kg/m^2
Plant diversity	Low	Greater than extensive	Greatest
Plant communities	Moss, sedum Herbs and grasses	Grass, herbs, and shrubs	Lawn herbs and shrubs
Uses	Ecological protection layer	Desgined green roof	Park-like garden
Cost	Low	Medium	Highest
Maintance	Minimal	Medium	Highest

4.4.2 Green Walls

A green wall is a vertical garden that can be installed on the exterior or interior of a building. This system can help purify the air, provide insulation, and create a visually appealing aesthetic. Vertical gardens are an innovative way to bring nature indoors and maximize green spaces in urban environments. They consist of plants and vegetation grown vertically on a wall or other vertical surface. Some of the benefits of vertical gardens in buildings are as follows [7]:

1. Improving air quality: Vertical gardens improve interior air quality and give tenants a better environment by filtering airborne toxins with plants.
2. Reducing urban heat island effect: The plants in vertical gardens absorb heat and release moisture through transpiration, reducing the temperature inside the building and improving the overall climate.
3. Enhancing aesthetics: Vertical gardens are visually appealing and provide a natural element to the interior or exterior of a building.
4. Providing insulation: Vertical gardens can also provide insulation to buildings, reducing energy consumption for heating and cooling.
5. Reducing noise pollution: The vegetation in vertical gardens can absorb and reduce noise, creating a more peaceful environment inside the building.
6. Increasing property value: Buildings with vertical gardens may have a higher property value and marketability due to their sustainability and aesthetic appeal.
7. Creating a positive impact on mental health: Studies have shown that exposure to green spaces and nature can reduce stress and improve overall mental health, making vertical gardens a valuable addition to building interiors.

FIGURE 4.5 Illustrative image of a vertical garden [8].

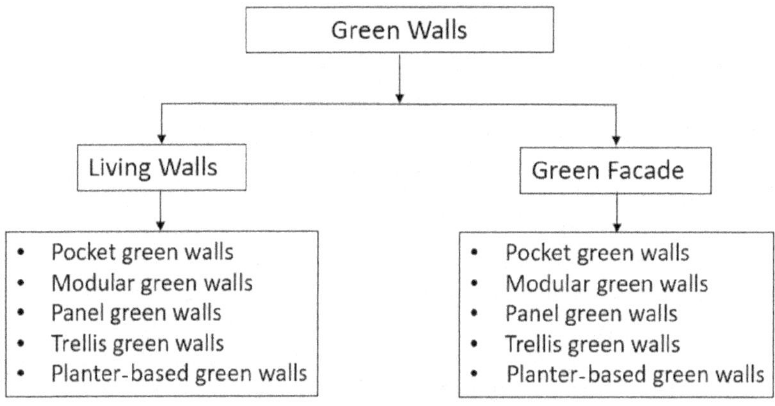

FIGURE 4.6 Classification of green walls.

Vertical gardens in buildings provide numerous benefits and are a sustainable and environmentally friendly way to maximize green spaces in urban environments.

4.4.2.1 Types of Green Walls

Green walls system can be divided into two categories based on their techniques and application. Living walls and green facades. Further, living walls and green facades can be divided into sub-categories depending on maintenance, desired plant diversity, and budget. Figure 4.7 shows living wall and green facade.

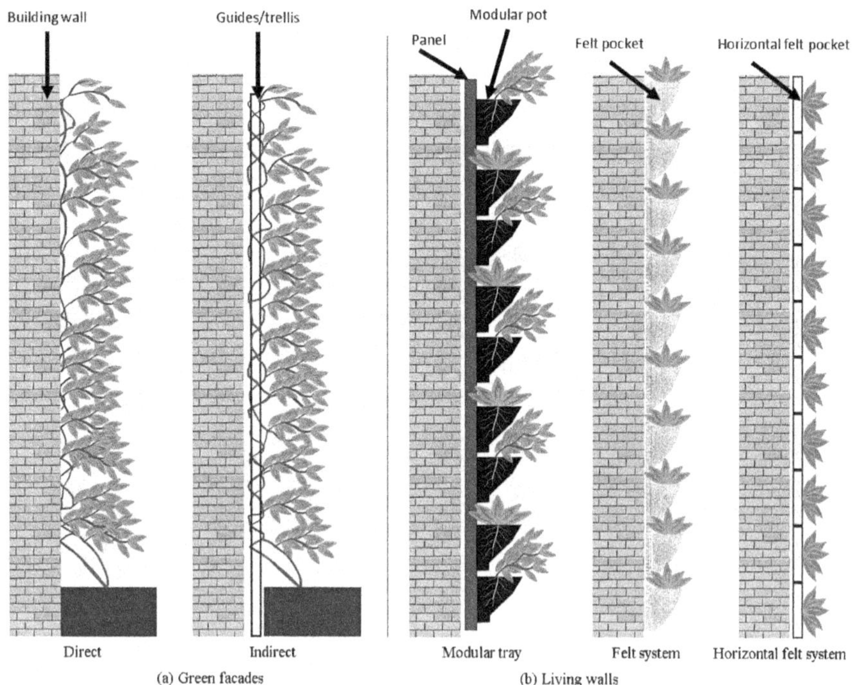

FIGURE 4.7 (a) Green facade and (b) living wall design [9].

Modular trellis panel System and Cable & wire rope net system. On the other hand, Living Wall techniques can be divided into three sub-categories: Modular living walls, biofilters, and Vegetated mat walls. There are several types of green walls, including:

4.4.2.1.1 Pocket Green Walls

Pocket green walls are similar to planter-based green walls but use pockets to hold the plants. The pockets can be made from various materials, including felt or canvas, and can be attached to a wall or freestanding. Pocket green walls are a type of vertical garden designed to be compact and portable. They consist of a panel or frame with pockets or compartments that can hold soil or a soilless growing medium and plants. These panels can be used indoors and outside and mounted on walls, fences, or other vertical surfaces. Green pocket walls are a popular solution for people who wish to add greenery to tiny spaces or give walls or other vertical surfaces visual flair.

They can also be utilized as decorative elements, room dividers, or privacy screens. Depending on the design and surroundings of the wall, different plants can be utilized to create green pocket walls. Succulents, herbs, and miniature-blooming plants are examples of common plants. Since the plants are growing vertically, maintenance of green pocket walls may be more complex than of conventional horizontal gardens. It may also call for more frequent feeding and watering. However, many pocket green walls incorporate integrated irrigation systems to make upkeep simpler.

FIGURE 4.8 Illustrative image of pocket green walls [10].

Environmentally friendly materials like recycled plastic felt or fabric are frequently used to create green pocket walls. They are a popular option for DIY gardeners or people wishing to establish a small-scale vertical garden without the money or labor of a more enormous, permanent installation because they are lightweight and straightforward.

4.4.2.1.2 Modular Green Walls

Modular green walls are pre-fabricated panels that can be easily installed on any surface. They come in various sizes and forms and are made to sustain a wide range of plant types. Vertical gardens or living walls are other names for modular green walls, a type of green infrastructure involving growing plants on a vertical surface. They consist of easy-to-assemble modules or panels mounted on fences, walls, or other structures. Urban regions with limited space frequently choose modular green walls because they enable the creation of green spaces in inaccessible locations. They can also offer several environmental advantages, such as better air quality, reduced noise, and thermal insulation. Many kinds of modular green walls are available, from soil-based systems that rely on conventional gardening methods to hydroponic systems that employ a nutrient-rich water solution to support plant development. Some modular green walls are also made self-sufficient with drainage and irrigation systems included to keep plants healthy. In a variety of contexts, including residences, workplaces, and public areas, modular green walls can be used [11]. The plant varieties can be modified to meet specific design requirements and vary depending on the region and desired appearance. For those looking to add more greenery to their surroundings, modular green walls are a popular alternative since they provide several aesthetically pleasing and environmentally beneficial advantages.

4.4.2.1.3 Panel Green Walls

Panel green walls, like modular green walls, are typically created to fit certain places. They are made of many materials, including wood, metal, or plastic, and can support a variety of plant types. Panel green walls or green wall systems are vertical gardens with panels or modules to encourage plant growth. These systems typically comprise numerous interconnected panels and can be installed on a wall or other vertical surface. Several irrigation and drainage systems for green panel walls are included, making them easy to install and maintain. With panels being offered in various sizes, shapes, and materials, they may also be altered to meet unique design requirements. Panel green walls have a variety of uses, which is one of their benefits. They can be used inside and outside, in private residences, workplaces, public areas, and commercial structures. Also, they can offer many environmental advantages like better air quality and less heat absorption. Panel green walls may accommodate a variety of plants, including herbs, flowers, and miniature trees [12]. Specific systems enable the incorporation of edible plants, making them a popular option for community agriculture and urban gardening initiatives. Anybody looking for a diverse and adaptable solution to add greenery to their surroundings frequently chooses green panel walls. They are generally simple to install and maintain and can offer a variety of environmental advantages as well as aesthetic appeal.

4.4.2.1.4 Trellis Green Walls

Trellis green walls are vertical gardens that support climbing plants using a trellis or lattice construction. The trellis can be fashioned in various ways and made from various materials, including wire, metal, and wood, to suit diverse surroundings and aesthetics. Typically, climbing or trailing plants like ivy, jasmine, or climbing roses create green trellis walls. A living green wall that offers shade, privacy, and aesthetic appeal can be made by training these plants to grow up and along the trellis. Trellis green walls can meet a space's precise measurements and can be utilized indoors and outdoors. However, they can also be used in commercial or public contexts, such as outside buildings or parks. They are popular in home settings such as balconies, patios, and gardens. Because the plants need constant trimming and training to ensure they properly grow along the trellis, maintaining green trellis walls can be more complex than maintaining other vertical gardens. Maintaining the health and vitality of the plants also requires fertilization and watering [13]. Green trellis walls can soften and make hard surfaces feel more inviting. They can be a lovely and valuable addition to any room. The decision will be based on the location's unique needs and constraints for the green wall installation. Each green wall design has its advantages and disadvantages.

4.4.2.1.5 *Plant-based Green Walls*

Vertical gardens that employ plants to cover and grow on walls or other vertical surfaces are called plant-based green walls. These can be found indoors and outdoors and are gaining popularity because they bring more vegetation into cities, enhance air quality, and provide areas with visual flair. Green walls can be created in various methods, but they typically include a supporting framework, a planting medium, and a watering and drainage system. With hydroponics or aeroponics, the plants can be raised either in soil or in a soil environment. They can be chosen based on beauty, upkeep needs, and environmental factors. In addition to lowering the heat island effect in cities and insulating buildings, green walls can improve air quality by removing pollutants. Also, they can enhance biodiversity, make spaces more welcoming and pleasant, and offer chances for urban agriculture. Ferns, succulents, and other low-maintenance species that may flourish in vertical growing circumstances are some of the common plants used for green walls. However, the specific setting and layout of the green wall will determine the plants to be used. A green wall's long-term health and success may necessitate professional installation and upkeep.

4.4.3 Green Facades

Buildings with green facades have outside vegetation. These plants are often planted on a unique support structure linked to the building's exterior, giving it a lush and attractive appearance. Green facades can improve air quality, lessen the impact of urban heat islands, provide insulation, and lower a building's overall energy usage, among other things. Due to the plants' ability to absorb and filter water, they can also aid in reducing the quantity of rainfall runoff from structures. Green facades can be fitted to various building types and architectural styles on both new and old construction. They can produce a particular pattern or image on the building's exterior using various plant species, from blooming vines to succulents. Soil-based and hydroponic green facades are the two primary varieties. A growing soil media is used in containers or pockets affixed to the building's exterior for soil-based green facades. Plants are grown in these pots and either manually watered or watered by an irrigation system. Soil-based green facades are employed for projects that are more noticeable and long-lasting. Hydroponic green facades use a system that allows plants to grow without soil using a nutrient-rich solution instead. The solution is delivered to the plants through drip irrigation or misting systems. Hydroponic green facades are often used in smaller installations or locations where soil-based systems may need to be more practical. Green facades can also be categorized by design, including panel, modular, and cable systems. Panel systems are pre-grown panels of plants attached to the building's facade. Modular systems use individual plant modules arranged in different patterns and configurations. Cable systems use wire cables to support the plants and can be used to create a more minimalistic design. By integrating these greenery systems within buildings, we can create more sustainable, healthy, and attractive living and working spaces that benefit both people and the environment.

4.4.3.1 Types of green facades

Different types of green facades exist, depending on where they are mounted and what plant is used. It often depends on the building itself which solutions are the most

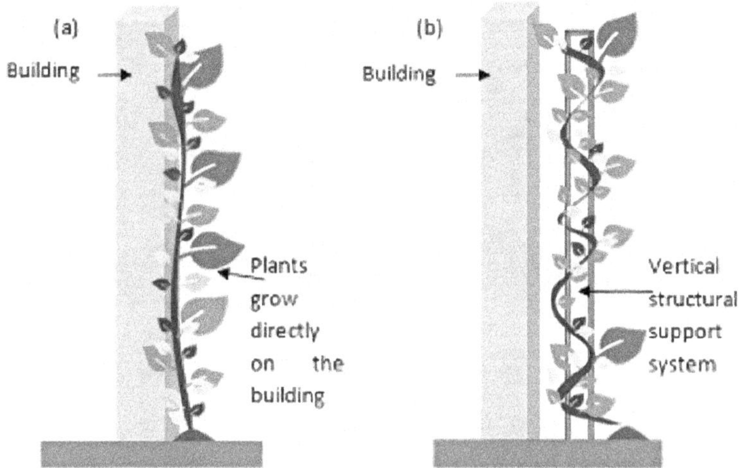

FIGURE 4.9 Illustrative image of green facade. (a) Traditional green facade. (b) Double-skin green facade [14].

suitable. Generally, factors like costs, maintenance, and underlying structure should be kept in mind.

4.4.3.1.1 Traditional green facades

The traditional green facade is a vertical garden used for centuries in various cultures worldwide. It typically involves using climbing plants trained to grow up the exterior of a building, either on a trellis or a wire mesh structure. The plants used in a traditional green facade are often chosen for their aesthetic qualities, such as colorful flowers or attractive foliage. However, they may also provide other benefits like shade, privacy, or insulation.

Conventional green facades can benefit buildings and their occupants in several ways. They can offer insulation for structures, lessen the urban heat island effect, and enhance a building's aesthetic appeal. A trellis or wire mesh framework is often installed on the building face before climbing plants are planted, making the construction and upkeep of a classic green facade very straightforward. They must be pruned regularly to maintain the plant's health and vigor. Ivy, wisteria, honeysuckle, and clematis are climbing plants frequently utilized in traditional green facades. These plants can help buildings and their occupants in various ways because they are well-suited to growing vertically.

4.4.3.1.2 Double-skin green facades

Buildings and their occupants can benefit from double-skin green facades in several ways. They might improve the air quality, lessen the impact of the urban heat island, and improve building insulation. A double-skin green facade demands particular knowledge and tools for installation and upkeep. Typically, the approach entails mounting a trellis or wire mesh framework on the building face, afterward covered in climbing plants.

FIGURE 4.10 Traditional green facade [15].

FIGURE 4.11 Double-skin green facades [15].

The various plant species are then planted on the inside side of the trellis after the modular panels have been put in place. The plants are watered using a drip or mist irrigation system, and regular pruning and upkeep may be necessary to maintain their health and vigor. The Hydrotech Wall Garden and Oxyvital Green Facade systems are two double-skin green facade systems. These solutions combine the advantages of a climbing plant with a modular green facade.

4.4.3.1.3 Continuous green facade

A vertical garden, known as a continuous green facade, covers the whole exterior of a building instead of being constructed in modular panels. A living wall system produces a continuous green surface on a building's front. A continuous green facade uses plants specifically chosen to flourish in the unique growth circumstances of the facade. They are often selected for their capacity to survive the challenging weather factors on a building's exterior, including high winds, glaring sunlight, and dry soil. Buildings and their occupants can gain various advantages from continuous green facades. They can enhance air quality, reduce the impact of urban heat islands, and provide building insulation. They can also contribute to a building's aesthetic appeal and make its occupants feel more at home. Continuous green facade installation and upkeep require specialized knowledge and tools. On the building face, a substrate or growing media is usually installed as part of the system before other plant species are added. The plants are watered using a drip or mist irrigation system, and regular pruning and upkeep may be necessary to maintain their health and vigor. The Green Façade and Bioskin systems are two instances of continuous green facade systems. These systems, which are frequently utilized in commercial and public settings, are created to seamlessly produce a green surface on the facade of a structure.

4.4.3.1.4 Modular green facade

Plants are cultivated in individual modules positioned all over the wall for a modular green facade. In the event of plant growth issues, it is simpler to substitute certain sections. Also, maintenance is less expensive because the facade hidden by the vegetation is easier to access. A vertical garden that may be erected on the outside of a building is a modular green facade. It is made up of several readily installed and maintained modular panels. The panels often have tiny pockets or cells for planting and are composed of lightweight materials like recycled plastic. The plants in a modular green facade are specifically chosen to flourish in the facade's unique growing circumstances. They are frequently selected for their capacity to withstand the challenging environmental factors on a building's facade, such as ferocious winds, glaring sunlight, and arid soil [16]. Buildings and those living in them can benefit from modular green facades in several ways. They can improve the air quality, lessen the impact of the urban heat island, and insulate buildings. The Fytowall, LivePanel, and Green screen systems are a few instances of modular green facade systems. These systems are available in various sizes and configurations to meet various structures and environments.

4.4.3.1.5 Direct green facade

A building facade with plants and vegetation is known as a "direct green facade" since the plants are grown right on the facade's surface. The plants are supported by a wire mesh or trellis system attached to the building and can be grown in soil or a hydroponic system. The direct green facade provides several advantages, including improving air quality, consuming less energy, lowering the impact of the urban heat island, and offering natural insulation. The plants used in the facade can be selected based on their adaptability to the local climate and aesthetic appeal, and they can be switched out seasonally for diversity. A direct green facade can significantly reduce a building's energy needs for cooling and heating because the plants serve as natural insulation and shading. Also, the plants remove carbon dioxide and other air pollutants, enhancing air quality and fostering a healthier atmosphere for residents [17]. Direct green facade installation and maintenance make present particular difficulties. The installation procedure can be challenging and needs thorough planning and engineering to guarantee that the facade is structurally sound and does not affect the structure. Since the plants require regular watering, pruning, and fertilization, maintenance can also be more difficult and expensive. The direct green facade is an innovative and sustainable way to improve buildings' environmental performance and make urban landscapes more appealing and sustainable. However, for it to succeed, careful planning and upkeep are necessary.

4.4.3.1.6 Indirect green facade

When a building has an indirect green facade, the plants are not fastened to the wall in any direct way. Mesh or wires are typically installed along the wall so the plants can climb and develop higher. A building facade with plants and flora is called an indirect green facade. The plants are cultivated on a separate structure that is positioned in front of the building, as opposed to a traditional green facade where they are planted directly on the front. In front of the building, the structure, which can be a trellis, wire mesh, or other support systems, allows the plants to grow vertically. Besides offering insulation and shade, the indirect green facade also reduces noise pollution, improves air quality, and improves the appearance of the surrounding region. The plants used in the facade can be selected based on their adaptability to the local climate and aesthetic appeal, and they can be switched out seasonally for diversity. The ability to be installed on existing structures makes an indirect green facade one of its key benefits and an excellent option for upgrading older buildings [18]. The installation procedure is often less invasive than a conventional green facade because the support structure is attached to the building facade rather than having plants grown directly on it. It also makes it simpler to maintain and, if required, replace the plants. The indirect green facade is a creative way to enhance a building's environmental performance while improving the urban surroundings' aesthetics and sustainability.

4.4.4 Indoor Plants

Any interior space can benefit greatly from having indoor plants because they offer aesthetic value and have many positive health effects. By releasing oxygen and absorbing carbon dioxide, plants enhance air quality and lower the number of dangerous poisons in the atmosphere. They can also aid in lowering stress, boosting productivity, and elevating mood. Regarding their function as a greenery system, indoor plants can be employed to produce a lush, green environment within the home that is comparable to outdoor gardens. They can be used as a backdrop or a focal point to enhance a room's decor. The area's lighting and humidity levels must be considered while selecting indoor plants as a greenery system. While some plants do better in low light than others, some do better in more sunshine. Like us, some plants prefer dry environments, while others need high humidity. Popular houseplants include peace lilies, pothos, snakes, and spiders [19]. These plants are simple to maintain and do well in various lighting and humidity levels.

Indoor plants can serve as an effective greenery system, providing aesthetic and health benefits while creating a more natural and inviting indoor environment.

4.4.5 Biophilic Design

The biophilic design approach aims to promote a sense of connection between people and nature by introducing natural elements into the built environment. It may have water features, natural lighting, and organic materials. An approach to architecture and interior design known as "biophilic design" aims to incorporate natural aspects into constructed surroundings. It involves utilizing organic materials, natural lighting, and, most significantly, plants and other vegetation. Biophilic design can offer several advantages to interior areas as a vegetation system. Biophilic design can enhance overall well-being, reduce stress and anxiety, and improve air quality by incorporating plants and other natural elements. According to research, outdoor exposure has been linked to improved cognition, creativity, and productivity. Plants and other natural

FIGURE 4.12 Building design using a biophilic approach [21].

features can improve a space's aesthetic appeal and be suitable for physical and mental well-being. They can enhance the environment by adding texture, color, and visual curiosity [20]. A few examples of biophilic design strategies include living walls, green roofs, indoor gardens, and organic elements like wood, stone, and water features.

These methods can be used indoors in commercial buildings, retail stores, medical facilities, or residential constructions. Incorporating plants into interior spaces using the biophilic design method is a novel and effective way to promote health and well-being while enhancing the surrounding region's aesthetic appeal.

4.4.6 Green Atriums

A building's green atrium is a space that is open to the elements of nature and vegetation. It can provide residents of buildings with a peaceful haven, improve the air quality, and bring in more natural light. Atriums featuring many plants and other natural elements, usually trees and plants, are called "green atriums." Atriums can serve as focal points for both visitors and inhabitants. They are often enormous, open spaces frequently placed in the center of a structure or near the entrance. Atriums with vegetation provide beauty to interior spaces. They can improve interior air quality, reduce noise, and regulate temperature and humidity. Indoor vegetation has also been shown to reduce stress and improve general well-being [22].

Green atriums can also lower the demand for artificial lighting and energy use by offering natural light. They can reduce energy costs and offer significant environmental benefits. Depending on the specific needs and goals of the area, there are many different ways to design a green atrium. Some may be built to support a variety of plant species, while certain atriums may concentrate on including trees or water features. The design of the atrium can also vary, ranging from more organic and naturalistic to modern and minimalist forms. Green atriums may successfully incorporate vegetation into interior spaces, providing occupants with several benefits and generating an aesthetically pleasing and stimulating environment.

4.4.7 Vertical Greenery Systems

Plants growing on vertical surfaces are typically referred to as vertical greenery systems (VGSs) [23]. In other words, this is the process through which various plant species, whether created by humans or occurring naturally within or outside of a building, can grow on vertical surfaces [24]. The VGS may be affixed to the building's wall surface or standing in front of it [25]. VGSs are basically any kinds of plants growing on any kinds of vertical building surfaces [26]. The VGSs can be categorized as extensive and intensive systems in accordance with the criteria of implementation and maintenance costs [27, 28]. There is yet another classification of VGSs based on plant species, construction techniques, and growing conditions [29]. The different literature studies identifies four kinds for vertical greenery systems (VGSs), including wall-climbing type, module type, hanging-down type, and tree-against-wall type, as shown in Figure 4.13.

Vertical greenery systems (VGSs) and tree-against-wall systems enhance greenery in architectural spaces similarly despite differing operational principles. Wall

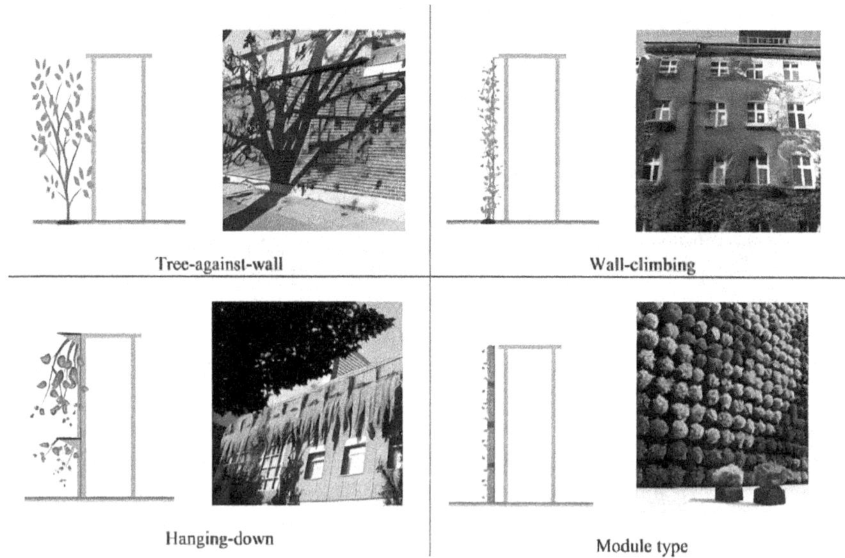

FIGURE 4.13 Different categories of vertical greenery systems [30].

climbers, which are common in traditional structures, use plants to cover wall surfaces or climb up trellises. While simple to deploy, this VGS requires time to cover the space with flora [31]. Hanging-down VGSs are commonly found on balconies or rooftops, quickly transforming them into rich green places. In VGSs, color, and aesthetic appeal may be considered, as well as the selection of diverse plant species. Module-type VGSs are a newer technology that provides benefits such as colorfulness, aesthetic variation, quick development, ease of modification, and adaptability for diverse plant varieties [31]. Since the early twentieth century, there has been an increase in research on VGSs, stressing their importance in urban contexts. The use of VGSs and green roofs has grown in popularity as people become more aware of the environmental impact of buildings on both the outside and inside environments [32]. Today, VGSs are intended to provide long-term solutions with important implications for society, the environment, and the economy [34]. Using sustainable materials alone is no longer sufficient in green building development. VGSs, the insulating characteristics of green roofs, tree shading effects, and urban heat reduction via evapotranspiration are all factors to consider in addition to sustainable materials [32]. VGSs play an essential role in "greening processes," helping to turn building envelopes into green spaces [32]. Both plants and VGSs provide benefits in urban areas by changing microclimates through shading, reflection, and absorption. Green spaces provide a direct cooling impact, and vertical garden systems are an innovative approach to introducing varied plant species into metropolitan settings [35]. Combining green roofs and VGSs is considered an excellent temperature management technique, while VGSs, as a novel approach to temperature control, require additional research and evaluation [41]. Attractive VGSs improve urban environments, increase biodiversity, reduce pollution, and provide economic benefits through energy savings and lower surface temperatures [45]. The ability of VGSs to accommodate many plant species with different attributes is a crucial advantage [46].

4.4.7.1 Significance of Vertical Greenery Systems

Economic, environmental, and social advantages of vertical greenery systems are among their three categories of benefits [47–50]. Vertical greenery systems (VGSs) function similarly to tree-against-wall systems, albeit with distinct operational principles. In traditional buildings, wall climbers often utilize plants to directly cover wall surfaces or climb trellises, albeit this method requires more time to fully green the area. Hanging-down type VGSs, commonly found on balconies or rooftops, rapidly transform buildings into lush green spaces. To enhance the aesthetic appeal of buildings, it may be necessary to incorporate various plant species and add color within VGSs. Research on VGSs in metropolitan areas has surged since the 20th century, highlighting their importance in urban settings. The consideration of VGSs and green roofs has intensified due to the evident environmental impact of buildings on both external surroundings and internal climates. Nowadays, VGSs are designed to provide long-term solutions with significant societal, environmental, and economic impacts. Sustainable building development now encompasses not only the use of eco-friendly materials but also the integration of VGSs, green roofs for insulation, tree shading, and urban heat reduction through evapotranspiration.

Both plants and VGSs offer advantages for urban environments by influencing microclimates through shading, reflection, and absorption. Research confirms the direct relationship between green spaces and temperature, demonstrating that even small green areas can effectively cool surroundings. Utilizing VGSs on underused building surfaces presents a strategic approach to incorporating diverse plant species into metropolitan areas. Optimal use of greenery systems in buildings involves combining green roofs with VGSs, leveraging their respective benefits in temperature management. While green roofs have been commonly used for temperature control, VGSs represent a novel approach requiring further research and evaluation. Attractive VGSs have the potential to enhance urban environments, promote biodiversity, reduce pollution, and yield economic benefits such as energy savings and surface temperature reduction. The versatility of VGSs lies in their ability to accommodate a variety of plants with different characteristics.

1. Environmental advantages: There are several environmental advantages to vertical greenery systems. For instance, plants used in vertical landscaping trap dust and purify the air [51, 52], acting as a form of natural air filtration. Additionally, according to photosynthesis, plants take in carbon dioxide and expel oxygen [53–55]. This improves air quality and lowers carbon dioxide emissions. The earth warms up because carbon dioxide blankets its surface like a blanket [56]. Additionally, it causes an increase in significant greenhouse gas growth in the atmosphere [57]. The potential of vertical greenery systems to regulate noise and their usage as barriers for noise abatement are two further significant environmental benefits [58, 59]. They can also lessen noise disturbance and sound reflection [48].

2. Economic advantages: Modern society is increasingly focusing on the economic advantages of vertical greenery systems. One method is to use vertical vegetation systems as window coverings [60]. Appropriate shade systems that are prepared by vertical vegetation systems have the features of increasing daylight and minimizing discomfort glare, and they finally result in decreased electricity consumption [61]. Additionally, they can behave as spongy surfaces and regulate storm waters [62]. Vertical greenery systems are appropriate for retrofitting projects such as eco-retrofitting, which aims to improve the human and environmental conditions and is more cost-effective than demolition and reconstruction [63]. Temperature reduction is a significant benefit of vertical greenery systems in addition to their many economic advantages [64]. The primary element of human comfort that is influenced by lifestyle is temperature [65]. Construction envelope design is influenced by a variety of elements, including function, social culture, aesthetics, environment, and technology [66], yet contemporary construction materials like concrete retain heat throughout the day more than rural settings [67]. Asphalt and concrete are hard, impermeable surfaces that absorb solar radiation and also radiate it back into the atmosphere [68]. The primary causes of high temperatures, particularly in metropolitan areas, are absorption and re-radiated solar radiation [69]. It results in urban heat island, which is a serious issue in cities and urban areas. Urban heat island is the maximum temperature difference between urban areas and rural areas [70, 71]. Utilizing vegetated spaces has a significant impact on lowering urban heat island because plants absorb shortwave radiation and diminish solar radiation from hard surfaces [72, 73].

   Additionally, they create a cooler atmosphere through transpiration, evaporation [72, 74–78], and the shadowing effects of plants [60, 79]. Accordingly, constructing vertical greenery systems as well as green roofs is an appropriate strategy to mitigate urban heat island effects in dense urban regions [67, 72, 80, 81]. There has been some research on the impacts of urban greenery and green roofs in lowering urban heat island [82–84], but more research is needed on the effects of vertical greenery systems on temperature reduction and urban heat island reduction. Additionally, the installation of a vertical greenery system lowers the temperature of a building or structure and lowers the urban heat island [85, 86]. Vertical greenery systems are appropriate systems for reducing cooling energy demand and enhancing building energy efficiency because of their capacity to store heat.
3. Social advantages: The use of vertical vegetation systems for societal advantages extends back to ancient times, and the famous Babylon Hanging Garden is one such them [48, 68, 87, 88]. Because connection to nature is physically fundamental, people utilized greenery in buildings and their living spaces in various ways for aesthetic purposes. It has been demonstrated that contact with nature has a psychological impact and improves human health and well-being [89] and that plants provide areas for relaxation and rest [90, 91]. Additionally, access to green spaces can reduce stress and

lessen obesity [92]. Humans consequently instinctively demand compound greenery in urban settings and switch from grey and soulless surfaces to green screens. All respondents found houses with green facades to be aesthetically pleasant, according to an online study that compared one house with no greenery to others with various greenery positions [90]. It supports the theory put forth by Blanc [93] that plants in urban areas and compounds near buildings draw more people than plants in gardens. Making people aware of the economic and environmental benefits of vertical greenery systems is crucial, in addition to their social advantages.

4.4.7.2 General Knowledge about the Advantages of Vertical Garden Systems

Studies on greenery themes frequently focus on respondents' aesthetic senses and their desire for greenery places, locations, or particular plant species [90, 94]. There aren't many studies on people's knowledge of the financial and environmental advantages of vertical greenery systems. A survey questionnaire was created in Singapore to understand people's understanding of the benefits of vertical greenery systems [88]. The responders who were chosen from various populations did not concur that vertical greenery systems have a beneficial impact on lowering cooling energy demand via insulating.

Additionally, they held the view that vertical greenery systems could not extend building lifespan, improve water quality, or regulate runoff [88]. According to the findings of a different study, in addition to the advantages of vertical greenery systems, people also do not have adequate knowledge of the advantages of green roof systems, and some individuals were unaware that they had a green roof system put on their building [95].

To use greenery on buildings, the general public needs to be educated about the use and advantages of these systems. Landlords and investors often request the installation of vertical greenery systems due to their initial expense, but in reality, doing so has very minimal costs and many advantages [96]. This is because there is a lack of public understanding about the economic and environmental benefits of these systems. People typically employ vertical greenery systems for their graceful qualities rather than their ability to reduce temperature and generate income. In order to use the most efficient elements for lowering temperature and cooling energy consumption, a new movement to utilize these systems based on their economic and environmental benefits is required.

REFERENCES

[1] Wang X, Gard W, Borska H, Ursem B, van de Kuilen JWG. Vertical greenery systems: from plants to trees with self-growing interconnections. Eur J Wood Prod 2020; 78(5): 1031–1043. doi: 10.1007/S00107-020-01583-0.

[2] www.buildingdesignindex.co.uk/entry/112172/BUILDING-DESIGN/Green-roofs-a-guidance-article (accessed 28 April 2023).

[3] Bustami RA, Belusko M, Ward J, Beecham S. Vertical greenery systems: A systematic review of research trends. Build Environ 2018; 146: 226–237. doi: 10.1016/J.BUILDENV.2018.09.045.

[4] https://in.pinterest.com/pin/856035841651814948 (accessed 29 April 2023).
[5] Wang L. Green-roofed Colorado home is buried into the earth to save energy. https://inhabitat.com/green-roofed-colorado-home-is-buried-into-the-earth-to-save-energy (accessed 01 April 2023).
[6] Czemiel Berndtsson J. Green roof performance towards management of runoff water quantity and quality: A review. Ecol Eng 2010; 36(4): 351–360. doi: 10.1016/J.ECOLENG.2009.12.014.
[7] Manso S, et al. Bioreceptivity evaluation of cementitious materials designed to stimulate biological growth. Sci Total Environ 2014; 481(1): 232–241. doi: 10.1016/J.SCITOTENV.2014.02.059.
[8] Vertical Gardening and its Impact. www.assetzproperty.com/blog/vertical-gardening-impact (accessed 20 April 2023).
[9] Safikhani T, Abdullah AM, Ossen DR, Baharvand M. A review of energy characteristic of vertical greenery systems. Renew Sustain Energy Rev 2014; 40: 450–462. https://doi.org/10.1016/j.rser.2014.07.166.
[10] Mok K. Easy-to-install living wall system uses felt pockets for plants. www.treehugger.com/plants-on-walls-florafelt-living-wall-kit-4857954 (accessed 6 April 2023).
[11] Modular green wall I vertiss plus the vertical garden. www.vertiss.net/vertiss-plus-le-module-vegetalise?lang=en (accessed 17 April 2023).
[12] El Menshawy AS, Mohamed AF, Fathy NM. A comparative study on green wall construction systems, case study: South valley campus of AASTMT. Case Stud Construct Mater 2022; 16: e00808. doi: 10.1016/J.CSCM.2021.E00808.
[13] Manso M, Castro-Gomes J. Green wall systems: A review of their characteristics. Renew Sustain Energy Rev 2015; 41: 863–871. doi: 10.1016/J.RSER.2014.07.203.
[14] Kromoser B, Ritt M, Spitzer A, Stangl R, Idam F. Design concept for a greened timber truss bridge in city area. Sustainability (Switzerland), 2020; 12(8): 3218. doi: 10.3390/SU12083218.
[15] Manso M, Castro-Gomes J. Green wall systems: A review of their characteristics. Renew Sustain Energy Rev 2015; 41: 863–871. doi: 10.1016/J.RSER.2014.07.203.
[16] Radić M, Dodig MB, Auer T. Green facades and living walls-A review establishing the classification of construction types and mapping the benefits. Sustainability (Switzerland) 2019; 11(17). doi: 10.3390/SU11174579.
[17] Radić MR, Brković Dodig M, Auer T. Sustainability green facades and living walls-a review establishing the classification of construction types and mapping the benefits 2019. doi: 10.3390/su11174579.
[18] Talhinhas P, Ferreira JC, Ferreira V, Soares AL, Espírito-Santo D, do Paço TA. In the search for sustainable vertical green systems: An innovative low-cost indirect green Façade Structure Using Portuguese Native Ivies and Cork. Sustainability 2023; 15: 5446. doi: 10.3390/SU15065446.
[19] Liu F, Yan L, Meng X, Zhang C. A review on indoor green plants employed to improve indoor environment. J Build Eng 2022; 53: 104542. doi: 10.1016/J.JOBE.2022.104542.
[20] Zhong W, Schröder T, Bekkering J. Biophilic design in architecture and its contributions to health, well-being, and sustainability: A critical review. Frontiers of Architectural Research 2022; 11(1): 114–141. doi: 10.1016/J.FOAR.2021.07.006.
[21] Biophilic Design in Architecture. www.archiscene.net/design/biophilic-design-in-architecture (accessed 28 April 2023).
[22] Drapella-Hermansdorfer A., Gierko A. Rediscovering the 'atrium effect' in terms of the European green deal's objectives: a case study. Buildings 2020; 10: 46. doi: 10.3390/BUILDINGS10030046.
[23] Shiah, K., & Kim, J. (2011). An investigation into the application of vertical garden at the new SUB atrium [R]. doi:http://dx.doi.org/10.14288/1.0108430
[24] Loh S. Living walls—A way to green the built environment. BEDP Environ. Des. Guide 2008; 1: 1–7.

[25] Afrin, Shahrina. Green Skyscraper: Integration of Plants into Skyscrapers KTH Architecture and the Built Environment. 2009. {Afrin2009 GreenSI,title={Green Skyscraper: Integration of Plants into Skyscrapers}, author={Shahrina Afrin},year={2009}, url={https://api.semanticscholar.org/CorpusID:107564367}
[26] Cheng CY, Cheung KKS, Chu LM. Thermal performance of a vegetated cladding system on facade walls. Build Environ 2010; 45(8): 1779–1787.
[27] Pérez G, Rincón L, Vila A, González JM, Cabeza LF. Green vertical systems for buildings as passive systems for energy savings. Apply Energy 2011; 88(12): 4854–4859.
[28] Cuce PM, Cuce E, Guclu T, Demirci V. Energy saving aspects of green facades: Current applications and challenges. Green Build Constr Econ 2021; 2(2): 1–11.
[29] Hao X, Xing Q, Long P, Lin Y, Hu J, Tan H. Influence of vertical greenery systems and green roofs on the indoor operative temperature of air-conditioned rooms. J Build Eng 2020; 31: 1013–1073.
[30] Safikhani T, Abdullah AM, Ossen DR, Baharvand M. A review of energy characteristic of vertical greenery systems. Renew Sustain Energy Rev 2014; 40: 450–462.
[31] Hao X, Xing Q, Long P, Lin Y, Hu J, Tan H. Influence of vertical greenery systems and green roofs on the indoor operative temperature of air-conditioned rooms. J Build Eng 2020; 31: 1013–1073.
[32] Ottelé M. The green building envelope. Vertical Greening, Delft, 2011.
[33] Mitterboeck M, Korjenic A, Analysis for improving the passive cooling of building's surroundings through the creation of green spaces in the urban built-up area. Energy Build 2017; 148: 166–181.
[34] Zuo J, Zhao ZY. Green building research–current status and future agenda: A review. Renew Sustain Energy Rev 2014; 30; 271–281.
[35] Bernatzky A. The contribution of tress and green spaces to a town climate. Energy Build 1982; 5(1): 1–10.
[36] Picot X. Thermal comfort in urban spaces: Impact of vegetation growth: Case study: Piazza della Scienza. Energy Build 2004; 36(4): 329–334.
[37] Honjo T, Takakura T. Simulation of thermal effects of urban green areas on their surrounding areas. Energy Build 1990; 15(3): 443–446.
[38] Wong NH, Yu C. Study of green areas and urban heat island in a tropical city. Habitat Int 2005; 29(3): 547–558.
[39] MacIvor JS. Green roofs as constructed ecosystems: native plant performance and insect diversity. 2010 http://dx.doi.org/10.1371/journal.pone.0009677
[40] Nyuk HW, Yok TP, Yu C. Study of thermal performance of extensive rooftop greenery systems in the tropical climate. Build Environ 2007; 42(1): 25–54.
[41] Mir M. Green facades and building structures. 2011. http://resolver.tudelft.nl/uuid:f262c218-8801-4425-818f-08726dde5a6c
[42] Perini K, Ottelé M, Haas K, Raiteri R. Greening the building envelope, façade greening and living wall systems. Open J Ecol 2011; 1(1): 1.
[43] Chiquet C, Dover JW, Mitchell P. Birds and the urban environment: The value of green walls. Urban Ecosyst 2013; 16(3): 453–462.
[44] Charoenkit S, Yiemwattana S. Living walls and their contribution to improved thermal comfort and carbon emission reduction: A review. Build Environ 2016; 105: 82–94.
[45] Bolton C, Rahman M, Armson D, Ennos A. Effectiveness of an ivy covering at in- sulating a building against the cold in Manchester, UK: A preliminary investigation. Build Environ 2014; 80: 32–35.
[46] Jim CY, Chen WY. Assessing the ecosystem service of air pollutant removal by urban trees in Guangzhou (China). J Environ Manag 2008; 88(4): 665–676.
[47] Jaafar B, Said I, Rasidi MH. Evaluating the impact of vertical greenery system on cooling effect on high rise buildings and surroundings: A review. In Proceedings of the 12th international conference on sustainable environment and architecture (Senvar). Indonesia; 2011, p. 8.

[48] Shiah K, Kim JW, Oldridge S. An investigation into the application of vertical garden at the New SUB Atrium. The University of British Columbia, Vancouver, 2011.

[49] Kontoleon KJ, Eumorfopoulou EA. The effect of the orientation and proportion of a plant-covered wall layer on the thermal performance of a building zone. Build Environ 2010; 45:1287–303.

[50] Ong BL. Green plot ratio: An ecological measure for architecture and urban planning. Landsc Urban Plan 2003; 63:197–211.

[51] Donahue J. An empirical analysis of the relationships between tree cover, air quality, and crime in urban areas [1491329]. United States–District of Columbia, Georgetown University, Georgetown, 2011.

[52] Amir A, Yeok F, Abdullah A, Rahman A. The most effective Malaysian legume plants as bio facade for building wall application. J Sustain Dev 2011; 4:103.

[53] Ismail A, Abdul Samad M H, Abdul Rahman A. Using green roof concept as a passive design technology to minimize the impact of global warming. Second international conference on built environment in developing countries (ICBEDC 2008), Malaysia, 2008, pp. 588–598.

[54] Darlington A B, Dat JF, Dixon MA. The bio filtration of indoor air: Air flux and temperature influences the removal of toluene, ethylbenzene, and xylene. Environ Sci Technol 2001; 35: 240–246.

[55] Peck SW. Green backs from green roofs: Forging a new industry in Canada. Canada Mortgage and Housing Corporation, Montreal, 1999.

[56] Aziz HA, Ismail Z. Design guideline for sustainable green roof system. IEEE, Lampur, 2011, pp. 198–203.

[57] Ismail A, Abdul Samad MH, Rahman AMA, Yeok FS. Cooling potentials and CO_2 uptake of ipomoea pes-caprae installed on the flat roof of a single storey residential building in Malaysia. Procedia – Soc BehavSci 2012; 35: 361–368.

[58] Van Renterghem T, Botteldooren D. Reducing the acoustical façade load from road traffic with green roofs. Build Environ 2009; 44:1081–1087.

[59] Wong NH, Kwang Tan AY, Tan PY, Chiang K, Wong NC. Acoustics evaluation of vertical greenery systems for building walls. Build Environ 2010; 45: 411–420.

[60] Bass B, Liu K, Baskaran B. Evaluating rooftop and vertical gardens as an adaptation strategy for urban areas. CCAF impacts and adaptation progress report, April1, 1999–March 31, 2003, pp. 1–106.

[61] Kim G, Lim HS, Lim TS, Schaefer L, Kim JT. Comparative advantage of an exterior shading device in thermal performance for residential buildings. Energy Build 2012; 46: 105–111.

[62] Lang S. Green roofs as an urban storm water best management practice for water quantity and quality in Florida and Virginia [3416693]. University of Florida, Florida, 2010.

[63] Birkeland J. Eco-retro fitting with building integrated living systems. In: Proceedings of the 3rd CIB international conference on Smart and Sustainable Built Environment (SASBE09). Aula Congress Centre, Delft University of Technology, Delft, 2009, pp. 1–9.

[64] Bennett J. Looking behind the green facade. Feature 2012; 28: 4.

[65] Ahmed MHB, Rashid R. Thermal performance of rooftop greenery system in tropical climate of Malaysia. In: Proceedings of the conference on technology & sustainability in the built environment, 2009, pp. 391–408.

[66] Oral GK, Yener AK, Bayazit NT. Building envelope design with the objective to ensure thermal, visual and acoustic comfort conditions. Build Environ 2004; 39: 281–287.

[67] Pompeii IIWC. Assessing urban heat island mitigation using green roofs: A hardware scale modeling approach. Shippensburg University, New York, 2010.

[68] Figueroa M. Green roof performance in Los Angeles, California [1454104]. University of Southern California, Los Angeles, 2008.

[69] Rizwan AM, Dennis LYC, Liu C. Are view on the generation, determination and mitigation of Urban Heat Island. J Environ Sci 2008; 20: 120–128.

[70] Kikegawa Y, Genchi Y, Kondo H, Hana kiK. Impacts of city-block-scale counter measures against urban heat-island phenomena upon a buildingg's energy-consumption for air-conditioning. Appl Energy 2006; 83: 649–668.
[71] Kolokotroni M, GiannitsarisI, Watkins R. The effect of the London urban heat island on building summer cooling demand and night ventilation strategies. Sol Energy 2006; 80: 383–392.
[72] Kleerekoper L, van Esch M, Salcedo TB. How to make a city climate-proof, addressing the urban heat island effect. Resour Conserv Recycl 2012; 64: 30–38.
[73] Wong NHY, Wong Chen, Tee Siu, Chung Calvin. Exploring the thermal benefits of plants in industrial areas with respect to the tropical climate. In: Proceedings of the 23rd conference on passive and low energy architecture. Geneva, 2006, p. 6.
[74] Alexandri E, Jones P. Temperature decreases in an urban canyon due to green walls and green roofs in diverse climates. Build Environ 2008; 43: 480–493.
[75] Takakura T, Kitade S, Goto E. Cooling effect of greenery cover over a building. Energy Build 2000; 31: 1–6.
[76] Taib N, Abdullah A, Fadzil S, Yeok F. Anassessment of thermal comfort and users' perceptions of landscape gardens in a high-rise office building. J Sustain Dev 2010; 3: 153.
[77] Arnold J. Using green roofs to mitigate the effects of solar energy on an unconditioned building in the southern United States [1502702]. Mississippi State University, Mississippi, 2011.
[78] Sheweka SM, Mohamed NM. Green façade sasa new sustainable approach towards climate change. Energy Procedia 2012; 18:507–520.
[79] Newton JJJ. Building green–a guide to using plants on roofs, walls and pavements. City Hall, The Queen's Walk, London SE12 AA: Greater London Authority, 2004.
[80] Pompeii IIWC, Hawkins TW. Assessing the impact of green roofs on urban heat island mitigation: a hardware scale modeling approach. Geogr Bull 2011; 52: 52–61.
[81] Hui SCM, Chan HM. Development of modular green roofs for high-density urban cities. In: Proceedings of the world green roof congress, 2008, pp. 17–18.
[82] Wong NH, YuC. Study of green are as and urban heat island in a tropical city. Habitat Int 2005; 29: 547–558.
[83] Oliveira S, Andrade H, Vaz T. The cooling effect of green spaces as a contribution to the mitigation of urban heat: a case study in Lisbon. Build Environ 2011; 46: 2186–2194.
[84] Kardinal Jusuf S, Wong NH, Hagen E, Anggoro R, Hong Y. The influence of land use on the urban heat island in Singapore. Habitat Int 2007; 31: 232–242.
[85] Wong NH, Tan AYK, Tan PY, Wong NC. Energy simulation of vertical greenery systems. Energy Build 2009; 41: 1401–1408.
[86] Eumorfopoulou EA, Kontoleon KJ. Experimental approach to the contribution of plant-covered walls to the thermal behavior of building envelopes. Build Environ 2009; 44: 1024–1038.
[87] Binabid J. Vertical garden: the study of vertical gardens and their benefits for low-rise buildings in moderate and hot climates [1476132]. University of Southern California, Los Angeles, 2010.
[88] Wong NH, Tan AYK, Tan PY, Sia A, Wong NC. Perception studies of vertical greenery systems in Singapore. J Urban Plan Dev–ASCE 2010; 136: 330–338.
[89] Gidlöf-Gunnarsson A, Öhrström E. Noise and well-being in urban residential environments: The potential role of perceived availability to nearby green areas. Landsc Urban Plan 2007; 83(2–3): 115–126. https://doi.org/10.1016/j.landurbplan.2007.03.003
[90] White EV, Gatersleben B. Greenery on residential buildings: Does it affect preferences and perceptions of beauty? J Environ Psychol 2011; 31(1): 89–98. https://doi.org/10.1016/j.jenvp.2010.11.002

[91] Papadakis G, Tsamis P, Kyritsis S. An experimental investigation of the effect of shading with plants for solar control of buildings. Energy Build 2001; 33: 831–836.
[92] Nielsen TS, Hansen KB. Do green areas affect health? Results from a Danish survey on the use of green areas and health indicators Health Place 2007; 13: 839–850.
[93] Blanc P. The vertical garden: in nature and the city. W.W. Norton & Company, New York, 2008.
[94] Vanden Berg AE, Koole SL, Vander Wulp NY. Environmental preference and restoration: (How) are they related? J Environ Psychol 2003; 23: 135–146.
[95] Yuen B, Hien WN. Resident perceptions and expectations of roof top gardens in Singapore. Landsc Urban Plan 2005; 73: 263–276.
[96] Bass B. Green roofs and green walls: potential energy savings in the winter. Environment Canada, UoT Adaption and Impacts Research Division, Toronto, 2007.

5 Bioclimatic Building Technology

Ankush Banyal, D. K. Singh, and Rajan Kumar

5.1 INTRODUCTION

The environmental footprints of the conventional construction industry and alarming situation of climatic change have raised concerns about sustainability. The continuous depletion of energy sources is one of the major contributory factors for the same. Further substantial energy consumption owing to growing population, urbanization, and income and lower energy efficiency and performances of current technology has intensified the issues of energy crises at a global scale.[1–3] As per reports published in Forbes, 40% of global energy is used by buildings, half of which is wasted due to inefficiency.[4] As per another published report, current building operations contribute 27% of greenhouse gas emissions, and it's imperative to decarbonize the buildings.[5] Human comfort is the primary driving force for all the development and growth that human civilization has undergone since industrial revolution. But if it continues at the same pace, it could lead to irreversible and irreparable damage to environmental sustainability and, consequently, to the existence of humankind. Therefore, to avert this unwelcoming scenario, a paradigm shift towards enhancing the energy performance of buildings has been noticed that resulted in the emergence of bioclimatic building technology which promises to address human comfort and environmental sustainability simultaneously.[6] The technology involves the interplay of bioclimatism and architecture, with the objective to address sustainability in modern buildings and providing thermal comfort to their occupants. The technology employs the strategic use of bioclimatic designs and globally accessible technology to enhance the energy performance of the buildings. However, the energy performance of the buildings depends not only properties of a building; building controls which are occupant dependent also act as a determinant. Thus, the main idea is to ameliorate the risks associated with sensory organs to provide thermal comfort to the occupants. Bioclimatic building technology involves using local climate data to design buildings that minimize the need for active thermal comforters, resulting in reduced energy consumption and lower carbon emissions.[7]

This book chapter will discuss the bioclimatic building technology in detail, including its basic concept, the practical utility of Givoni's bioclimatic chart, and the utilization of local climate data to extend the thermal comfort. All the relevant natural strategies like passive solar heating, passive solar cooling, cooling with thermal mass, evaporative cooling, and cooling through ventilation have been discussed in detail with practical utility. The importance of vernacular architecture and its

DOI: 10.1201/9781003415695-5

inclusion in modern architecture has also been discussed. The chapter also gives a quick overview of recent approaches in the field of bioclimatic building technology with the objective of highlighting bioclimatic architecture's integration in the current scenario. Further, LEED (Leadership in Energy Efficient Design) certification based on building monitoring and performance has also been briefly presented in this chapter.

5.2 BASIC CONCEPT OF BIOCLIMATIC BUILDING TECHNOLOGY

The concept of bioclimatic building technology has been deemed to be originated from the basic principles of vernacular and traditional buildings across the world. The evolutionary process and theory of bioclimatism in architecture has been discussed in detail recently by Nguyen and Reiter, who highlighted the role of the theory of natural evolution in the emergence of innovative eco-friendly designs leading to green buildings.[8] Further, local culture, religion, and traditions also influence the vernacular architecture. To introduce the bioclimatic concept to architecture, a multidisciplinary approach is needed to be followed, as defined by Olgyay: a) *define the measure and aim of requirements for human comfort*, b) *review the existing climatic conditions*, and c) *attain a rational architectural technological solution.*[9] Thus, knowledge of biological sciences to measure human comfort, meteorological sciences to understand the climatic impact and existing climatic conditions, and engineering sciences to develop such an architecture that brings meteorological sciences and engineering sciences to the same platform to harmonize biological sciences.[8,9] The main aim of BBT is to provide human thermal comfort both indoors and outdoors by utilizing solar energy and other relevant natural environmental resources. The thermal comfort of a person is a subjective parameter evaluated by dry bulb temperature, clothing level, metabolic rate, air velocity, humidity, and mean temperature and defined by person's response to the surrounding environment and scale of satisfaction that he gets from the thermal environment.[10] To achieve thermal comfort, the monthly local climate is assessed and analyzed through the Givoni bioclimatic chart[11] (Figure 5.1) and the Mahoney table to strategize suitable bioclimatic designs.[12]

Bioclimatic designs can also be strategized by Mahoney tables, designed by British architect C. Mahoney. As the name suggests, Mahoney tables are a set of some reference tables that help to analyze the meteorological data to set some indices to draw framework guidelines for the construction of climate-friendly architecture.[12]

Setiawan has explained the use of Mahoney table to strategize bioclimatic models in a very detailed manner and explained that the charting of geographical coordinates of the targeted location is the first task, followed by the charting of average extremes of temperature and humidity and then of rainfall and wind. Sorting of selected locations into humidity groups is important to understand the interactions with the temperature extremes to diagnose the limit of thermal stress. Next step is to identify indicators of stress to come up with thermal stress indices, on the basis of which recommendations are framed and suggestions are delivered.[13] Thus, it is now clear that BBT involves deeper insights into the microclimate of the targeted location before designing any of the interior, exterior, or outdoor sections of the buildings to make the best use of natural solar energy and environmental sources with a purpose

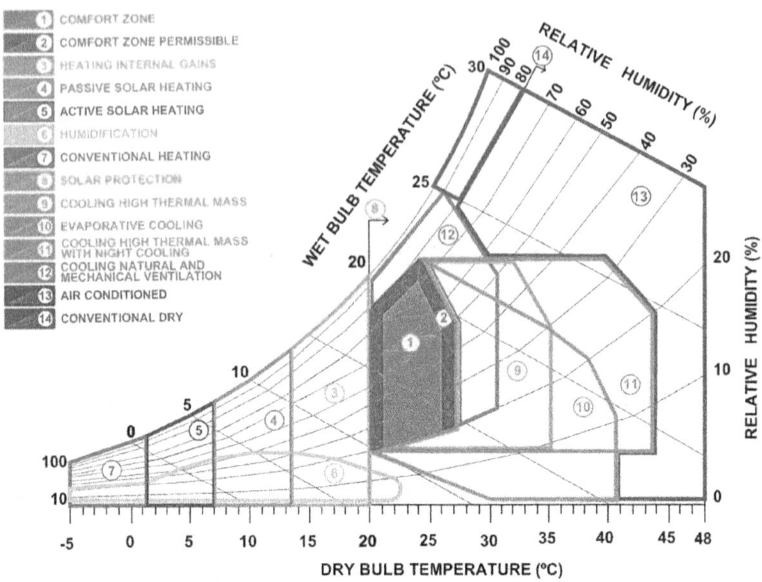

FIGURE 5.1 Givoni's psychometric bioclimatic chart.

to reduce energy consumption with maximum comfort to its residents. These environmental sources are utilized through passive solar systems for heating, cooling, and lighting the buildings, which is a basic element of bioclimatic design. However, there are some fundamental principles of bioclimatic design to have a holistic insight of BBT as proper enveloping of the building to insulate it in extreme weather; smart orientation of buildings along its openings to capture enough solar energy to utilize it as a primary source of heating in summer and light throughout the day around the year; adequate shading of the buildings to protect from the summer sun and dissipation of collected heat to the surroundings using natural means; and the designing of exterior spaces of buildings to improve microclimatic environment as a whole.[14] In order to fully appreciate the BBT, we should first understand the basic concepts of thermal load and provisions for reducing the thermal load of buildings.

5.3 THERMAL LOAD

In order to appreciate bioclimatic technologies, one must understand the scope of energy conservation in buildings. The total energy consumption, on the basis of application, in the buildings can be categorized into three components: to maintain thermal comfort, to maintain illumination, and to operate appliances (electrical and electronic devices such as ACs, heaters, computers).

Amongst the aforementioned three categories, *thermal comfort* commands the majority of energy consumption in modern buildings. Hence, the greatest scope of energy conservation lies in this category. We will restrict our discussion to the measures for reducing energy consumption in this category only. The broad categorization of the techniques for reducing thermal load while maintaining thermal comfort has been illustrated in Figure 5.2.

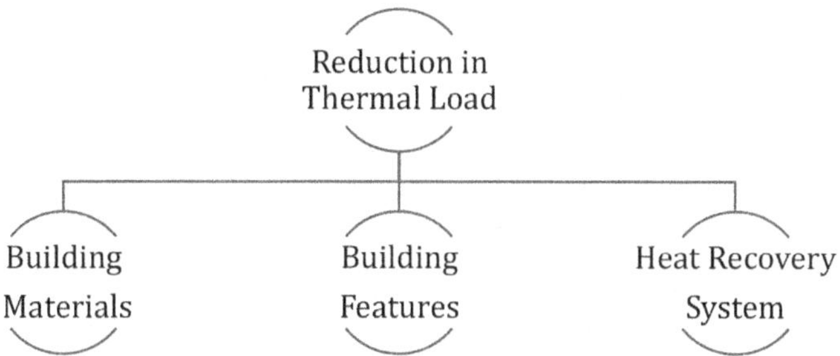

FIGURE 5.2 General approach for a reduction in thermal load.

In cold climates, the outflow of heat is prevented, whereas in hot climates inflow of heat is prevented. In the same spirit, more exposure to the sun is desirable in cold climates, and less exposure is desirable in hot climates. Therefore, the underlying principles of energy conservation are the same. A combined approach has been explained without any special reference to any specific conditions.

5.3.1 Improvements in Building Materials and Design

5.3.1.1 Insulation

Insulation of any material is done by the introduction of a material which is a poor conductor of heat. By doing so, there is an increase in the thermal resistance of the material or building, thereby reducing the heat transfer rate. As per Fourier's Law of heat conduction, the amount of heat transferred through (via conduction) the material bears a direct relation with the thermal conductivity of the material.

$$Q = -k\,A\frac{dT}{dx}$$

Q = Rate of heat transfer through the material by conduction (W)
k = Thermal conductivity of the material (W/m-K)
A = Area of the cross section normal to the direction of flow of heat
$\frac{dT}{dx}$ = Temperature gradient in the direction of the flow of heat

Insulating materials have low thermal conductivity and inhibit heat transfer. If the temperature of the surroundings is greater than that of the inside, the maximization in terms of time lag appears when the insulating material is applied on the outer side of the building material. Applying an insulation layer will not only contribute to the thermal comfort of the residents but also curb energy consumption.[15] The selection of the insulating material is dependent upon the composition, thermal conductivity, operating range, durability, ease of application, etc. The location of the insulating layer and thermal mass layer in the wall is very important and is a decisive factor. For instance, a layer of heavy insulation exterior to the thermal mass layer results in more damping of load fluctuation and a smaller peak load of the inner surface heat flow.[16] In a hot, dry climate, it should be on the outer side of the building envelope,[17] as this gives little fluctuation.[18] The application of a higher amount of thermal mass on the inner side is also considered beneficial to improve thermal comfort.[19] However, one study suggested that during intermittent heating, the application of thermal insulation on the inner side has the ability to reduce 40% of energy consumption.[20] Further, the occupant's behaviour (passive or active) also affects the efficacy of external and thermal insulation. For instance, if the behaviour is passive (no moveable shadings and closed window), internal thermal insulation is effective in terms of energy consumption; however, with active behaviour (moveable shadings and night ventilation), external thermal insulation efficiency is more.[21]

5.3.1.2 Air Cavities

The working principle of air cavities is similar to that of insulation. Air is a bad conductor of heat, and the air entrapped between two layers of the building's wall or roof acts as a barrier to heat transfer. Such a technique is effective only until no convection sets up in the cavity, which would facilitate the heat transfer, thereby defeating the purpose. Therefore, the thickness of the air cavity is a very important factor which governs its effectiveness.[22]

It has been found that the thickness of the air cavity beyond 50 mm facilitates heat transfer due to the movement of the trapped air. Therefore, cavities larger than 50 mm are generally avoided. However, in cases where greater thickness of the air cavity is required for achieving the required insulation levels, partition of the main large air cavity can be provided as an alternative.[23] The thermal resistance offered by the air cavities depends upon the thickness, location, and orientation of such cavities. Further, the most economical cavity configuration depends on the thermal emissivity and the insulation material used.[23] In addition, the air cavities between the insulation boards result in an additional cooling effect, as described by Halik et al.[24]

5.3.1.3 Improved Glazing

The glazing of a building allows the entry of not only heat but also natural light. There are times when the type and size of the window become major design parameters, as it may become the largest source of heat transfer. Ordinary windows having one layer allow sunlight into the building; however, adding further layers of glass, typically two or three layers, namely double-glazed and triple-glazed windows, respectively, results in a drastic reduction in heat transfer. These multiple layers of glazing work on the same principle as air cavities. Inert gases, mainly argon and krypton, can also be used in place of air. These inert gases have lower thermal conductivity than air. Being heavier than air, convection/movement would not set easily in these inert gases as compared to air. Moisture is often a problem on the interior size of the cold single-glazed glass. The visual problem due to condensation is not much, as it evaporates during the day since the glass receives heat during the day.[25] Nowadays, the concept of smart glazing employing the use of thermochromic, electrochromic, photochromic, and gasochromic materials is in trend with an objective of reducing energy consumption to lower the temperature by dynamic regulation of glazing transmittance, absorptance, and reflectance.[26]

5.3.1.4 Coatings

Heat transfer through glazing occurs via all three modes, that is, conduction, convection, and radiation. Special coatings having low values of emissivity are used to reduce heat transfer by radiation, especially for *long-wavelength* radiation. The value of transmissivity should be high simultaneously in order to facilitate illumination inside the space.

Such coatings result in a reduction in long-wavelength radiation from the glazed surfaces. The coatings can either be coated on the surface of the glazing or can be applied in the form of thin films. These coatings get deteriorated on the exposed surface, thereby reducing the performance of the coatings over time. Therefore, the coatings are generally applied to the first available protected surface of glass in the heat flow direction.

In the case of cold climates (heat flow from inside to outside is to be prevented), the coating is applied on the outer surface of the inner glazing. However, in hot climates (heat flow from outside to inside is to be prevented), the coating is applied on the inner surface of the outer glazing. Besides this, high transmission–low emissivity, selective transmission–low emissivity, and low transmission–low emissivity products can be used depending upon the location of the building. High transmission–low emissivity products are best suited for heating-dominated areas, particularly to south-facing windows. Selective transmission–low emissivity products are suited for locations having both summer cooling and winter heating. The low emissivity reduces heat loss during the winter, and the selective property allows the natural light during the day but blocks the summer heat. Similar effects can be achieved by combining a low-emissivity coating with a spectrally selective tinted glass.

Nowadays, thin-layer nanocomposite coatings like zirconium oxide and aluminium oxide are being used as insulation to decrease the rate of heat transfer through the building envelope. Such specialized coatings prevent energy loss effectively by acting as a thermal barrier, even at small thicknesses.[27]

5.3.1.5 Seals

Moisture is often a problem on the interior size of the cold single-glazed glass. The visual problem due to condensation is not much, as it evaporates during the day since the glass receives heat during the day. However, the moisture between the layers of the glazing is a problem. The simple way of achieving this is the proper sealing of the windows so as to avoid the entry of moisture between the glazing.[28] In cases where seals are to be provided as improved building material as a stand-alone measure for reducing the thermal load, either only secondary seals (single sealed) or both primary and secondary seals (dual sealed) can be used as per requirement. Multi-glazed windows/fenestration are sealed on the sides, which provides structural strength along with restriction of the ingress of moisture into the inside of the glazing. A further modification results in terms of the usage of the edge seal, which consists of the spacer (metal as well as non-metal) along with a desiccant and a sealant[29] (Figure 5.3).

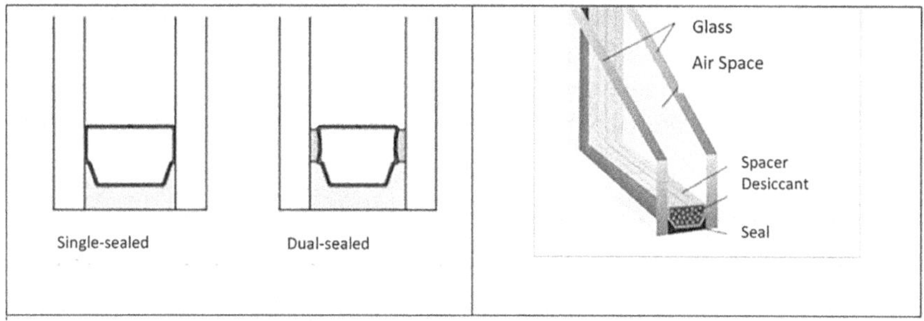

FIGURE 5.3 Single and dual seals (left) and components of double glazing (right).

5.3.2 Modifications in the Building Features

After discussing the methods of reduction in thermal load by incorporating changes in the building materials and components, further enhancement of energy conservation can be achieved by controlling air infiltration into the building, which causes unwanted air changes between the building envelope and the outside environment through openings such as cracks, clearances, fenestration, etc. Therefore, deciphering the parameters on which air infiltration depends is of significant importance.

Air infiltration between the building envelope and the outside depends on the following factors:

- The meteorological factors such as average wind speed, prevailing direction of the wind, seasonal and daily variation in wind speed and direction, etc.
- The shape of the building/house, the nature of the surroundings, and local obstacles such as nearby buildings, trees, hills, etc., as these factors influence the pressure distribution between the interior and exterior of the building.
- The airflow characteristics of all opening and flow paths.
- The temperature difference between the outside and inside of the building.

The gains/losses due to air infiltration are proportional to the number of air changes and the volume of the building. The volume of air changes is given by:

$$V = A \times L \times (\Delta P)^{n}$$

Here,

'A' stands for joint/opening discharge coefficient.

'L' stands for joint/opening length.

'ΔP' stands for pressure difference.

'n' exponent stands for pressure difference.

The pressure difference is primarily dependent upon the flow of air with respect to the building and secondarily on the temperature difference. The difference in the velocity of wind between the *windward side* (facing the prevailing wind) and *leeward side* (facing away from the prevailing wind) results in a pressure difference across the building. This effect is known as the *wind effect*.[30] The difference in temperature creates a difference in the density of the air. The less dense air (high temperature) rises up, allowing the relatively cool air to come into the lower side of the building, thereby creating a pressure difference. This effect is known as the *stack effect* or *buoyancy effect*.[31,32]

The infiltration of outside air can be reduced by making collective efforts by changing shape, orientation, etc., which can be reduced to the broader phrase of 'modifying external features of the building'.

5.3.3 Heat Recovery Systems

The heat demand of a building can be reduced by possibly recovering the waste heat. A heat recovery system in an arrangement that operates like a heat exchanger. In buildings, the heat recovery system preheats the incoming air with the help of

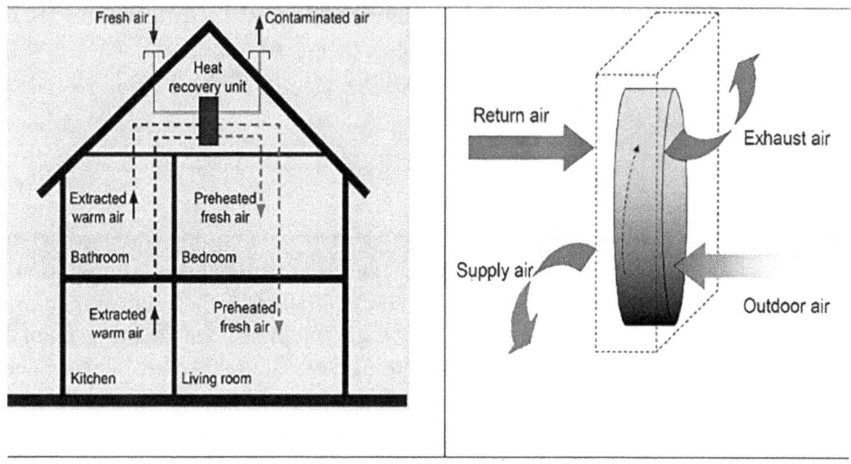

FIGURE 5.4 Heat recovery system (left) and regeneration (right).

outgoing/recycled waste heat. Heat recovery systems have the ability to recover heat in the form of sensible heat as well as latent heat; therefore, these systems can cater to the requirement of fresh air supply and indoor air humidity for the building. About 60–95% of waste energy can be recovered through these systems.[33] Figure 5.4[34] illustrates the components of a heat exchanger system. These systems are of four types: rotary wheel, fixed plate, heat pipe, and run around. All of four types have been briefly reviewed by Xu et al.,[35] where the basic concepts and functionalities of these recovery systems have been described by the authors.

5.4 BIOCLIMATIC BUILDING TECHNOLOGIES

The main emphasis of bioclimatic building technologies is on *local climate, thermal comfort and passive design technologies*, with the objective to improve energy efficiency. The bioclimatic building technologies, when implemented, move the conditions from outside the zone of thermal comfort to the zone of thermal comfort, as mentioned in the Givoni diagram (Figure 5.2).[11] Most of the time, it is achieved by passive methods, as they do not consume any energy. However, low-energy-consuming strategies can also be used in cases when passive designs do not yield the required outcome.

5.4.1 Comfort Zone and Permissible Comfort Zone

As displayed in Figure 5.1 and marked as label ‘1’, *the comfort zone* depicts the ideal condition for thermal comfort for the human body assuming light clothing, little activity, and having no energy requirements for thermal comfort.[11,36] The temperature in this zone ranges from 21 °C to 26 °C, whereas relativity humidity ranges between 20% and 70%. This zone requires no interventions. However, the *permissible comfort zone* is an extended zone wherein there are finite energy requirements. As displayed in Figure 5.1 and marked as label '2', it is an area bounded by temperature ranging from 20 °C to 27 °C, relativity humidity ranging between 20% and 80%, and the intersection of lines having coordinates of (temp. 24 °C, relativity humidity 80%) and (temp. 27 °C, relativity humidity 50%).[11,37,38]

5.4.2 Heating Internal Gains

Internal heat gain is an important parameter while calculating the cooling load and in turn is significant for the bioclimatic calculation of the building. In addition to the contribution of external factors, due to the metabolism of the *occupants* (human beings), there is heat gain in the building.[39] The amount of heat gain depends upon the level of activity and other factors such as clothing, sex, etc. In the case of space cooling, the usage of artificial lighting in the conditioned space is often required, as the use of sunlight brings in solar heat gain, which shall increase the total energy load on the air conditioner. All *lighting devices* consume electricity, which is converted into heat. The heat transfer to the building conditioned space takes place by convection and radiation. This radiation is absorbed by the various surfaces such as walls, floors, furniture, etc. and is re-released by these surfaces. There is a significant delay in the time of switching on the lighting appliance and re-radiation of the heat absorbed by the surface. There are *electric equipment and appliances* such as laptops, computers, kitchen appliances, etc. installed in the building space which add heat gain to the building in the form of convection and radiation.[40] The internal heat gains from the aforementioned parameters (occupants, lighting devices, electric equipment, appliances) can be calculated as follows:

$$Q_{s,o} = (\text{Number of people}) \times (\text{Sensible heat gain per person}) \times \text{CLF}$$
$$Q_{l,o} = (\text{Number of people}) \times (\text{Latent heat gain per person})$$
$$Q_{s,li} = (\text{Wattage}) \times (\text{Usage factor}) \times \text{CLF}$$
$$Q_{s,a} = (\text{Wattage}) \times (\text{Usage factor}) \times \text{CLF}$$

Where, Q means heat gain of the conditioned space.

First subscript 'l' represents latent, and 's' represents sensible.

Second subscript *'o', 'li', 'a'* represent occupants, lighting, and appliances, respectively.

CLF represents the cooling load factor, which can be looked up in the relevant ASHRAE handbooks.

5.4.3 Passive Solar Heating

The basic idea of passive heating is to increase the net heat gain of the building without the use of energy-consuming traditional methods that depend upon the use of fossil fuels or electricity. This is particularly usable for buildings in cold climates where the objective is to keep the conditioned space warmer than the surroundings.[41]

Passive heating techniques hinge on the three elements for achieving thermal comfort: collecting naturally available energy, storing the collected energy, and utilizing (distributing and conserving) the stored energy (Figure 5.5).

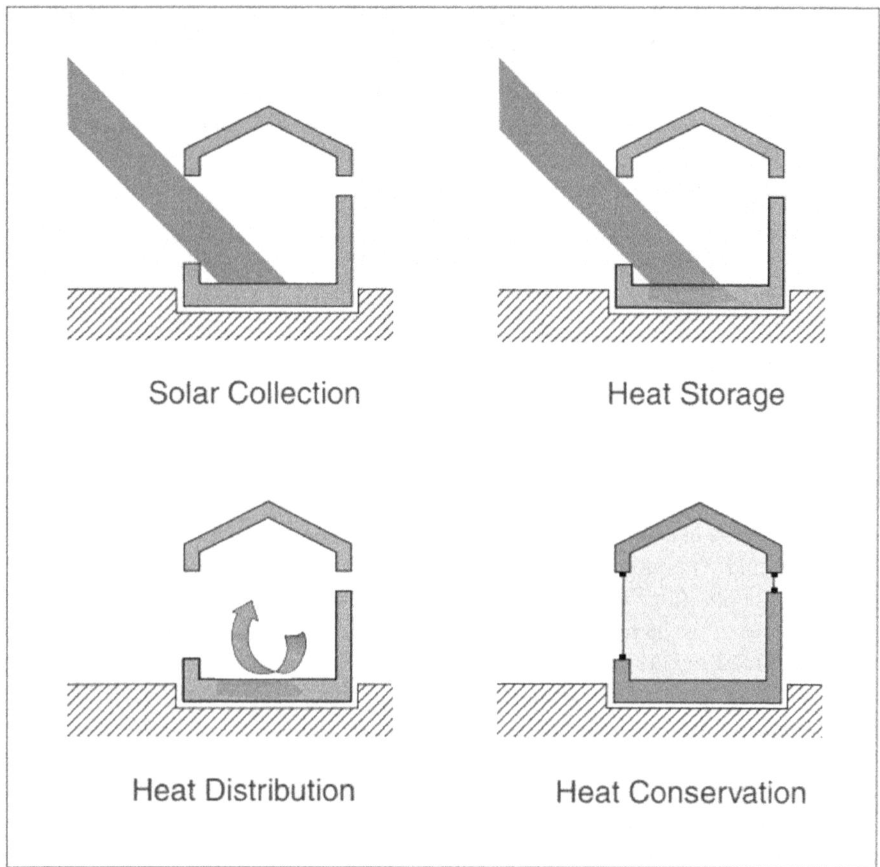

FIGURE 5.5 Steps for achieving thermal comfort.

The largest available source of energy for use is solar energy; therefore, it is the preferred source of energy for passive heating. The performance of passive solar heating devices hinges upon maximizing the exposure of solar radiation, thereby collecting maximum solar radiation. Consequently, the heat losses from the building are to be minimized, thereby achieving a greater value of net heat gain. Passive heating methods can be classified into direct and indirect methods.[42,43]

5.4.3.1 Direct

In these methods, solar energy is utilized or directed directly into the building. A few of the methods by which it can be achieved are as follows.

5.4.3.1.1 Orientation

The orientation of a building is one of the most important parameters of the bioclimatic building and decides the amount of solar radiation that can be collected for heating purposes. In the northern hemisphere, the amount of solar radiation received

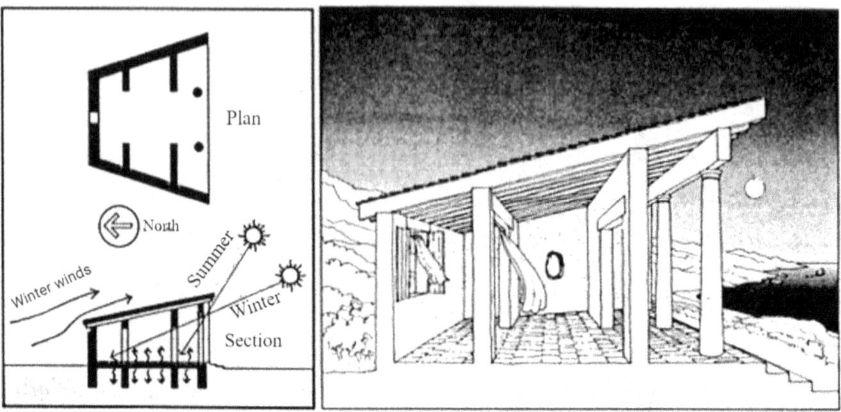

FIGURE 5.6 The house of Socrates.

by a building on the southern side is larger than that on any other side or orientation of the building. Therefore, it is desirable to have the longest axis of the building lie in an east–west direction, with glazed windows concentrated on the southern side.[44] For all the latitudes in the northern hemisphere, the south side of the building receives maximum solar radiation. Therefore, the shape of the building should be such that it meets the requirement for maximum exposure to solar radiation in winter; however, proper care should be taken for the utility in the summers. The orientation should not increase heat gain in the summer. For example, consider the House of Socrates, as shown in Figure 5.6. During winters, the sun goes southwards, and the altitude is not high, so the vertical glazing allows solar radiation to enter the house; however, the use of an appropriate length of extension avoids the undesirable radiation of the summer sun, which goes at a higher altitude.[45]

5.4.3.1.2 Glazing and Reflectors

The direct heat gain through glazing is simple. The solar radiation is transmitted through the windows or glazing. The amount of solar radiation received at a window opening is fixed by the season, weather conditions, and geometry of the building with respect to the sun. Although the intensity of direct solar radiation cannot be increased, irradiation received by the surrounding surfaces can be reflected into the opening, thereby increasing the effective collector area of the window. Clerestory windows, as shown in Figure 5.7, can also be used for reflecting solar radiation and allowing sunlight into the building space. The reflectivity of a few surfaces is given in Table 5.1.[46]

5.4.3.2 Indirect Methods

Unlike in direct heat gain, in these methods, the sun's energy is introduced into the building space in an indirect method.

5.4.3.2.1 Mass Wall

In this method, the heat from solar radiation is directly taken as a result of conduction through the wall and subsequent convection and radiation remitted by the thermal wall. A thermal mass wall or simply mass wall is made of concrete or masonry

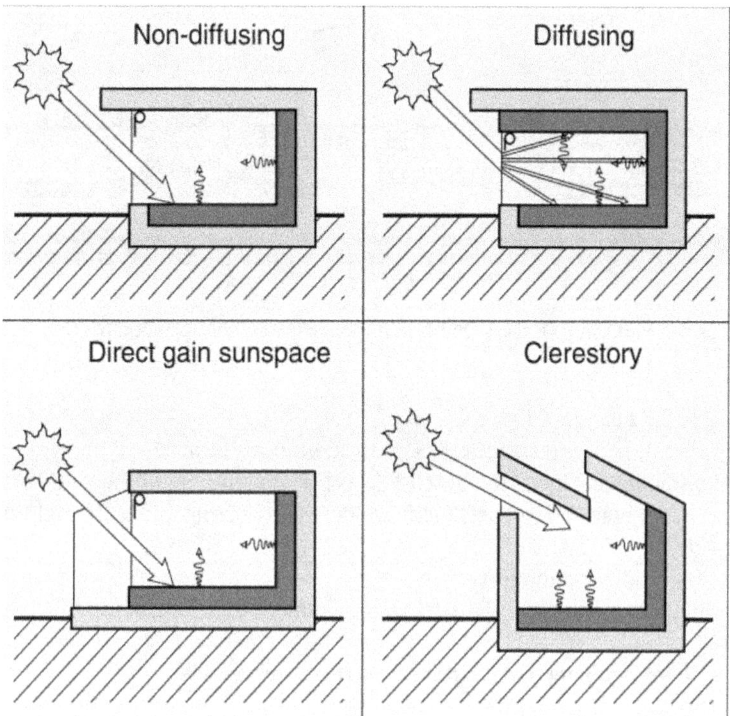

FIGURE 5.7 Various types of heat gains into the building space.

TABLE 5.1
Reflectivity of a Few Surfaces

Material	Reflectivity (% of incident radiation) at 27 °C
White fire clay	12
Asbestos	11
Cork	20
Wood	14
Porcelain	05
Concrete	10
Aluminium	92
Graphite	40

facing south, suitably blackened and glazed. The greatest advantage of this method is that it reduces the temperature fluctuations in the room air, which is the disadvantage of the direct heat gain method. This is also known as a Trombe wall without vents (Figure 5.8).[43,47]

$$Q = U_L \times A \times [\frac{(\alpha\tau)I}{ho} + (T_a - T_i)]$$

$$\text{And } \frac{1}{U_L} = \frac{1}{h_o} + \frac{L}{k} + \frac{1}{h_i}$$

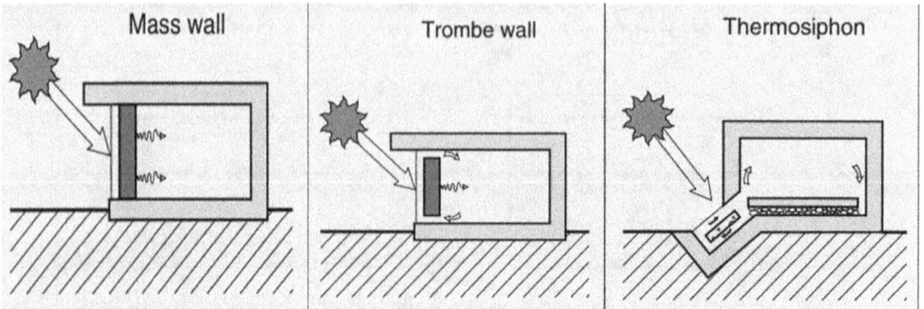

FIGURE 5.8 Mass wall (left), Trombe wall (centre), and thermosiphon (right).

Where Q is the heat gain into the space,

U_L is overall heat transfer coefficient from ambient to inside room,

τ is the transmissivity of the glazing, α is the absorptivity of the wall surface,

h_o and h_i are convection coefficient are between glazing and wall, and the inside, respectively,

'I' is solar radiation intensity,

L is thickness of the wall, k is thermal conductivity of the wall, A is the area of wall, and

T_a and T_i are ambient and inside temperatures respectively

5.4.3.2.2 Trombe Wall

This method is a modification of a mass wall. It has openings in the mass wall towards the top and the bottom to allow the self-sustaining distribution of heat in the building. In this system, the thermal storage lies just behind the glazed area. The openings of vents allows for convective heat exchange from the mass wall to the room, which is the conditioned space.[47] Normally, the ratio of the opening area to the wall area may be in the range of 0.01–0.03. If the thermal storage of the wall between the glazing and the living space is replaced by the water stored in drums (drums stacked over one another), then the modified system is known as a water wall. Such a system is more effective than concrete because the heat storage capacity of water is almost double as compared to concrete. The disadvantage of such a modified Trombe wall is illumination since water drums will be blocking the entry of the sunlight.[41]

5.4.3.2.3 Thermosiphon

A thermosiphon is essentially a solar energy collector for heating the air, usually fitted to the south wall of the building for supplying warm air to the building or conditioned space. The movement of air takes place due to a density difference which gets generated due to the local heating of air beneath and/or inside the panel. A thermosiphon is installed at the level below the space to be heated. After getting heated, air flows into the building space and replaces the old air. During summer, the collector can be disconnected to avoid overheating the room air. A thermosiphon can also be installed on sloped roofs or on south-facing walls.[48,49]

5.4.4 Passive Solar Cooling

In order to achieve natural cooling in the building, it is paramount to reduce unnecessary thermal loads that interact with the building. There are two main interactions with the building: a) the exterior loads due to the climate and b) the internal loads due to people, appliances, lighting, etc. Various strategies have been outlined herein in order to achieve natural cooling.[50]

5.4.4.1 Orientation

The interaction of solar radiation with the building is the source of maximum heat gain inside the building space. The natural way to cool a building is to minimize the incidence of solar radiation. One of the methods is the proper orientation of the building. The maximum solar radiation is interrupted by the roof, followed by the east and west walls. Therefore, it is desirable that the building is oriented with the longest walls facing north and south. The east and west windows are the sources of maximum solar heat gain and should either be avoided or reduced in size. It is also advisable to place unconditioned spaces such as garages and closets on east and west sides. Orientation to take advantage of prevailing breezes in warm and humid climates and prevention of hot winds in hot and dry climates is also important.[50,51]

5.4.4.2 Shade by Neighbouring Structures

Buildings in close proximity can be planned such that they shade each other. The amount and effectiveness of the shading, however, depends on the types of building clusters. Martin and March have classified the building clusters into three categories, that is, pavilions, streets, and courts. ‘Pavilions’ are isolated buildings, single or in clusters, surrounded by open spaces. ‘Streets’ are long building blocks arranged in parallel rows, separated by actual streets in open spaces. ‘Courts’ are defined as open spaces surrounded by buildings on all sides. The courts are created by the conjunction of repeating L-shaped figures. The area in all three sites remains the same; however, the ‘courts’ cluster occupies 75% of the site space, the ‘streets’ occupy 50% of the site space, and the ‘pavilions’ occupy 25% of the site space. It makes their respective floor space index values in the proportion of 3:2:1.[52,53]

5.4.4.3 Shade by Vegetation

The process of shading by trees and vegetation is a very effective and environmentally friendly method of cooling the ambient air and shielding the building from solar radiation. The solar radiation absorbed by the leaves of the vegetation is mainly utilized for photosynthesis. A part of solar radiation results in evaporative heat losses and partly is stored as heat by the fluids in the plants and trees. The best place to plant shady trees is to be decided by observing which part of the fenestration admits the most sunshine during the peak hours during the hottest months. Usually, east- and west-oriented walls and windows receive 50% more sunshine than the north and south-oriented walls and windows. Trees should be planted at positions determined by lines from the centres of the windows on the west or east walls towards the position of the sun at the designated time. On the south side only, deciduous trees should be planted. The precise distance should be decided on the availability of space and size of trees.[54,55]

5.4.4.4 Shading by Overhangs and Louvres

Roller blinds, curtains louvres (fixed or adjustable), and overhangs can be used to provide shading into an opening such as doors, windows etc. Moveable blinds and curtains block the transmission of solar radiation through the fenestration of east and west walls. In hot and dry climates, the temperature of ambient air is greater than the room air; therefore, such devices help in reducing the heat gain. The horizontal louvres, with spacing, parallel to the wall allow air circulation and reduce heat gain by conduction. Louvres hung against a solid overhang prevent the low sun angles, but the comfort of view is compromised. The criteria for shading on the basis of climate zones are listed in Table 5.2.[56,57]

5.4.4.5 Reflecting Surfaces

The external surfaces of a building can be painted with such colours which could maximize the reflection of solar radiation and minimize the absorption of solar radiation. In such cases, the heat transfer into the building decreases considerably. Table 5.3 shows that 'whitewash' has lower reflectivity and high emissivity as compared to aluminium, which results in its lower temperature upon exposure to solar radiation.[58]

TABLE 5.2
Criteria for Shading on the Basis of Climate Zones

Climatic zone	Shading need
Hot and dry	Comprehensively throughout the year
Warm and humid	Comprehensively throughout the year with effective ventilation
Temperate	Comprehensively throughout the year during sunlight time only
Cold and cloudy	No need
Cold and sunny	During summers alone
Composite	During summers alone

TABLE 5.3
Reflectivity and Emissivity of Different Materials

Material	Reflectivity	Emissivity
Aluminium foil	0.95	0.05
Polished aluminium	0.80	0.05
Aluminium paint	0.50	0.50
Galvanized steel sheet	0.75	0.25
Whitewash new	0.88	0.90
Grey colour, light	0.60	0.90
Grey colour, dark	0.30	0.90
Red brick	0.40	0.90
Glass	0.08	0.90
Green colour, light	0.60	0.90

5.4.5 Cooling with Thermal Mass

Insulating the building with material having low thermal conductivity is a well-known and most practised aspect of energy-efficient buildings. In hot climate conditions, weak coupling between the thermal mass and the external heat source, that is, solar radiation, is ensured; however, strong coupling with the inside of the building is ensured so that the internal gains can be absorbed easily. Along with the coupling of thermal mass, the provision of air cavities in the walls or the empty space between the roof and the ceiling results in the reduction of the solar heat gain.[16,59]

5.4.5.1 Cooling with Thermal Mass and Nocturnal Renovation

The emission of longwave radiation takes place continuously from the building during the day and night. During the daytime, the absorbed solar radiation at the surface outweighs the cooling effect produced by the emission of longwave radiation. However, radiant cooling can be obtained during the night hours, and therefore it is known as nocturnal renovation. Radiative cooling often takes place with convection and evaporation. In humid climates, evaporation is limited; therefore, radiative cooling plays a vital role in total heat loss from wet surfaces exposed to the sky.[60] The major problems during the design of such a system are as follows:

- The first problem is the estimation of the meteorological potential for radiant loss or cooling potential. It includes the effective sky temperature, emissivity, and convective heat exchange.
- The second problem is the technique for augmenting the net radiative loss from the radiating surface and reducing the convective gains.
- The third problem is the utilization of the cooling effect produced by the nocturnal radiant loss, that is, the way it is used to provide thermal comfort.

Direct cooling of the thermal mass in a building element such as a roof can be achieved by shielding it from the direct incident radiation and hot ambient air during the day. It shall result in lowering heat gain by radiation and convection from hot ambient air.[43,61] This can be achieved by the following technologies.

5.4.5.1.1 Moveable Elements

The application of moveable elements such as moveable canvas having insulating material with high value of reflectivity on the outer surface permits the protection of the roof from solar radiation and convective gain during the daytime. During the night time, it can be removed to expose the roof to the sky to facilitate the nocturnal radiant loss.[61]

5.4.5.1.2 Vegetation and Earthen Pots

The provision of vegetation on the roof results in the effective shading of the roof. The shading, as discussed earlier, results in a decrease in the heat gain of the building. The earthen pots on the roof facilitate radiant cooling by effectively increasing the surface area; however, it limits the use of the roof for practical purposes by the occupants.[62]

5.4.6 Evaporative Cooling

During evaporation, water changes its state from liquid to vapour. When water and unsaturated air come in contact, heat and moisture transfer takes place. The sensible heat of the air is transferred to water and gets converted into the latent heat of vaporization. This results in lowering the dry bulb temperature of air and increasing the humidity content. The wet bulb temperature remains constant throughout the process. The decrease in air temperature is proportional to the change in the amount of water vapour contained in it. It is termed as adiabatic cooling since no external heat to the system is involved in this process, and the evaporation occurs by the transfer of the existing sensible heat of the air. This process continues proceeds until the stage of equalization of temperatures and vapour pressures of the water and air is reached.[63] The various process involving evaporative cooling can be categorized as outlined in the following section.

5.4.6.1 Outdoor Air Cooling

The air temperature in the shade below the trees is 3 °C to 4 °C lower than the ambient temperature. The free flow of air by the proper vegetation into the building will result in the cooling of the ambient air and its consequent flow into the building. There are many ways such as fountains and artificial sprinklers where a film of water is exposed to the outside air of the building. The cool and humid air flows into the building. This form of cooling is preferred in arid zones where the relative humidity is low and undesirable. Evaporative cooling by means of vegetation can also be done. Vegetation (plants, trees) exterior to the building and on the roof helps in augmenting the outdoor air cooling, which in turn results in reducing the temperature of the conditioned space.[64]

5.4.6.2 Indoor Air Cooling

Vegetation can be used as a means for indoor evaporative cooling. The sensible heat in the air is absorbed by the vegetation (plants inside the conditioned space), resulting in natural evaporative cooling inside the building. This phenomenon can be augmented by the regular watering of the plants inside the building. Indoor air cooling is also provided by active power and requires energy consumption. Indoor air cooling can be provided by either increasing the humidity or keeping it at its original level.[65,66] Accordingly, the process of indoor air cooling can be categorized into the following categories.

5.4.6.2.1 Direct Evaporative Cooling

When the room air is in direct contact with a water surface, as shown in Figure 5.9, the evaporation of water into the air increases the humidity. This process is called direct evaporative cooling. The resultant cool air is directed into the conditioned space. Several designs of direct evaporative cooling are in vogue, some of which are: drip-type evaporative coolers, spray-type evaporative coolers, rotary pad evaporative coolers, and textile mill evaporative coolers. All these systems are useful for rooms and apartments. The evaporative cooling pads used in these systems suffer from clogging and sagging over a period of time. The cooled air produced is extremely humid. The high rate of airflow and a large number of air changes, which are necessary for

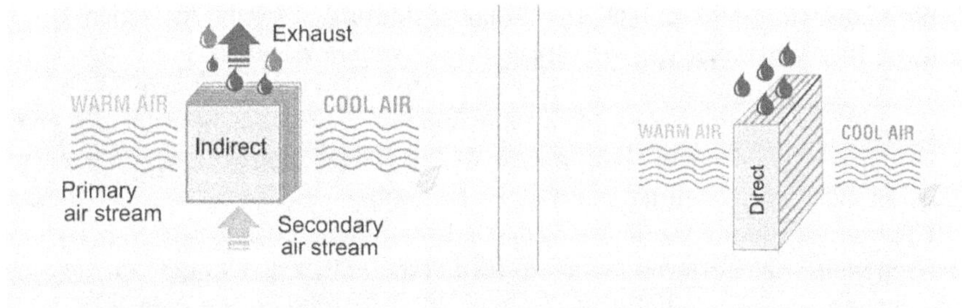

FIGURE 5.9 Evaporative cooling: indirect (left) and direct (right).

effective cooling, cause large variations in the airspeed and associated thermal sensation within the conditioned space.[67]

5.4.6.2.2 *Indirect Evaporative Cooling*

This is when the room air does not come into direct contact with the water. In this process, the evaporation occurs separately, and air/water is cooled without any increase in the humidity. The resultant cool air/water is then used to cool the room air. Different types of indirect evaporative cooling systems are in vogue such as simple dry surfaces, regenerative dry surfaces, plate-type heat exchangers, Pennington heat storage wheels, indirect evaporative radiant cooling, and two-stage indirect evaporative cooling.[66]

Indirect types of systems overcome the defect in direct cooling systems, as the air does not come in direct contact with water. Therefore, the problem of excessive humidity is solved automatically. The number of air changes also gets reduced. The only disadvantage of such system is that suitable filters are to be used to avoid the accumulation of dust and wet surfaces that are prone to scale formation.[68]

5.4.6.3 Building Surface Cooling

The process of cooling of building surfaces by the evaporation of water provides a heat sink for the room air for the dissipation of heat. A film of water is maintained over the surface of the building element, which brings down its temperature below the wet bulb temperature of the ambient air. The roof surface acts as a means of heat transmission from inside of the conditioned space to the ambient air without having any effect on the humidity of the air present in the room. Roof surface or roof water evaporative cooling involves maintaining a uniform thin film of water over on the roof of the building. This helps the roof to achieve a value of temperature much lower than that of any other part of the building. The roof evaporation process can be very effective in hot and dry and also in warm and humid climate zones.

5.4.7 Cooling through Ventilation

One of the most effective ways of providing cooling to non-conditioned buildings is by the means of ventilating the building space with cool ambient air. Ventilation facilitates heat transfer rather loss from the inside of the building by replacing the internal air with cool ambient air, whereas the movement of air decreases the effective temperature by augmenting convective and evaporative cooling of the body.

Traditionally, in humid regions, the design is dominated by the movement of cool ambient. In inland areas, the wind speed is relatively low, and one has to opt for provisions such as designing and constructing on stilts or at times on treetops. This convective cooling can be achieved by a) creating natural movement of the air, known as natural ventilation, or b) use of power-driven devices such as fans, etc., known as forced or mechanical ventilation.[36,69]

The heat transfer by ventilation is dependent on the exchange rate, the difference between interior and exterior temperature, and thermal capacity of the medium (i.e., air). Heat transfer by ventilation (Q_v) can be approximated as:

$$Q_v = (N \times V \times \Delta T) / 3$$

Where N is air changes per hour, V is the volume of conditioned space in m^3, and ΔT is the temperature difference between the inside and outside of the building or conditioned space.

In order to achieve cooling by indoor natural ventilation, a few of the methods have been explained as follows:

5.4.7.1 Windows

Windows play an important role in inducing indoor ventilation due to wind forces. The various parameters that affect the ventilation are *climate, wind direction, area of fenestration, size of inlet and outlet of the openings, volume of the room, shading devices, etc.* Air moves from a zone of high pressure to a zone of low pressure if openings are made on the walls of the respective zones in a building. If the inlet and outlet are placed at different heights, as shown in Figure 5.10, the air flows from the inlet to the outlet due to the density difference created by the upward movement of warm air. In order to attain sensible air movement, it is essential to provide cross-ventilation. Single-sided ventilation will provide air movement to a very shallow depth of the building. An alternative would be to provide exhaust for the air via a ridge or chimney on the leeward side of the chimney. Louvres can be provided for protection against rain and for the prevention of the direct entry of sunlight. Along with this, the presence of a veranda on the windward or leeward side influences the air motion of the room. However, windows have a certain set of limitations. It may not be always possible to locate the openings on the windward side due to reasons such as site restriction, privacy, solar radiation, etc.[70,71]

5.4.7.2 Wind Tower

The principle of wind towers can be understood with the help of Figure 5.11. The hot ambient air enters the tower through the openings in the tower and is cooled when it comes in contact with the cool tower and thus becomes heavier and sinks down. When an inlet is provided to the rooms with an outlet on the other side, there is a draft of cool air. After a whole day of heat exchange, the wind tower becomes warm in the evening. During the night, the reverse happens, the cooler ambient air comes in contact with the bottom of the tower through the rooms; it gets heated up by the warm surface of the wind tower and begins to rise due to buoyancy, and thus airflow is maintained in the reverse direction. This system works effectively in hot and dry

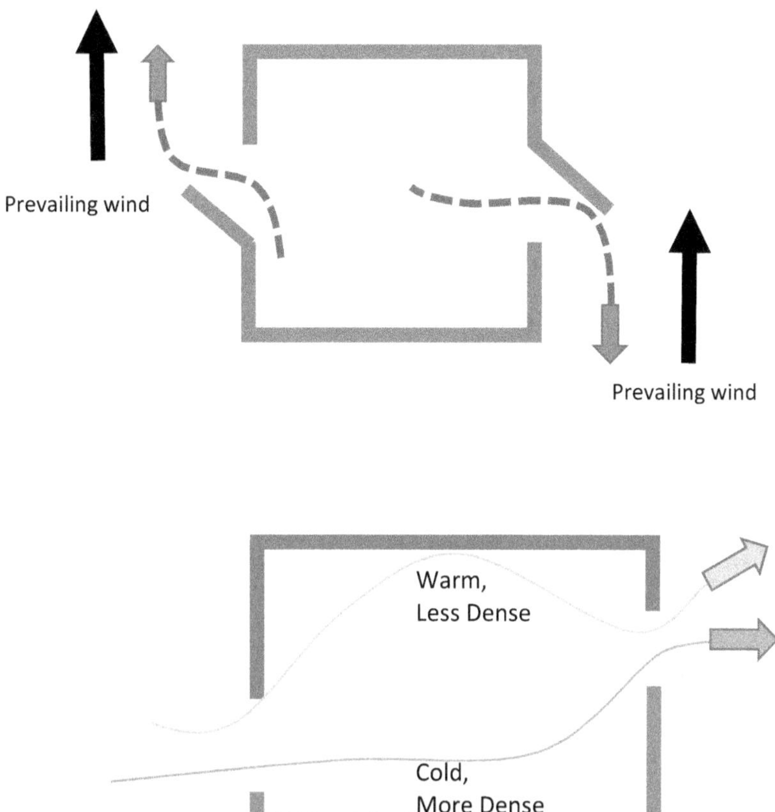

FIGURE 5.10 Ventilation by cross-ventilation (upper) and different height windows (lower).

types of climates, where the daily variations in the temperatures are high. In such climates, there exist high temperatures during the day time and low temperatures during the night time. The openings of the wind tower are provided in the direction of the wind, and outlets on the leeward side thereby take advantage of the pressure difference created by wind speed and direction. Normally, the outlets have thrice the area as compared to the inlet to offer better efficiency. The traditional architecture of the Middle East, having hot and dry climates, has wind towers as a common frequent design feature in the buildings. The concept of a wind tower is suited for single individual dwelling units but not suited for multistorey buildings. In dense urban areas, the wind tower has to be sufficiently high so as to catch enough air. Also, the surface of the wind tower collects dust over the course of time, and heat exchange between the wind tower and the air gets reduced.[72–75]

5.4.7.3 Courtyard Effect

Due to the incident radiation in the courtyard, the air in the courtyard becomes warmer and rises up. To replace it, cool air from the ground level flows through openings of the room, thus producing the airflow. During the night, the process is

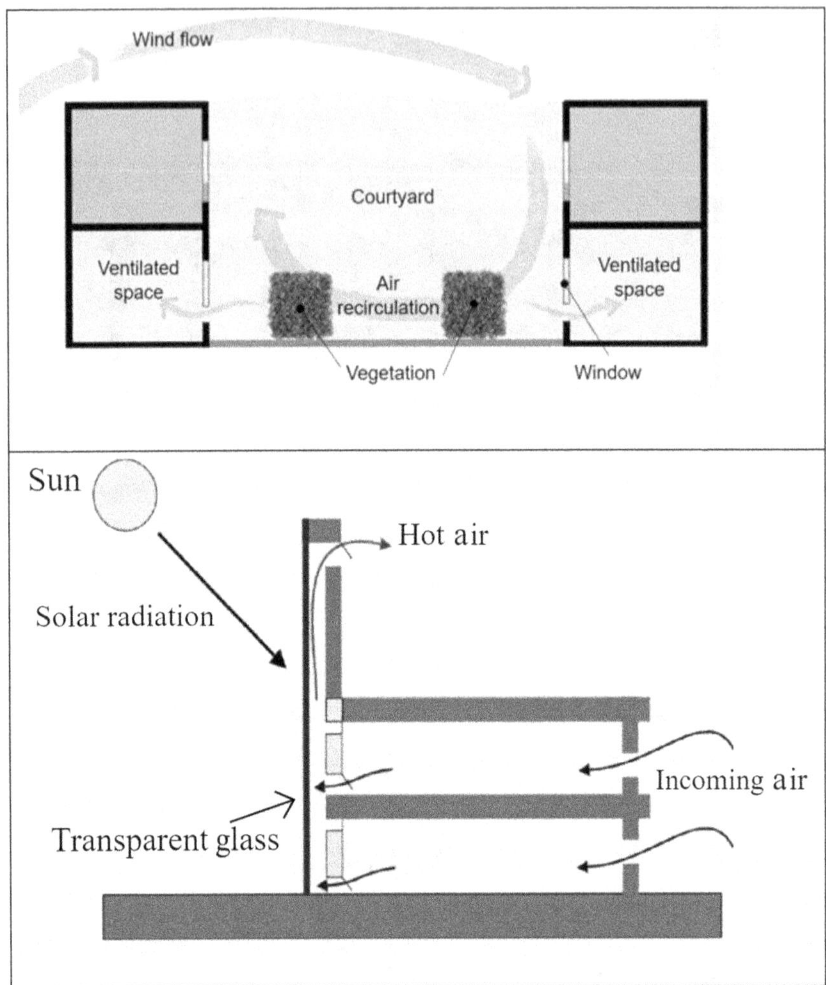

FIGURE 5.11 Courtyard (upper) and wind tower (lower).

reversed. As the warm surface gets cooled by convection and radiation, a stage is reached when its temperature equals the dry bulb temperature of the ambient air. Further cooling by radiation continues as the night sky temperature is lower than the ambient temperature. If the net heat transfer reduces the roof surface temperature to the wet bulb temperature of the surrounding air, condensation of the atmospheric moisture takes place on the roof, and the heat gained due to condensation limits further cooling. If the roof is sloped towards an internal courtyard, the cooler air sinks into the court and enters the living spaces through the low-level openings. The efficiency of the courtyard effect can be augmented by introducing an evaporative cooling system in the cool courtyard. The concept of a courtyard can be applied in a warm and humid climate. It should be ensured that the courtyard gets sufficient radiation to produce the necessary draft through the interior. When the courtyard is allowed to receive intense solar radiation, a much heat is conducted and radiated into the rooms as against the induced draft of air, which may be counterproductive. The intense radiation in the courtyard also produces intense glare indoors.[76–78]

5.4.8 Active Solar Devices

Along with passive utilization for lighting and heating, we can harvest and utilize solar energy in the form of many active solar devices. These devices include photovoltaics and solar thermal systems. Photovoltaic (PV) cells utilize incident solar energy and convert it into electricity. Photovoltaic cells are merged into solar panels which can further be installed over the south-facing roof in the form of a matrix. The inclination angle of the solar panel depends upon the geographical location and latitude of the building. The electricity produced by the solar panels can be further used to run the electrical appliances. In a completely *off-grid system*, the electricity produced during the daytime is stored in the batteries. This stored electricity in the batteries is further utilized during the night time or whenever the need arises. There are also systems that are connected to the national electricity grid, known *as on-grid systems*. Such systems don't have the provision of any energy storage. The electricity generated by the solar panels is used for internal use, and any excess production of electricity is fed to the grid. During the night time, electricity supply from the grid is utilized to meet the electrical requirements of the building. Solar panels have an estimated life of 20 to 25 years. The economics of design integration of solar panels with buildings offers interesting possibilities. The installation cost of solar panels can be compensated against roof or facade elements. These solar panels can also act as shading devices reducing solar heat gain, and semi-transparent systems have the potential to replace glazing. Solar thermal systems are devices which utilize incident radiation to heat air and are known as solar air heaters (SAHs) or solar water heaters (SWHs). In a domestic building,[79–83] a SWH setup consists of collector installed on a south-facing building element, the fluid which is generally water that circulates through the collector tubes, a water tank for storage, and, at times, an active device operating on electricity to heat the water during non-sunshine hours. SAHs generally find applications to dry grains, fruits, vegetables, and seafood at moderate temperatures.[84] The thermal and thermohydraulic performance of the solar air heaters is enhanced by using roughness-producing elements on the heat absorbing side (absorber plate) of the collector. The roughness-producing elements include pin fins, packed beds, fins with perforations, ribs (V, W, intermittent V type), and baffles. The intermittent nature of solar radiation can be overcome with the use of phase change materials (PCMs).[85] PCMs, as the name suggests, absorb the heat during the day time by changing from one phase to another in the form of latent heat, and during the intermittency of solar radiation, they release the stored latent heat, thereby undergoing several repetitive cycles of freezing and melting. A few examples of PCMs include paraffins, fatty acids, salt hydrates, etc. In order to enhance the effectiveness of PCMs, nanoparticles are used as enhancers. These nanomaterials include SiO_2, Al_2O_3, graphene, carbon nanotubes, etc.[86]

5.5 RECENT APPROACHES TO BIOCLIMATIC ARCHITECTURE

Bioclimatic architecture is the heart of bioclimatic building technology that completely and closely integrates human beings with nature. Recent approaches to bioclimatic architecture focus on integrating sustainable design principles and technologies to create energy-efficient and environmentally friendly buildings. These approaches prioritize the use of natural resources and climatic conditions to minimize energy consumption and maximize occupant comfort. Here are some key aspects of recent approaches in bioclimatic architecture.

5.5.1 Adoption of Vernacular Architecture

Vernacular architecture refers to traditional structures of well-tried forms made of indigenous local materials[87] and has the adaptability to local climate without the need for any additional energy-consuming devices.[88] These types of structures came into existence when human beings experienced the need for shelter, and they started to utilize natural resources for the same and passed the knowledge from generation to generation.[89] As the ecological footprints of current conventional architecture are very high and employ the utilization of climate control devices, vernacular architectural strategies work as a base model to them to get insight into the behaviour of a building in the local environment with an objective to reduce energy consumption. The culture, climate, and topography of the location play a decisive role in determining the adoption of vernacular materials.[90] In this context, Alrashed and co-workers found that structures/buildings constructed of adobe building materials in vernacular designs like 'mushrabiyah' consumed 8% lower annual electricity in almost all locations of Ghana.[90] Similarly, vernacular building materials like earth/laterite may also be adopted in modern architecture.[91] Further, several instances of the adoption of multiple vernacular designs in modern conventional architecture for passive designs are there, including roof ponds, domes, air vents, cooling towers, etc., besides the use of adobe and rammed earth construction, as well as wattle and daub.[90,92,93] Though the strategies are location- and climate-specific, they can be exported to other locations.[6]

5.5.2 Inclusion of Bioclimatic Architecture in Urban Planning

The integration of bioclimatic architecture with urbanization is imperative for green development and to have harmony with the local environment and make the best use of it. Since urban planning refers to cheap road and rail transport and specialized land-use zoning, it is closely associated with higher energy consumption. Thus, urban planning, as a more holistic way to integrate energy, environmental, and wider social considerations, is the recent approach for which considerations of site climate and nature in existing settlements as well as in new ones are pre-requisite. While integrating urbanization with green technology, planning the build morphology and orientation plays a crucial role in the provision of thermal comfort. Strategic access to solar radiation and shelter from winds and rains can be achieved through the integration of bioclimatic structures and orientations. Further, solar and wind access can also be strategically accessed through street orientations. For example, to achieve solar shading in southern latitudes, western orientation is not advisable, as evening sun possesses a lower altitude, and air temperatures tend to be high at this time of day.[94] Similarly, in arid areas, vegetation is imperative to obtain improved macro-climatic indices like temperature and relative humidity conditions for better human well-being.[88,95]

5.5.3 Renewable Energy Integration

Recent bioclimatic architecture approaches increasingly integrate renewable energy technologies into building design. This includes incorporating solar photovoltaic systems, wind turbines, and geothermal systems to generate clean energy on site; reducing reliance on fossil fuels; and minimizing the environmental impact of the building.[96,97]

5.5.4 Water Conservation and Rainwater Harvesting

Bioclimatic architecture aims to reduce water consumption through various strategies such as efficient plumbing fixtures, greywater recycling systems, and rainwater harvesting. These approaches promote sustainable water management and reduce the strain on local water resources.[98]

5.5.5 Green Building Materials

The use of environmentally friendly and sustainable materials is a key consideration in bioclimatic architecture. This includes selecting materials with low embodied energy, recycled content, and non-toxic properties. Additionally, incorporating locally sourced materials reduces transportation energy and supports local economies.[99]

5.5.6 Building Performance Monitoring and Optimization

Advanced monitoring systems and data analytics enable real-time measurement and optimization of building performance. This allows for the continuous evaluation of energy usage, indoor air quality, and thermal comfort, facilitating adjustments and improvements to enhance overall building efficiency. The performance appraisal of a building and its certification as bioclimatic is conducted by a building certification program known as LEED (Leadership in Energy and Environmental Design) devised by the United States Green Building Council (USGBC) with an objective to develop a "*consensus-based, market-driven rating system to accelerate the development and implementation of green building practices and to encourage market transformation towards sustainable design.*"[100] LEED certification is available for both residential and cooperative buildings and projects. Though the system is voluntary, it is internationally accepted, and to get the same, building projects are required to satisfy a checklist of prerequisites and earn points. These earned points further decide their eligibility for the level of certification. Four levels of certifications, certified, silver, gold, and platinum, can be accredited to buildings projecting depending upon the earned points. For instance, with 40–49 points, 'certified' can be achieved.

Similarly, certifications of 'silver' with 50–59 points, 'gold' with 60–79 points, and 'platinum' with 80 points and above can be achieved accordingly. Thus, the LEED system is credit-based, where a project earns more points by constructing a more efficient, environmentally friendly bioclimatic building.[101,102]

Biophilic design principles are gaining prominence in bioclimatic architecture, aiming to create connections between occupants and nature within the built environment. Integrating natural elements such as green walls, indoor plants, and views of nature enhances occupant well-being, productivity, and satisfaction.

5.6 CONCLUSIONS

This chapter has discussed the development of bioclimatic building technologies. These technologies can be applied across the world, and similar strategies can work in various areas with similar exterior climates. Actual engineering projects should attempt to reduce the energy demand as much as possible in respect of the available meteorological and climatic conditions. However, it is necessary to place greater importance on climatic comfort as a fundamental engineering objective. One must put effort to achieve this objective without the use of devices that consume electricity and reduce the ecological footprints. Awareness and importance of bioclimatic technologies and vernacular architecture should spread across all sections of society. The requirement of an appropriate comfort zone with the minimum use of energy-consuming cooling and heating devices should be the way forward. To meet this objective, awareness must be disseminated about the environmental impacts of energy consumption and the consequences on climate change and humankind. Additionally, energy-efficient techniques must be adopted across the globe similar to those which are already being adopted in many parts of the world to establish energy-efficient and comfortable architecture. Therefore, we can say that the bioclimatic approach will play an increasingly important role in sustainable holistic development.

REFERENCES

1. Widera B. Comparative analysis of user comfort and thermal performance of six types of vernacular dwellings as the first step towards climate resilient, sustainable and bioclimatic architecture in western sub-Saharan Africa. *Renewable and Sustainable Energy Reviews*. 2021;140:110736. doi:10.1016/j.rser.2021.110736
2. Pajek L, Košir M. Can building energy performance be predicted by a bioclimatic potential analysis? Case study of the Alpine-Adriatic region. *Energy and Buildings*. 2017;139:160–173. doi:10.1016/j.enbuild.2017.01.035
3. Khambadkone NK, Jain R. A bioclimatic analysis tool for investigation of the potential of passive cooling and heating strategies in a composite Indian climate. *Building and Environment*. 2017;123:469–493. doi:10.1016/j.buildenv.2017.07.023
4. Hopping D. Siemens Smart infrastructure brandvoice: energy crisis: how buildings can drive the energy transition. www.forbes.com/sites/siemens-smart-infrastructure/2023/02/17/energy-crisis-how-buildings-can-drive-the-energy-transition/. *Forbes*. Published online 2023. Accessed April 14, 2023. www.forbes.com/sites/siemens-smart-infrastructure/2023/02/17/energy-crisis-how-buildings-can-drive-the-energy-transition/

5. Kiessling T. Siemens smart infrastructure brandvoice: decarbonization: the future green economy of the building industry. *Forbes*. Published online 2023. Accessed April 14, 2023. www.forbes.com/sites/siemens-smart-infrastructure/2023/03/15/decarbonization-the-future-green-economy-of-the-building-industry/
6. Manzano-Agugliaro F, Montoya FG, Sabio-Ortega A, García-Cruz A. Review of bioclimatic architecture strategies for achieving thermal comfort. *Renewable and Sustainable Energy Reviews*. 2015;49:736–755. doi:10.1016/j.rser.2015.04.095
7. Barea G, Victoria Mercado M, Filippín C, Monteoliva JM, Villalba A. New paradigms in bioclimatic design toward climatic change in arid environments. *Energy and Buildings*. 2022;266:112100. doi:10.1016/j.enbuild.2022.112100
8. Nguyen AT, Reiter S. Bioclimatism in architecture: An evolutionary perspective. *International Journal of Design & Nature and Ecodynamics*. 2017;12:16–29. doi:10.2495/DNE-V12-N1-16-29
9. Olgyay V. *Design with Climate: Bioclimatic Approach to Architectural Regionalism*. Princeton University Press; 1963.
10. American National Standards Institute. *Thermal Environmental Conditions for Human Occupancy*. Vol 55. American Society of Heating, Refrigerating and Air-Conditioning Engineers; 2004.
11. Givoni B. Comfort, climate analysis and building design guidelines. *Energy and Buildings*. 1992;18(1):11–23. doi:10.1016/0378-7788(92)90047-K
12. Elshafei G. Bioclimatic design strategies recommendations for thermal comfort using mahoney tables in hot desert bioclimatic region. *Journal of Urban Research*. 2021;39(1):59–74. doi:10.21608/jur.2021.39201.1019
13. Setiawan MF. Building design recommendation for thermal comfort in cities on the island of Java, Indonesia. *IOP Conf Ser: Earth Environ Sci*. 2022;969(1):012068. doi:10.1088/1755-1315/969/1/012068
14. CRES. Bioclimatic design and passive solar systems. Center for renewable energy sources and saving. *Pikermi Attiki Greece*. www.cres.gr/kape/energeia_politis/energeia_politis_bioclimatic_eng.htm. Published 2023. Accessed April 15, 2023. www.cres.gr/kape/energeia_politis/energeia_politis_bioclimatic_eng.htm
15. Long L, Ye H. The roles of thermal insulation and heat storage in the energy performance of the wall materials: a simulation study. *Scientific Reports*. 2016;6(1):24181. doi:10.1038/srep24181
16. Al-Sanea SA, Zedan MF, Al-Hussain SN. Effect of thermal mass on performance of insulated building walls and the concept of energy savings potential. *Applied Energy*. 2012;89(1):430–442. doi:https://doi.org/10.1016/j.apenergy.2011.08.009
17. Eben Saleh MA. Thermal insulation of buildings in a newly built environment of a hot dry climate: The Saudi Arabian experience. *International Journal of Ambient Energy*. 1990;11(3):157–168. doi:10.1080/01430750.1990.9675170
18. Ozel M. Effect of insulation location on dynamic heat-transfer characteristics of building external walls and optimization of insulation thickness. *Energy and Buildings*. 2014;72:288–295. doi:https://doi.org/10.1016/j.enbuild.2013.11.015
19. Verbeke S, Audenaert A. Thermal inertia in buildings: A review of impacts across climate and building use. *Renewable and Sustainable Energy Reviews*. 2018;82:2300–2318. doi:10.1016/j.rser.2017.08.083
20. Bojić MLj, Loveday DL. The influence on building thermal behavior of the insulation/masonry distribution in a three-layered construction. *Energy and Buildings*. 1997;26(2):153–157. doi:10.1016/S0378-7788(96)01029-8
21. Kolaitis DI, Malliotakis E, Kontogeorgos DA, Mandilaras I, Katsourinis DI, Founti MA. Comparative assessment of internal and external thermal insulation systems for energy efficient retrofitting of residential buildings. *Energy and Buildings*. 2013;64:123–131. doi:10.1016/j.enbuild.2013.04.004

22. Husain D, Shukla S, Umrao VK, Prakash R. Thermal Load Reduction with Green Building Envelope. *Open Journal of Energy Efficiency*. 2017;6(3):112–127. doi:10.4236/ojee.2017.63009
23. Bekkouche SMEA, Benouaz T, Mohamed Kamal C, Hamdani M, Yaïche R, Benamrane N. Thermal resistances of air in cavity walls and their effect upon the thermal insulation performance. *International Journal of Energy and Environment*. 2013;4(3):459–466.
24. Hallik J, Klõšeiko P, Piir R, Kalamees T. Numerical analysis of additional heat loss induced by air cavities between insulation boards due to non-ideality. *Journal of Building Engineering*. 2022;60:105221. doi:10.1016/j.jobe.2022.105221
25. Adhikari P. A review of glazing materials and its prospects in buildings of Kathmandu Valley. In: *Proceedings of IOE Graduate Conference*. Kathmandu; 2014.
26. Su L, Fraaß M, Wondraczek L. Design guidelines for thermal comfort and energy consumption of triple glazed fluidic windows on building level. *Advanced Sustainable Systems*. 2021;5(2):2000194. doi:10.1002/adsu.202000194
27. Azemati A, Rahimian Koloor SS, Khorasanizadeh H, Sheikhzadeh G, Hadavand BS, Eldessouki M. Thermal evaluation of a room coated by thin urethane nanocomposite layer coating for energy-saving efficiency in building applications. *Case Studies in Thermal Engineering*. 2023;43:102688. doi:10.1016/j.csite.2022.102688
28. Back Y, Bach PM, Jasper-Tönnies A, Rauch W, Kleidorfer M. A rapid fine-scale approach to modelling urban bioclimatic conditions. *Science of the Total Environment*. 2021;756:143732. doi:10.1016/j.scitotenv.2020.143732
29. Van Den Bergh S, Hart R, Jelle BP, Gustavsen A. Window spacers and edge seals in insulating glass units: A state-of-the-art review and future perspectives. *Energy and Buildings*. 2013;58:263–280. doi:10.1016/j.enbuild.2012.10.006
30. Mironova J. Wind impact on low-rise buildings when placing high-rises into the existing development. *IOP Conference Series: Materials Science and Engineering*. 2020;890(1):012055. doi:10.1088/1757-899X/890/1/012055
31. Moosavi L, Mahyuddin N, Ab Ghafar N, Azzam Ismail M. Thermal performance of atria: an overview of natural ventilation effective designs. *Renewable and Sustainable Energy Reviews*. 2014;34:654–670. doi:10.1016/j.rser.2014.02.035
32. Lomas KJ. Architectural design of an advanced naturally ventilated building form. *Energy and Buildings*. 2007;39(2):166–181.
33. Cuce PM, Riffat S. A comprehensive review of heat recovery systems for building applications. *Renewable and Sustainable Energy Reviews*. 2015;47:665–682.
34. Mardiana-Idayu A, Riffat S. Review on heat recovery technologies for building applications. *Renewable and Sustainable Energy Reviews*. 2012;16(2):1241–1255.
35. Xu Q, Riffat S, Zhang S. Review of heat recovery technologies for building applications. *Energies*. 2019;12(7):1285. doi:10.3390/en12071285
36. Bhamare DK, Rathod MK, Banerjee J. Evaluation of cooling potential of passive strategies using bioclimatic approach for different Indian climatic zones. *Journal of Building Engineering*. 2020;31:101356. doi:10.1016/j.jobe.2020.101356
37. Teitelbaum E, Miller C, Meggers F. Highway to the comfort zone: history of the psychrometric chart. *Buildings*. 2023;13(3):797. doi:10.3390/buildings13030797
38. Pereira PF da C, Broday EE. Determination of thermal comfort zones through comparative analysis between different characterization methods of thermally dissatisfied people. *Buildings*. 2021;11(8):320. doi:10.3390/buildings11080320
39. Lapinskiene V, Motuzienė V, Džiugaitė-Tumėnienė R, Mikučionienė R. *Impact of Internal Heat Gains on Building's Energy Performance*. VGTU Press; 2017. doi:10.3846/enviro.2017.265
40. Liang R, Ding W, Zandi Y, Rahimi A, Pourkhorshidi S, Khadimallah MA. Buildings' internal heat gains prediction using artificial intelligence methods. *Energy and Buildings*. 2022;258:111794. doi:10.1016/j.enbuild.2021.111794

41. Chan HY, Riffat SB, Zhu J. Review of passive solar heating and cooling technologies. *Renewable and Sustainable Energy Reviews*. 2010;14(2):781.
42. Nanda AK, Panigrahi CK. A state-of-the-art review of solar passive building system for heating or cooling purpose. *Front Energy*. 2016;10(3):347–354. doi:10.1007/s11708-016-0403-0
43. Gupta N, Tiwari GN. Review of passive heating/cooling systems of buildings. *Energy Science & Engineering*. 2016;4(5):305–333. doi:10.1002/ese3.129
44. Yüksek I, Karadayi TT, Yüksek I, Karadayi TT. Energy-efficient building design in the context of building life cycle. In: *Energy Efficient Buildings*. IntechOpen; 2017. doi:10.5772/66670
45. Sinha A. Building orientation as the primary design consideration for climate responsive architecture in urban areas. *Architecture and Urban Planning*. 2020;16:32–40.
46. Ismaeil EMH, Sobaih AEE. High-performance glazing for enhancing sustainable environment in arid region's healthcare projects. *Buildings*. 2023;13(5):1243. doi:10.3390/buildings13051243
47. Abbas EF, Al-abady A, Raja V, AL-bonsrulah HAZ, Al-Bahrani M. Effect of air gap depth on Trombe wall system using computational fluid dynamics. *International Journal of Low-Carbon Technologies*. 2022;17:941–949. doi:10.1093/ijlct/ctac063
48. Ranatunga DBJ. *Thermosyphon Flow in Solar Collectors*. Thesis. UNSW Sydney; 1978. doi:10.26190/unsworks/5283
49. Fantozzi F, Filipeschi S, Mameli M, et al. An innovative enhanced wall to reduce the energy demand in buildings. *Journal of Physics: Conference Series*. 2017;796(1):012043. doi:10.1088/1742-6596/796/1/012043
50. Almasri RA, Abu-Hamdeh NH, Esmaeil KK, Suyambazhahan S. Thermal solar sorption cooling systems—A review of principle, technology, and applications. *Alexandria Engineering Journal*. 2022;61(1):367–402. doi:10.1016/j.aej.2021.06.005
51. Yu CR, Guo HS, Wang QC, Chang RD. Revealing the impacts of passive cooling techniques on building energy performance: A residential case in Hong Kong. *Applied Sciences*. 2020;10(12):4188. doi:10.3390/app10124188
52. Abdel Aziz DM. Effects of tree shading on buildingâ's energy consumption. *J Architure Engineering Technology*. 2014;03(04). doi:10.4172/2168-9717.1000135
53. Tsiros I. Assessment and energy implications of street air temperature cooling by shade tress in Athens (Greece) under extremely hot weather conditions. *Renewable Energy*. 2010;35:1866–1869. doi:10.1016/j.renene.2009.12.021
54. US EPA O. Using trees and vegetation to reduce heat islands. Published June 17, 2014. Accessed May 29, 2023. www.epa.gov/heatislands/using-trees-and-vegetation-reduce-heat-islands
55. Zeng Y. Research on building greening and shading in hot summer and warm winter region. *IOP Conference Series: Materials Science and Engineering*. 2018;394(3):032019. doi:10.1088/1757-899X/394/3/032019
56. Sadeghi R. Effect of louvres, overhang, and sidefins shading patterns on energy consumption in an office building. 2019. Accessed May 29, 2023. https://papers.ssrn.com/abstract=3927195
57. Idchabani R, El Ganaoui M, Sick F. Analysis of exterior shading by overhangs and fins in hot climate. *Energy Procedia*. 2017;139:379–384. doi:10.1016/j.egypro.2017.11.225
58. Sen S, Khazanovich L. Limited application of reflective surfaces can mitigate urban heat pollution. *Nature Communications*. 2021;12:3491. doi:10.1038/s41467-021-23634-7
59. Avcı AB, Beyhan ŞG. Revealing the climate-responsive strategies of traditional houses of Urla, İzmir. *International Journal of Sustainable Building Tech*. 2023;14(1):18–34. doi:10.22712/susb.20230003
60. Toldi T de, Craig S, Sushama L. Internal thermal mass for passive cooling and ventilation: adaptive comfort limits, ideal quantities, embodied carbon. 2022;3(1):42–67. doi:10.5334/bc.156

61. Kuczyński T, Staszczuk A, Gortych M, Stryjski R. Effect of thermal mass, night ventilation and window shading on summer thermal comfort of buildings in a temperate climate. *Building and Environment*. 2021;204:108126. doi:10.1016/j.buildenv.2021.108126
62. Kamal MA. An overview of passive cooling techniques in buildings: Design concepts and architectural interventions. *Acta Technica Napocensis: Civil Engineering & Architecture*. 2012;55:84–97.
63. Cui X, Yang X, Sun Y, et al. Energy efficient indirect evaporative air cooling. In: *Advanced Cooling Technologies and Applications*. IntechOpen; 2018. doi:10.5772/intechopen.79223
64. Abdullah MS, Baharun A, Siti Halipah I, Azlan W, Abidin WAWZ. Bioclimatic home cooling design for acceptable thermal comfort in Malaysian climate. *Malaysian Construction Research Journal*. 2018;25:1–12.
65. Dong J, Lan H, Liu Y, Wang X, Yu C. Indoor environment of nearly zero energy residential buildings with conventional air conditioning in hot-summer and cold-winter zone. *Energy and Built Environment*. 2022;3(2):129–138. doi:10.1016/j.enbenv.2020.12.001
66. Tewari VP, Verma RK, von Gadow K. Climate change effects in the Western Himalayan ecosystems of India: Evidence and strategies. *Forest Ecosystems*. 2017;4(1):13. doi:10.1186/s40663-017-0100-4
67. Kapilan N, Isloor AM, Karinka S. A comprehensive review on evaporative cooling systems. *Results in Engineering*. 2023;18:101059. doi:10.1016/j.rineng.2023.101059
68. Amer O, Boukhanouf R, Ibrahim H. A review of evaporative cooling technologies. *International Journal of Environmental Science and Development*. 2015;6:111. doi:10.7763/IJESD.2015.V6.571
69. Al-Shamkhee D, Al-Aasam AB, Al-Waeli AHA, Abusaibaa GY, Moria H. Passive cooling techniques for ventilation: An updated review. *Renew Energy Environ Sustain*. 2022;7:23. doi:10.1051/rees/2022011
70. Gan G. A parametric study of Trombe walls for passive cooling of buildings. *Energy and Buildings*. 1998;27(1):37–43. doi:10.1016/S0378-7788(97)00024-8
71. Bugenings LA, Kamari A. Bioclimatic architecture strategies in denmark: A review of current and future directions. *Buildings*. 2022;12(2):224. doi:10.3390/buildings12020224
72. Sadeghi H, Kalantar V. Performance analysis of a wind tower in combination with an underground channel. *Sustainable Cities and Society*. 2018;37:427–437. doi:10.1016/j.scs.2017.12.002
73. Bouchahm Y, Bourbia F, Belhamri A. Performance analysis and improvement of the use of wind tower in hot dry climate. *Renewable Energy*. 2011;36(3):898–906. doi:10.1016/j.renene.2010.08.030
74. Bouchahm Y, Bourbia F, Belhamri A. Performance analysis and improvement of the use of wind tower in hot dry climate. *Renewable Energy*. 2010;36. doi:10.1016/j.renene.2010.08.030
75. Rabeharivelo R, Kavraz M, Aygün C. Thermal comfort in classrooms considering a traditional wind tower in Trabzon through simulation. *Build Simul*. 2022;15(3):401–418. doi:10.1007/s12273-021-0804-9
76. He C, Tian W, Shao Z. Impacts of courtyard envelope design on energy performance in the hot summer–cold winter region of China. *Buildings*. 2022;12(2):173. doi:10.3390/buildings12020173
77. Chinedu Alozie G. Environmental architecture: Courtyard as element of sustainable energy efficient building development. 2020;20:56–61. doi:10.5829/idosi.aejaes.2020.56.61
78. Sánchez de la Flor FJ, Ruiz-Pardo Á, Diz-Mellado E, Rivera-Gómez C, Galán-Marín C. Assessing the impact of courtyards in cooling energy demand in buildings. *Journal of Cleaner Production*. 2021;320:128742. doi:10.1016/j.jclepro.2021.128742

79. Vassiliades C, Kalogirou S, Michael A, Savvides A. A roadmap for the integration of active solar systems into buildings. *Applied Sciences*. 2019;9(12):2462. doi:10.3390/app9122462
80. Buonomano A, Palombo A, Forzano C, et al. Building integration of active solar energy systems: A review of geometrical and architectural characteristics. *Renewable and Sustainable Energy Reviews*. 2022;164. doi:10.1016/j.rser.2022.112482
81. Zaręba A, Krzemińska A, Kozik R, Adynkiewicz-Piragas M, Kristiánová K. Passive and active solar systems in eco-architecture and eco-urban planning. *Applied Sciences*. 2022;12(6):3095. doi:10.3390/app12063095
82. Chen Y, Chen Z, Wang D, et al. Co-optimization of passive building and active solar heating system based on the objective of minimum carbon emissions. *Energy*. 2023;275:127401. doi:10.1016/j.energy.2023.127401
83. Noaman DS, Moneer SA, Megahed NA, El-Ghafour SA. Integration of active solar cooling technology into passively designed facade in hot climates. *Journal of Building Engineering*. 2022;56:104658. doi:10.1016/j.jobe.2022.104658
84. Chand S, Chand P, Kumar Ghritlahre H. Thermal performance enhancement of solar air heater using louvered fins collector. *Solar Energy*. 2022;239:10–24. doi:10.1016/j.solener.2022.04.046
85. Palacio M, Ramírez C, Carmona M, Cortés C. Effect of phase-change materials in the performance of a solar air heater. *Solar Energy*. 2022;247:385–396. doi:10.1016/j.solener.2022.10.046
86. Radhakrishnan N, Thomas S, Sobhan CB. Characterization of thermophysical properties of nano-enhanced organic phase change materials using T-history method. *J Therm Anal Calorim*. 2020;140(5):2471–2484. doi:10.1007/s10973-019-08976-1
87. Salman M. Sustainability and vernacular architecture: rethinking what identity is. In: Hmood K, ed. *Urban and Architectural Heritage Conservation within Sustainability*. IntechOpen; 2019. doi:10.5772/intechopen.82025
88. Cañas I, Martín S. Recovery of Spanish vernacular construction as a model of bioclimatic architecture. *Building and Environment*. 2004;39(12):1477–1495. doi:10.1016/j.buildenv.2004.04.007
89. Jayasudha P, Dhanasekaran M, Devadas MD, Ramachandran N. A study on sustainable design principles: a case study of a vernacular dwelling in Thanjavur region of Tamil Nadu, India. *Indian Journal of Traditional Knowledge*. 2004;13:762–770.
90. Alrashed F, Asif M, Burek S. The role of vernacular construction techniques and materials for developing zero-energy homes in various desert climates. *Buildings*. 2017;7(1):17. doi:10.3390/buildings7010017
91. Costa C, Cerqueira Â, Rocha F, Velosa A. The sustainability of adobe construction: past to future. *International Journal of Architectural Heritage*. 2019;13(5):639–647.
92. Agyekum K, Kissi E, Danku JC. Professionals' views of vernacular building materials and techniques for green building delivery in Ghana. *Scientific African*. 2020;8:e00424. doi:10.1016/j.sciaf.2020.e00424
93. Niroumand H, Zain MFM, Jamil M. A guideline for assessing of critical parameters on Earth architecture and Earth buildings as a sustainable architecture in various countries. *Renewable and Sustainable Energy Reviews*. 2013;28:130–165. doi:10.1016/j.rser.2013.07.020
94. Goulding JR, Lewis JO. Bioclimatic architecture. Published online 1997. https://www.yumpu.com/en/document/view/18718283/download-ucd-energy-research-group-university-college-dublin
95. Kurbán A, Papparelli A, Cúnsulo M, Montilla E. Shading by urban forests in arid ecosystems. *Architectural Science Review*. 2007;50(2):122–129. doi:10.3763/asre.2007.5018

96. Kumar VSKVH Amit Vilas Sant, Arun, ed. *Renewable Energy Integration with Building Energy Systems: A Modelling Approach*. CRC Press; 2022. doi:10.1201/9781003211587
97. Yüksek İ, Karadağ İ, Yüksek İ, Karadağ İ. Use of renewable energy in buildings. In: *Renewable Energy—Technologies and Applications*. IntechOpen; 2021. doi:10.5772/intechopen.93571
98. Şahin Nİ, Manioğlu G. Water conservation through rainwater harvesting using different building forms in different climatic regions. *Sustainable Cities and Society*. 2019;44:367–377. doi:10.1016/j.scs.2018.10.010
99. Sharma N. Sustainable building material for green building construction, conservation and refurbishing. *MATTER: International Journal of Science and Technology*. 2020;29:5343–5350.
100. Green building council. LEED rating system. *U.S. Green Building Council*. Published 2023. Accessed May 29, 2023. www.usgbc.org/leed
101. U.S. Green Building Council. What is LEED certification? *U.S. Green Building Council*. Published April 7, 2022. Accessed May 29, 2023. https://support.usgbc.org/hc/en-us/articles/4404406912403-What-is-LEED-certification-
102. RTS. What is LEED certification & steps for getting a certification. *RTS. Recycle Track Systems*. Published 2023. Accessed May 29, 2023. www.rts.com/resources/guides/what-is-leed-certification/

6 Responsive Building Components and Systems

Bhartendu Mani Tripathi and Shailendra Kumar Shukla

6.1 INTRODUCTION

A new age in the building business has begun due to the spectacular confluence of technology and construction in recent years. Innovative building materials and components are revolutionizing how we design, create, and use buildings. With embedded sensors, sophisticated control systems, and connected networks, these cutting-edge materials can respond intelligently to environmental changes, optimize energy use, increase occupant comfort, and boost building performance. Responsive building components and systems refer to architectural elements and technologies that adapt and respond to various environmental conditions, user needs, and changing circumstances. These components and systems are designed to increase buildings' comfort, efficiency, and functionality while promoting sustainability and overall good health. They comprise new materials, sensors, and control systems to enable dynamic and intelligent responses. The concept of responsiveness in building design emerged with a rise in the focus on energy efficiency, sustainability, and occupant comfort. Traditional constructions usually rely on static parts and systems that cannot be altered to suit different situations, leading to energy loss and poor performance. Responsive building systems and components, which incorporate automation, flexibility, and adaptation into architectural design, help to overcome these limitations.

Smart materials are transforming the future of architecture and urban development, from self-healing concrete that seals its cracks to dynamic glazing systems that dynamically change transparency based on sunshine intensity. The capacity of intelligent building materials to increase energy efficiency is one of its main advantages. They can adjust to shifting environmental factors like temperature, humidity, and air quality, resulting in a healthier interior atmosphere and are more conducive to productivity. For instance, intelligent insulation materials can alter their thermal qualities to maintain ideal temperatures, while intelligent windows can automatically tint to prevent glare and heat absorption. The uses for intelligent building materials and components are countless. These materials present enormous prospects for environmentally friendly, energy-efficient, and intelligent construction, from private residences to commercial buildings, from skyscrapers to smart cities. Buildings can

DOI: 10.1201/9781003415695-6

become more dynamic, responsive, and adaptive thanks to their seamless integration with other innovative technology and systems. Innovative building materials have enormous potential but come with new considerations and difficulties. To enable the successful deployment and operation of these technologies, data privacy, cybersecurity, interoperability, and long-term durability concerns must be addressed appropriately. Architects, engineers, contractors, and policymakers must stay current on the most recent developments and opportunities provided by intelligent building materials and components as we enter this exciting new era of innovative construction. This knowledge will not only influence how our built environment develops in the future, but it will also help ensure a sustainable and resilient future for future generations. This chapter provides a brief introduction to modified facades; smart building materials such as insulation, phase change materials, recycled materials, sustainable wood- and glass-based materials; water-efficient fixtures; and components that have emerged as game-changing elements that integrate technology into the very structure of our built environment. These materials feature inbuilt sensors, state-of-the-art control systems, and connected networks, which enable them to respond autonomously to environmental changes, maximize energy use, and enhance occupant comfort. This chapter also discusses the benefits and uses of smart building materials, including their potential to create sustainable, intelligent structures, boost energy efficiency, and improve occupant well-being.

6.2 ADAPTIVE FACADES

Exterior building envelopes with dynamic response and adaptation capabilities are called adaptive facades. They use a variety of architectural elements, materials, and technology to enhance a building's energy efficiency, daylighting, thermal comfort, and visual comfort. Depending on outside parameters, including sunshine intensity, glare, and privacy requirements, adaptive facades can change how transparent their glazing or shading parts are. With the help of this feature, natural lighting may be maximized while heat gain or loss is reduced. Adaptive facades can regulate how much sunlight enters the structure by including movable shading elements like blinds, louvers, or electrochromic glass. It aids in lowering glare and solar heat gain while maintaining a cozy indoor climate. Adaptive facades control heat transfer via the building envelope using insulation materials, phase change materials, or movable insulation panels. As a result of these elements' ability to adjust to temperature changes outside, less mechanical heating or cooling is required, improving energy efficiency. Some adaptive facades have ventilation systems that can control the airflow through the building envelope. They allow for natural air exchange, temperature control, and ventilation, which reduces reliance on mechanical HVAC systems and enhances indoor air quality. Solar panels and other energy-harvesting technology can be included in adaptive facade designs. By doing so, they can produce electricity from green energy sources, lowering the building's reliance on outside power and promoting sustainability. Adaptive facades frequently contain sensors, actuators, and automation systems that continuously monitor and react to the outside environment. It makes it possible for the facades to adapt dynamically in response to variables like temperature, occupancy, and user preferences. Improved energy efficiency, less reliance on mechanical systems, higher occupant comfort, more natural daylighting, and the potential for energy cost savings are all advantages of adapted facades. Because they can alter their look dynamically, these facades also help a building's aesthetics and visual appeal. This results in a dynamic and responsive architectural statement.[1] While adaptive facades have many benefits, they also come with architectural obstacles, such as integrating intricate control systems, maintaining them, and considering the cost. Adaptive facades, which offer sustainable and responsive solutions for buildings, are becoming increasingly common in contemporary architecture because of developments in materials, technology, and design methodologies.

6.3 INSULATION MATERIALS

Insulation materials are used to improve the thermal performance of buildings by reducing the amount of energy needed for heating and cooling. Materials like cellulose, rock wool, and spray foam insulation are some examples of smart insulation materials that can improve energy efficiency in buildings. When it comes to green buildings, selecting the right insulation material can be crucial in achieving energy efficiency and reducing environmental impact. Some insulation materials commonly used in green buildings are summarized and shown in Figure 6.1.

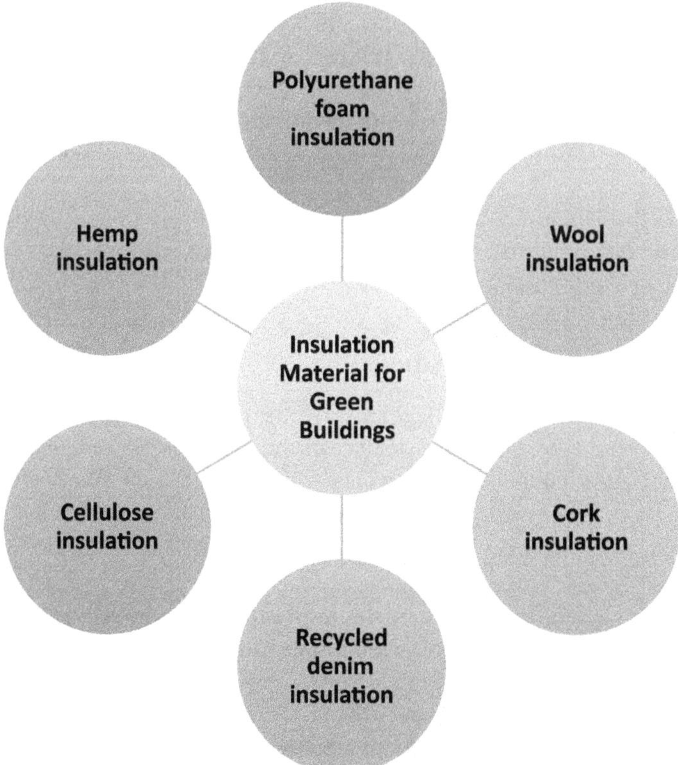

FIGURE 6.1 Different types of insulation used for green building technology.

6.3.1 Cellulose Insulation

This insulation is environmentally responsible because it is created from recycled newspapers and other paper goods. It is reasonably priced and has good thermal and acoustic qualities. Due to its affordability and environmental friendliness, cellulose insulation is a common building material. It is frequently used for new construction and remodeling of older buildings.[2] Here are a few essential advantages and things to consider while employing cellulose insulation in buildings.

Benefits

1. Effective thermal insulation is provided by cellulose insulation, keeping buildings cool in the summer and warm in the winter.
2. Because cellulose insulation is dense and fibrous, less sound is transmitted through walls and ceilings.
3. Cellulose insulation is created from recycled materials, lowering waste production and environmental harm.
4. Cellulose insulation is made fire retardant by the chemicals used to treat it, adding an extra layer of security for buildings.

Considerations

1. If improperly built, cellulose insulation can absorb moisture, resulting in mold and other problems. In buildings with cellulose insulation, ensuring sufficient ventilation and moisture control are in place is critical.
2. Cellulose insulation tends to settle with time and becomes less efficient as insulation. Ensuring the insulation is fitted correctly is crucial to reduce settling.
3. Foam insulation reduces air leakage more effectively than cellulose insulation, so it could be essential to utilize additional materials or methods to reduce air infiltration.

6.3.2 Recycled Denim Insulation

Cotton insulation, commonly called recycled denim insulation, is a relatively new type created from used cotton and denim fabrics. It is a viable and environmentally friendly choice for building insulation. The following are some essential advantages and factors to consider while using recycled denim insulation in buildings.

Benefits

1. Buildings are effectively thermally insulated using recycled denim insulation, which keeps them cool in the summer and warm in the winter.
2. Denim insulation efficiently reduces sound transmission through walls and ceilings because it is dense and fibrous.
3. Since recycled denim insulation is created from post-consumer waste, it has a lower waste output and fewer adverse environmental effects.

4. To meet fire safety requirements for structures, recycled denim insulation is treated with chemicals that are fire resistant.

Considerations

1. Recycled denim insulation, like other insulating material, can absorb moisture if improperly put in, resulting in mold and other problems. Buildings with recycled denim insulation must ensure sufficient ventilation and moisture management.
2. Recycled denim insulation may settle with time and become less efficient as insulation. Ensuring the insulation is fitted correctly is crucial to reduce settling.
3. Compared to more conventional insulation materials like fiberglass or cellulose, recycled denim insulation might be more expensive.

6.3.3 Wool Insulation

Wool, a natural, renewable, and biodegradable substance, is used more frequently as insulation. It is fire resistant, has excellent thermal and acoustic qualities, and is mold- and mildew-resistant. Wool insulation for building insulation is both energy- and environmentally friendly. It is a renewable resource created from natural wood fibers, usually from sheep. Compared to other insulation kinds, wool insulation provides several benefits. Due to its outstanding thermal insulation capabilities, it keeps buildings cool in the summer and warm in the winter.

Additionally, it is very breathable, letting moisture escape and lowering the risk of mold and wetness. Wool insulation is also hypoallergenic and non-toxic and does not emit poisonous chemicals or gases. The sustainability of wool insulation is another advantage. Unlike many other insulation materials, wool is a renewable resource that can be produced with little environmental harm. It is also biodegradable, which means it will naturally decompose over time, minimizing waste and contamination. Several types of wool insulation include batts, rolls, and loose fill. It can be utilized in new and old structures' walls, floors, and roofs. There may be better solutions for some projects, though, as wool insulation can be more expensive than other insulation options. In conclusion, wool insulation is highly efficient and environmentally friendly for insulating buildings. Its numerous advantages make it a well-liked choice for individuals trying to increase their residential or business structures' environmental performance and energy efficiency.

6.3.4 Hemp Insulation

Building insulation made from hemp is sustainable and eco-friendly. It is created from the industrial hemp plant, which is grown exclusively for its seeds and fibers. Figure 6.2 is an example of the advantages of hemp insulation has over other types of insulation. It can help keep buildings warm in the winter and cool in the summer since it has high thermal insulation capabilities. Additionally, it is very breathable,

FIGURE 6.2 Detail of the structure of hemp.[5]

allowing moisture to evaporate and preventing mold and mildew growth. Hemp insulation is non-toxic, secure to handle, and devoid of contaminants or harmful substances. Because of this, it is the perfect option for those who are allergic to synthetic materials or want to lessen their exposure to toxins at home or work. Hemp insulation is also a renewable resource that can be grown naturally without the use of chemicals or herbicides. It is indicated that it has a negligibly negative environmental impact and can lessen our reliance on non-renewable resources. There are several different types of hemp insulation, including batts, rolls, and loose fill. It can be utilized in new and old structures' walls, floors, and roofs. In addition to being reasonably simple to install, hemp insulation is a preferred option for DIY enthusiasts.[3,4] Overall, hemp insulation is a sustainable and effective choice for insulating buildings. Its many benefits make it a popular option for those looking to improve the energy efficiency and environmental performance of their homes or commercial buildings.

6.3.5 Cork Insulation

Cork insulation is created from the cork oak tree's bark. Building insulation applications often use natural, renewable, and ecological cork insulation. Compared to conventional insulating materials like fiberglass or mineral wool, cork insulation offers several benefits. First, cork insulation has a high R-value per inch of thickness, making it a superior thermal insulator.[6] In order to lower energy expenses and increase comfort, it can assist in keeping buildings cool in the summer and warm in the winter. Second, cork insulation can aid in reducing noise transmission between rooms or from outside because it is an excellent acoustic insulator. It can benefit noisy structures, including apartment buildings or those near busy roads or railroads. In addition to being smoke- and fume-free when exposed to fire, cork insulation is also fire resistant and safer than certain alternative insulation materials in the case of a fire. Cork insulation provides excellent thermal, acoustic, and fire resistance properties and is suitable for the environment. Cork oak trees are harvested for their bark without harming the tree every 9 to 12 years.[7] Because of this, cork insulation is a sustainable and renewable material whose production has no adverse environmental effects. Cork insulation is a flexible, eco-friendly, and effective substance widely used in building projects.

6.3.6 Polyurethane Foam Insulation

A common form of insulation material used in buildings is polyurethane foam insulation. It is produced by combining two liquid chemicals, which, when mixed, react to form a foam that expands and solidifies, serving as a thermal barrier to help prevent heat transfer. There are many advantages to using polyurethane foam insulation in structures. First, it insulates effectively per inch of thickness because of its high R-value, which could help reduce heating and cooling costs.[8] Additionally, polyurethane foam insulation can be used to seal fractures and openings, preventing air penetration and enhancing the insulation's insulating properties. Another advantage of polyurethane foam insulation is its capacity to act as a sound barrier and reduce noise transmission between spaces or from outside. It may be beneficial for noisy buildings near highways or airports.

Additionally, water-resistant polyurethane foam insulation can aid in preventing moisture intrusion, which can promote the formation of mold and mildew. Additionally, it is fire resistant and can aid in containing the spread of flames in the event of a fire. Using polyurethane foam insulation in buildings could have significant drawbacks.

1. It could cost more than conventional insulating materials like fiberglass or cellulose.
2. Improper installation could result in harmful fumes being released, endangering the health of building occupants.
3. Because it comprises petrochemicals and is not biodegradable, polyurethane foam insulation is not considered an environmentally beneficial choice.

However, some producers use recycled resources to make polyurethane foam insulation, and some businesses are attempting to provide more environmentally friendly substitutes.[9] All things considered, polyurethane foam insulation is a flexible and efficient insulating material that can offer a variety of advantages for structures, as demonstrated in Figure 6.3. However, it is crucial to consider both the benefits and drawbacks when selecting an insulation material for a particular project. Considering aspects like environmental effects, energy efficiency, and health and safety is crucial

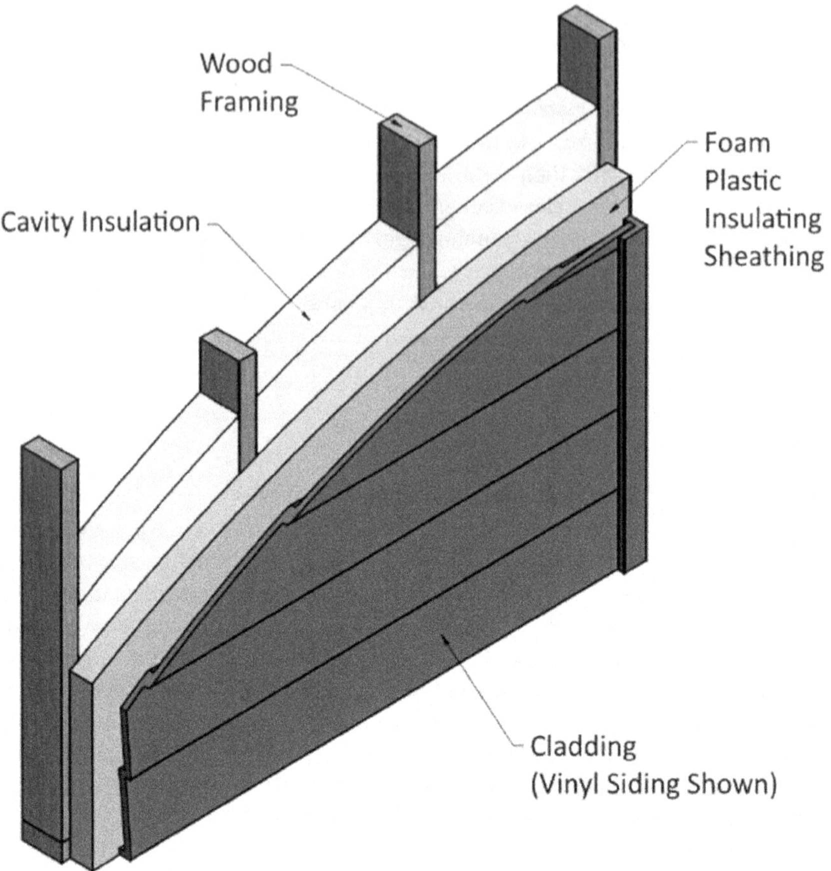

FIGURE 6.3 Wood-framed wall insulation.[10]

when choosing insulation materials for green buildings. A combination of these materials may be employed to achieve the necessary level of energy efficiency while reducing environmental effects.

6.4 PHASE CHANGE MATERIALS

Phase change materials (PCMs) are substances that absorb and release heat energy as they change from a solid to a liquid or vice versa. They are used in the design of green buildings to raise thermal comfort and increase energy efficiency. PCMs are frequently used in building envelope components like plasterboard and insulation in green structures.

These materials can be infused with PCMs to provide thermal bulk. By absorbing excess heat during the day and releasing it at night, thermal mass reduces indoor temperatures. PCMs are also utilized in HVAC systems in green buildings. PCMs can be used in thermal storage systems or as chilled ceilings to reduce the energy needed to cool and heat a building. When used in chilled ceilings, PCMs can offer cooling without using energy by absorbing heat during the day and releasing it at night. Here are a few uses for PCMs in environmentally friendly constructions.

6.4.1 Thermal Energy Storage

PCM thermal energy storage systems are becoming increasingly popular for use in buildings and other applications where there is a need to store and release thermal energy over a period of time. Lowering a building's or system's energy usage is one of the critical benefits of employing PCM thermal energy storage. During high energy demand, such as during the daytime when the sun is shining, PCMs can absorb additional heat and store it. In order to maintain a comfortable temperature in the building during times of low energy demand, such as at night or in overcast weather, the PCM can then release the heat stored during such times.[11] Lessening the need for heating and cooling systems will reduce energy expenses and greenhouse gas emissions. Enhancing thermal comfort in buildings is another benefit of PCM thermal energy storage. PCMs can assist in maintaining a more constant temperature in the structure by absorbing extra heat during the day and releasing it at night, lessening the need for artificial heating and cooling. It may give inhabitants a more comfortable interior environment, enhancing their well-being and productivity. PCM thermal energy storage systems are also incredibly dependable and require very little upkeep. PCM systems do not need moving components or pumps, unlike other thermal energy storage systems like water-based systems. Hence, there is less chance of a mechanical breakdown or failure. Using PCM thermal energy storage systems is not easy, however. One area for improvement is that the system's design must be thoroughly thought out to guarantee that the PCM can absorb and release heat at the proper periods. Another area for improvement is the cost of PCMs, mainly if they are designed specifically for a given purpose. In general, PCM thermal energy storage systems have several benefits for structures and other applications that require thermal energy storage. It is expected that PCM thermal energy storage will gain popularity as a method for lowering energy use and enhancing thermal comfort as technology advances and costs continue to decline.

6.4.2 Passive Cooling

In addition to being employed as a passive cooling system in buildings, phase change materials (PCMs) can also be used to maintain a suitable indoor temperature without air conditioning. A PCM is typically embedded into building materials such as walls or ceilings in a passive cooling system that employs PCMs. The PCM absorbs heat from the internal environment and melts during the day when the outside temperature is higher than the preferred indoor temperature, storing thermal energy. The PCM then releases the thermal energy it has stored. It solidifies it to chill the interior environment at night when the outside temperature is lower than the desired indoor temperature. Reducing the need for air conditioning, which may be costly to install and run, is one benefit of employing PCM as a passive cooling system. Lower energy costs and fewer greenhouse gas emissions may result.

Furthermore, PCM-based passive cooling systems have a minimal carbon footprint because they do not require energy inputs like electricity or gasoline. Improved thermal comfort in buildings is a benefit of employing PCMs as a passive cooling system. PCMs can create a more comfortable interior environment for occupants by keeping a constant indoor temperature, enhancing their well-being and productivity. However, employing as PCM as a passive cooling method has significant drawbacks. One restriction is that the PCM's ability to collect and release heat at the appropriate periods depends on the system's design.

Additionally, the efficacy of PCMs as a passive cooling system may be limited in hot and humid environments, when the cooling effects may need to be enhanced. In hot climates where maintaining a comfortable interior temperature without air conditioning is essential, PCMs can be a valuable material for passive cooling systems in buildings. However, the system must be implemented carefully to guarantee its effectiveness.

6.4.3 Radiant Heating and Cooling

Phase change materials may be used in buildings' radiant heating and cooling systems. Radiant heating and cooling systems require a network of pipes to circulate heated or cooled water around a building and radiate heat or coolness into the surrounding region. Using PCMs can improve these systems' capacity to store thermal energy and reduce energy usage. To absorb and retain heat from the flowing water, PCMs can be used in the floor or ceiling material of a radiant heating system.[12] As the temperature in the room drops, the thermal energy held in the PCM is released, increasing the heat in the space. In a radiant cooling system that uses it, the PCM can be included in the ceiling material, which absorbs and stores heat from the surrounding air. As the temperature in the room rises, the thermal energy held in the PCM is released, further cooling the area. Using PCMs, radiant heating and cooling systems can be made more effective, their energy consumption can be reduced, and their thermal comfort can be improved. PCMs can help with energy conservation, reducing greenhouse gas emissions and reducing the need for additional heating or cooling. During times of low energy demand, thermal energy is stored, and during times of high energy demand, thermal energy is released. PCMs must be appropriately planned and implemented in radiant heating and cooling systems in order to guarantee the system's functionality and effectiveness. The type and quantity of PCM used, as well as the size, layout, and type of the radiant heating and cooling system can all impact how well the system functions. PCMs may significantly increase buildings' radiant heating and cooling, improving thermal comfort and energy effectiveness. However, careful consideration must go into both the system's design and its implementation to ensure that the system is effective and successful.

6.4.4 Building Envelope

The "building envelope" refers to the external physical barriers that separate a structure's interior from the outside environment. Building envelopes ensure tenant health and safety, manage energy consumption, and maintains comfortable interior temperatures.[13] PCMs can be utilized in building envelopes to help regulate indoor temperatures by storing and releasing thermal energy. For instance, PCMs can be applied to insulation materials in walls, roofs, or floors to gather and release heat as indoor temperatures change. When indoor temperatures rise, the PCM absorbs additional heat to prevent the building from overheating. When the temperature inside dips, the PCM releases heat that has been stored, keeping the area warm. By lowering the energy needed for heating and cooling, PCMs in building envelopes can minimize energy costs and greenhouse gas emissions. In addition to improving thermal comfort and occupant well-being, PCMs may maintain a steady indoor temperature. Buildings can benefit from PCMs' improved thermal comfort, energy efficiency, and indoor air quality while reducing their vulnerability to moisture damage. By absorbing additional moisture, PCMs can assist in reducing the development of mold and other hazardous contaminants.

Additionally, some PCM variants emit few volatile organic chemicals and are non-toxic, which can improve indoor air quality. However, PCMs must be used very carefully for building envelopes to be practical and effective. The kind and quantity of PCM may influence the performance of the system used, its location within the building envelope, the building's temperature and energy requirements, and other factors. PCMs can improve energy efficiency, thermal comfort, indoor air quality, and overall occupant well-being when employed in building envelopes. By storing and releasing thermal energy, PCMs can help maintain indoor temperatures, reduce energy consumption, and improve a building's overall performance.

6.4.5 Solar Thermal Storage

In order to store the thermal energy produced by solar panels during the day so that they can be utilized to heat or deliver hot water to buildings when there is little or no solar radiation, phase change materials can be employed as solar thermal storage materials. During periods of high solar radiation, such as during the day, the solar collectors heat the PCM, causing it to melt and store thermal energy. When there is no solar radiation, such as at night or on cloudy days, the PCM hardens and releases the thermal energy that has been stored. Buildings can then use this thermal energy for heating or hot water. PCMs in solar thermal storage systems can improve the efficiency of solar systems by permitting the storage of thermal energy generated during high solar radiation for usage during low solar radiation times. They can increase the total amount of energy produced by the solar system and reduce the need for backup heating systems, conserving energy and reducing greenhouse gas emissions. For solar thermal storage systems to be effective and valuable, PCMs must be created and used correctly. The performance of the system can be affected by some variables, including the kind and quantity of PCM utilized, the temperature range of the PCM, and the thermal conductivity of the storage medium. Overall, solar systems can be more efficient and effective by integrating PCMs in solar thermal storage systems, resulting in energy savings and lower greenhouse gas emissions. PCMs can increase the system's overall energy output and offer dependable heating and hot water in buildings by storing thermal energy produced by solar systems.

6.5 RECYCLED MATERIALS

Utilizing recycled materials, like recovered wood or recycled steel, helps lessen the adverse environmental effects of construction materials. A sustainable practice that can significantly lessen the building industry's environmental impact is using recycled materials in green structures. Here are some strategies for incorporating recycled materials into environmentally friendly structures.

6.5.1 Recycled Steel

Recycled steel is a popular building material that lessens the environmental impact of new steel production. Recycled steel is a multipurpose building material used for structural framing, roofing, and cladding. Steel is a durable, versatile, and often used material in construction; recycling steel can help lessen the sector's environmental effects. Recycling metal trash involves melting it down and creating new steel products. Recycling steel utilizes less energy and produces fewer greenhouse gases than creating new steel from scratch.

Additionally, recycling steel can lengthen the useful life of current steel products, save natural resources, and reduce the amount of garbage dumped in landfills. There are several ways to use recycled steel in buildings. It might be used for the structural framing of a building, such as columns, beams, and trusses. Recycled steel can also be used to build architectural features like stairways, handrails, and lifts, as well as for roofing and cladding. Building with recycled steel has benefits for the economy and environment. As recycled steel is often less expensive than new steel, using it might be more cost-effective than using new steel. Utilizing recycled steel might also lessen the risk of price swings related to using new steel. Recycled steel use in buildings has its challenges, however. For instance, the source and recycling method might impact the consistency and quality of recycled steel. The necessity to guarantee that the steel fits the specifications for strength and durability might provide difficulties for producers and builders. Overall, using recycled steel in construction can benefit both the environment and the economy. The building sector may aid in lowering greenhouse gas emissions, protecting natural resources, and decreasing landfill trash by employing recycled steel. Special attention must be given to the quality and uniformity of the material in order to ensure that recycled steel satisfies the standards for strength and durability in building applications.

6.5.2 Recycled Concrete

Concrete is another common building material that can be made with recycled materials, such as crushed concrete or fly ash. Many construction projects can benefit from using recycled concrete as a building material. Since concrete is a material that is frequently used in construction, recycling it can help to lessen the industry's adverse environmental effects. In order to create a material that may be utilized as an aggregate in new concrete, recycling concrete entails crushing and grading old concrete. Compared to creating fresh concrete from scratch, reusing concrete uses less energy and emits fewer greenhouse gases. Recycling concrete can also lessen waste sent to landfills and preserve natural resources. Recycled concrete can be applied in a variety of ways in buildings. For instance, it can be utilized as an aggregate in fresh concrete and create precast concrete products like blocks. Recycled concrete can also be used as a base for roads and walkways and as a filler material in construction projects. There are many benefits to using recycled concrete in the construction sector. Using recycled concrete, for instance, can help reduce the environmental impact of the building sector by reducing greenhouse gas emissions and conserving natural resources. Additionally, using recycled concrete could be more economical than new concrete because it is frequently lower in price. However, there are several disadvantages to using recycled concrete in construction. For instance, the uniformity and quality of recycled concrete are influenced by the source and recycling process. Help may be required for producers and builders who need to ensure the concrete meets the requirements for strength and durability. Employing recycled concrete in buildings can benefit the economy and the environment. The construction industry may help reduce greenhouse gas emissions, safeguard natural resources, and reduce landfill waste by employing recycled concrete.

6.5.3 Reclaimed Wood

Reclaimed wood is a typical choice for recycled materials in applications involving building and construction. It is taken out of dilapidated or damaged old buildings, barns, and other structures. Reclaimed wood can have positive environmental consequences while giving a building project a unique appearance. Reclaimed wood can reduce the need for new wood, helping to protect natural resources and diminish the damaging effects of the construction industry on the environment. Additionally, old wood can be recovered and reused, avoiding landfill disposal and reducing waste and greenhouse gas emissions. Reclaimed wood can be used for various purposes in construction, including flooring, wall cladding, structural framing, and decorative features. The processing and preservation of salvaged wood must be prioritized to preserve its durability and structural integrity. There are many benefits to using reclaimed wood in buildings. It might, for instance, offer a distinctive aesthetic that is impossible to accomplish with brand-new materials. Reclaimed wood may provide warmth and character to a construction project.

Furthermore, using reclaimed wood instead of new wood may be more cost-effective because it is frequently less expensive than new wood. Recycled wood is used in buildings, but there are also some disadvantages. For instance, the quality and condition of the wood may vary depending on its origin and previous use. Builders and producers who must guarantee the wood meets the requirements for strength and durability might require assistance. Recycled wood may be beautiful and environmentally friendly when used in construction. Utilizing recovered wood in the construction industry could lower the demand for new wood, safeguard natural resources, and reduce waste.

6.5.4 Recycled Glass

Recycled glass is a flexible and environmentally friendly material that may be applied in construction and building. Glass recycling is melting down the used glass and molding it into new items, aiding in resource conservation and minimizing waste. Recycled glass can be applied in a variety of ways in construction. It can be utilized as an aggregate in producing glass blocks, countertops, and tiles, for instance, as well as concrete and asphalt. Additionally, recycled glass can be utilized as a filler material in building projects and as a decorative element in landscaping. Recycled glass is advantageous for several reasons when it comes to construction. Utilizing recycled glass, for instance, can assist in lowering the need for new raw materials, which can help save natural resources and lessen the construction industry's environmental effects. Utilizing recovered glass can also lessen waste sent to landfills and greenhouse gas emissions. The texture and color of recycled glass can add visual interest to a building project, giving it a distinctive and appealing appearance. Because recycled glass is frequently less expensive than new glass, it can also be more affordable. However, employing recycled glass in a building can be challenging. For instance, the source and recycling method might impact the consistency and quality of recycled glass. Builders and manufacturers that need to ensure the glass fits the criteria for strength and durability may need help. Using recycled glass in construction may be attractive and good for the environment. The building sector can contribute to resource conservation, waste reduction, and greenhouse gas emission reduction by employing recycled glass. However, careful consideration must be given to the quality and consistency of recycled glass to ensure that it meets the required standards for strength and durability in construction applications.

6.5.5 Recycled Insulation

Recycled insulation refers to insulation materials that are made from recycled materials. These materials can include recycled glass, cotton, denim, plastic bottles, and other materials that have been diverted from the waste stream. Insulation products manufactured from recycled materials are referred to as recycled insulation. These materials can include waste that has been diverted from the recycling process, such as recycled glass, cotton, denim, and plastic bottles. Recycled insulation can provide several advantages in construction.

1. It aids in lowering the volume of garbage dumped in landfills.
2. Exploring alternative materials or utilizing recycled materials for insulation can help lessen the environmental impact associated with building construction and renovation projects.
3. By using fewer materials that contain dangerous chemicals, recycled insulation can create a more sustainable and healthier interior atmosphere.

There are many places where recycled insulation can be used, including attics, floors, and roofs. The ability of recycled insulation to resist heat flow is measured by its R-value, which varies depending on the type of material utilized. For instance, post-industrial and post-consumer denim diverted from the trash stream makes recycled denim insulation. It is a chemical-free, natural, and environmentally friendly insulation material that is safe for people and the environment. An efficient insulator for walls, attics, and floors, recycled denim insulation has a high R-value. Cellulose insulation, created from recycled paper goods, including newspaper, cardboard, and other materials, is another type of recycled insulation. It is a fire-retardant-treated insulating material that is environmentally responsible and long-lasting. Employing recycled insulation in buildings can positively affect the environment, economy, and human health. Cellulose insulation has a high R-value and is appropriate for use in walls, attics, and floors. It can contribute to waste reduction, resource conservation, and creating a more sustainable interior environment.

6.5.6 Recycled Plastic

The flexible and resource-efficient recycled plastic material can be used in architecture and construction. Plastic recycling is the process of gathering, sanitizing, and processing used plastic products so they can be transformed into new products like building materials. Buildings can use recycled plastic in a variety of ways. For instance, it can replace traditional wood and concrete with plastic lumber, decking, and fencing. Recycling plastic allows for producing insulation, water-resistant barriers, and roofing materials. For several reasons, recycled plastic is helpful in construction. Utilizing recycled plastic, for example, can help reduce the need for new raw materials, which can help conserve natural resources and diminish the environmental effects of the building sector. Utilizing recovered plastic can help reduce landfill waste and greenhouse gas emissions. Recycled plastic has many advantages over virgin plastic, including longevity and resistance to rot, insects, and moisture. Building with recycled plastic materials can also save money over time because they often need less upkeep and last longer than conventional materials. However, employing recycled plastic in buildings can be challenging. For instance, the source and recycling method might impact the consistency and quality of recycled plastic. Builders and manufacturers that need to ensure the plastic fits the criteria for strength and durability may need help. Overall, using recycled plastic in construction can benefit both the environment and the economy. By employing recycled plastic, the construction sector can contribute to resource conservation, waste reduction, and greenhouse gas emission reduction. However, careful consideration must be given to the quality and consistency of recycled plastic to ensure that it meets the required standards for strength and durability in construction applications. Incorporating recycled materials into green buildings is an important sustainable practice that can significantly reduce the environmental impact of the building industry. By using recycled steel, concrete, wood, glass, insulation, and plastic in building designs, architects and builders can create sustainable buildings that are environmentally responsible and cost-effective.

6.6 SUSTAINABLE WOOD PRODUCTS

Wood has been a staple of human architecture throughout history because it is a natural and regenerative building element. Bamboo, salvaged wood, and FSC-certified wood are examples of sustainable wood products that help lessen the environmental impact of building materials and encourage good forestry practices. Green structures, which seek to minimize the use of non-renewable resources and lessen the environmental effect of construction, must include sustainable wood products. When using wood for construction, the following processes may affect the carbon balance.[14]

1. Switching from fossil fuels to alternative materials in manufacturing.
2. Avoiding emissions from cement production.
3. Storing carbon in wood products and forests.
4. Preventing fossil fuel emissions due to the use of biomass.
5. Influencing carbon dynamics in landfills.

These green buildings frequently use the following sustainable wood products.

6.6.1 Certified Wood

The term "certified wood" refers to wood grown and prepared according to strict social and environmental criteria. The Sustainable Forestry Initiative (SFI), the Programme for the Endorsement of Forest Certification (PEFC), and other independent organizations set these standards. Wood must fulfill requirements that guarantee its sustainable and ethical logging has been conducted before it can be deemed certified. It contains standards for forest management practices such as preserving biodiversity and wildlife habitats, maintaining soil and water quality, and having a minimal effect on nearby settlements. The building, furniture, and paper industries are just a few uses for certified wood. It is possible to reduce deforestation and promote sustainable forest management practices by using certified wood in construction projects to help ensure that the wood comes from sustainably managed forests. Utilizing certified wood can have economic and social advantages and environmental advantages. In addition to supporting fair labor practices and guaranteeing that employees are appropriately treated, certified wood can assist regional communities that rely on the forestry sector for their livelihoods. Utilizing certified wood can assist in encouraging sustainable forestry methods, lessen deforestation, and benefit regional people. As for construction materials and other wood goods, it might offer a more ecological and ethical alternative.

6.6.2 Reclaimed Wood

Reclaimed wood is a fantastic way to reduce the damaging effects of construction on the environment. Old houses, barns, bridges, and other structures that have been demolished or no longer used have been used to make reclaimed wood. Instead of being wasted, the timber is carefully removed and utilized in future construction or repair projects. Here are a few benefits of using reclaimed wood in construction.

1. By utilizing salvaged wood, less new wood is required, protecting forests and reducing the amount of wood waste disposed of in landfills.
2. It is impossible to replicate the specific characteristics of reclaimed wood with fresh wood, such as knots, nail holes, and grain patterns. As a result, buildings have a distinctive, rustic aspect.
3. Reclaimed wood is typically made from old-growth wood, which is denser and more resilient than fresh wood. Because of this, it is less prone to twist, bend, or rot, making it the ideal material for sturdy buildings and furniture.

4. Although reclaimed wood may initially cost more than new wood, it can ultimately be more affordable. Because reclaimed wood is sturdy, it will require less maintenance or replacement over time.
5. Reclaimed wood has a lower carbon footprint than new wood and requires less energy and fewer resources to produce because it has previously been harvested and treated.

Reclaimed wood is a distinctively characterful, long-lasting, cost-effective, and environmentally responsible building material with a low carbon impact. It is a terrific option for architects and designers who want to create a unique and environmentally friendly place.

6.6.3 Engineered Wood

Adhesives join wood strands or fibers to create engineered wood products like plywood and particleboard. These goods are frequently manufactured from small, quickly expanding trees that may be harvested sustainably. Compared to solid wood products, engineered wood products are also sturdier and more resilient. Here are a few advantages of engineered wood.

1. Engineered wood is frequently less expensive than solid wood, making it a popular option for builders and designers on a tight budget.
2. Engineered wood is more stable and dependable than solid wood for structural purposes because it is less likely to flex, shrink, or swell.
3. Engineered wood is versatile and can be produced in various sizes, shapes, and thicknesses to meet various construction and furniture-making needs.
4. Engineered wood is frequently created from waste wood products like sawdust or wood chips. As a result, it is a sustainable option that lowers waste.
5. Because engineered wood is produced in a controlled setting, its quality and other characteristics are maintained throughout the product.
6. Engineered wood can help to lower labor costs and installation time since it is simpler to cut, shape, and install than solid wood.
7. Engineered wood is aesthetically pleasing. Because it can be designed to resemble natural wood with a more uniform appearance, it is a popular option for modern and contemporary designs.

Alternatives to typical solid wood that are more affordable, adaptable, and sustainable include engineered wood. It is stable, consistent, and simple to use and can be finished to seem like solid wood for a more attractive result.

6.6.4 Bamboo and Cork

Fast-growing bamboo may be harvested sustainably and used in several construction projects. Bamboo is an excellent substitute for traditional wood items because it is sturdy and long-lasting. Cork can be extracted without destroying trees to create a sustainable wood product. Every nine years, the bark of the cork oak tree is removed to manufacture cork goods, allowing the tree to keep growing and producing more bark. In addition to supporting responsible forest management, using sustainable wood products in green buildings helps lessen construction's adverse environmental effects. Selecting sustainable wood products can contribute to a more sustainable future for everyone.

6.7 GLASS-BASED MATERIALS

Using glass-based materials in green buildings can offer a range of benefits, including energy efficiency, natural lighting, and improved indoor air quality. The different glass-based materials for green buildings are classified in Figure 6.4.

6.7.1 Smart Glazing with Micro-Mirrors

A new era for the intelligent window industry has begun with the development of a revolutionary glass with a thin covering of micro-mirrors that ranges in thickness from 0.15 to 0.2 millimeters. In order to improve air conditioning and natural lighting, they wish to manage sunlight. As a result, the building's heating and cooling costs will decrease. A high-precision laser is used to cut micro-mirrors inserted in a polymer film sandwiched between the layers of double-glazed windows. This system consists of a one-dimensional array of parabolic reflecting surfaces and an additional array of secondary reflective surfaces. As discussed herein, switchable or electrochromic glass, commonly called smart glazing with micro-mirrors, provides some advantages over conventional glass.

1. By regulating the amount of light and heat that enters a building, smart glazing with micro-mirrors can help lower energy use. The glass may alternate between transparent and colored modes by changing the micro-mirror orientation, eliminating the need for artificial heating, cooling, and lighting.
2. By decreasing glare and excessive brightness, which can be uncomfortable and strain the eyes, smart glazing with micro-mirrors creates a more

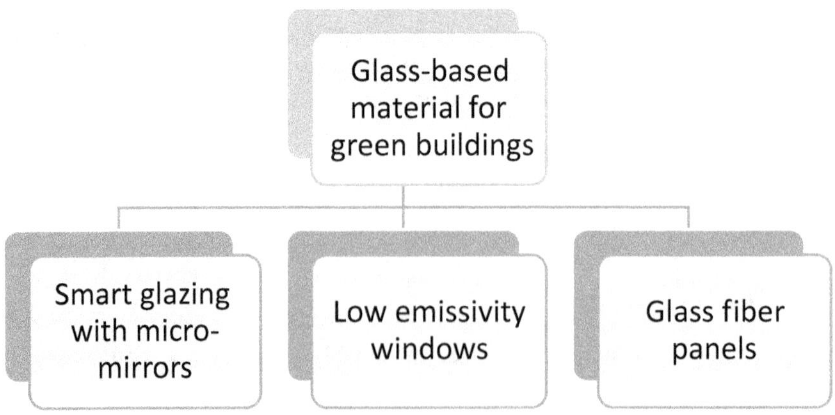

FIGURE 6.4 Different types of glass-based materials for green buildings.

comfortable environment. A constant temperature throughout the day is also beneficial.

3. Instead of curtains or blinds, privacy zones can be created by combining smart glass with tiny mirrors. By changing the direction of the micro-mirrors, the glass may go from transparent to opaque mode to provide privacy.
4. Smart glass with tiny mirrors gives structures a contemporary, futuristic appearance. It can also be used as a design element by incorporating it into architectural elements like facades, skylights, and partitions.
5. A building's security can be improved by using smart glass with micro-mirrors to create a shatterproof barrier. The glass is unbroken and keeps debris from entering the building in the case of a break-in or natural disaster.

6.7.2 Low-Emissivity Windows

In addition to reducing heat loss, low-emissivity (low-E) windows lower the quantity of UV and infrared light entering a structure. These windows can increase indoor comfort, lower energy expenditures for heating and cooling, and improve energy efficiency. Low-E windows are a crucial element in the construction of green buildings. They are specifically made to lessen heat gain in the summer and heat loss in the winter, which can significantly reduce the energy needed for heating and cooling. Figure 6.5 illustrates how low-emissivity windows operate. These low-E window advantages and green building considerations are listed as follows.

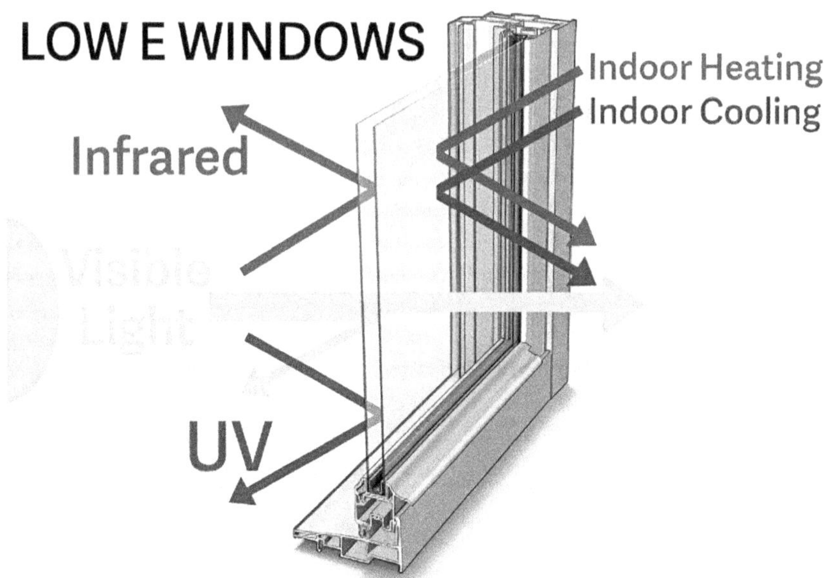

FIGURE 6.5 Illustrative image of low-emissivity windows.[16]

Low-E windows' benefits

1. Low-E windows are an energy-efficient option for green buildings because they contain a specific coating that reflects infrared radiation, which can considerably minimize heat gain and loss.[15]
2. By lowering temperature swings and minimizing draughts, low-E windows can contribute to maintaining a more agreeable indoor climate.
3. Low-E windows can lessen a building's carbon footprint by using less energy for heating and cooling.

Disadvantages of low-E windows

1. Cost: Low-E windows can be more expensive than traditional windows, although the energy savings over time can offset the initial cost.
2. A building's windows' orientation can impact low-E coatings' effectiveness. For example, north-facing windows may benefit less from low-E coatings than south-facing windows.
3. Glare and color distortion: Low-E coatings can sometimes cause glare and color distortion, which may affect the visual comfort of building occupants.

In addition to low-E coatings, other factors that can affect the energy efficiency of windows include window framing materials, glazing materials, and window placement. When designing green buildings, it is essential to consider all these factors and select the most appropriate window system for the building's specific needs and goals.

6.7.3 Glass Fiber Panels

A group of Spanish researchers developed a prototype fiberglass prefabricated facade that could optimize solar energy by absorbing and releasing it into the room for heating surroundings, as well as precisely increase the heat insulation, to assist minimize the energy consumption of buildings. A Merida construction used a mixture of glass fiber and organic binders as the "pilot model" for the prefabricated panel.[17] The performance of the fiberglass wall was evaluated in three key areas: corrosion resistance, water and wind resistance, thermal insulation, and sound insulation. Glass fiber panels, commonly referred to as fiberglass panels, have various benefits, such as[18]

1. Glass fiber panels are highly robust and resistant to abrasion, weather, and temperature variations. They are an excellent option for outdoor and industrial applications because they are rot, rust, and corrosion resistant.
2. Because glass fiber panels are light, handling and installation are straightforward. They are also less expensive to ship and more environmentally friendly because they are simple to transport.
3. Glass fiber panels are flexible and can fit several shapes and sizes, making them appropriate for various applications. Additionally, they can be painted or coated to match any color or style, giving designers more design freedom.

4. Because glass fiber panels are highly resistant to deterioration and damage, they require very little upkeep. In the long term, this can save time and money because there will not need to be as many repairs or replacements.
5. Because glass fiber panels are great insulators, they can help keep buildings warmer and cooler in the winter, which can help cut down on energy expenses.

A cost-efficient, low-maintenance, and highly durable choice for various industrial and commercial applications is provided by glass fiber panels.

6.8 SMART GLASS

This glass can switch between transparent and opaque modes, depending on the level of light exposure. It can help reduce cooling costs by reducing solar heat gain. Smart glass, also known as switchable or electrochromic glass, is a technology that can help make green buildings more energy efficient by reducing energy consumption for heating, cooling, and lighting. Smart glass can change its opacity or transparency in response to external stimuli such as light, heat, or electricity. An illustrative image of smart glass is shown in Figure 6.6. Here are some ways smart glass can be helpful for green buildings.

1. Windows use less energy. Smart glass windows contain tints that can be automatically altered to reflect or absorb sunlight, minimizing heat absorption in the building and the need for air conditioning. By minimizing the energy used for cooling, smart glass windows can help lower a building's overall energy use and carbon impact.
2. Smart glass may also maximize the amount of natural light in a structure. Smart glass may change color to balance natural and artificial lighting in a space using sensors to determine how much daylight is present. It lowers the electricity demand.

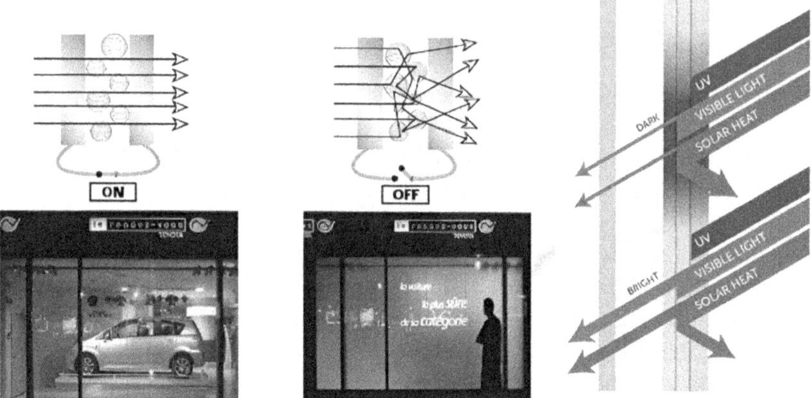

FIGURE 6.6 Illustrative image of smart glass.[19,20]

3. Smart glass can also be utilized to improve building privacy. The color of the glass can be altered to give residents privacy while preserving natural light. Smart glass can produce distinctive architectural designs that improve the attractiveness and appeal of a structure.

Advanced buildings can become more aesthetically pleasing, pleasant, and energy efficient thanks to smart glass technologies. The smart glass will continue to be a crucial tool for building designers and architects as the need for sustainable building materials rises.[21]

6.9 WATER-EFFICIENT FIXTURES

Low-flow toilets, showerheads, and faucets can help save water when used in buildings. Green buildings must have water-efficient fixtures since they can save utility costs, reduce water use, and protect limited natural resources. Here are some instances of water-saving fixtures utilized in green structures.

6.9.1 Low-Flow Toilets

Low-flow toilets flush with a lot less water than regular toilets do. They typically use less than 1.6 gallons per flush (gpf) instead of older toilets, which use 3.5 to 7 gpf. Low-flow toilet installations can reduce water use by up to 60%.[22] Reducing water use can save money on water bills while preserving water resources. There are both pressure-assisted and gravity-fed low-flow toilets available. Gravity-fed toilets use the force of gravity to flush waste away, whereas pressure-assisted toilets use a burst of pressurized air to push waste down the drain. Low-flow toilets might not be as effective in eliminating waste as standard toilets. Many modern low-flow toilets are constructed with broader trap ways and more efficient flushing mechanisms to guarantee they are as effective as regular toilets. In general, using low-flow toilets is a fantastic method to conserve water, reduce water costs, and quickly eliminate waste.

6.9.2 Low-Flow Showerheads

Low-flow showerheads use less water than standard showerheads while providing a pleasant and comfortable showering experience. These showerheads are made to be more cost-effective and water-efficient without sacrificing comfort. Low-flow showerheads have a maximum water flow restriction of 1.5 gallons per minute (gpm). They can provide a pleasant shower experience using much less water than traditional showerheads, which generally flow at 2.5 gpm or more.[23] Low-flow showerheads restrict the amount of water that passes through the showerhead, making the water flow more effective and efficient. Many low-flow showerheads have adjustable settings that change how much water flows through them. Concern should be expressed about the possibility of a weaker stream of water from low-flow showerheads than from regular showerheads. Aerating technology, which combines air and water to generate a more robust and energizing showering experience, is used by many modern low-flow showerheads. Low-flow showerheads are a great way to save water and reduce water bills while providing a pleasant and comfortable showering experience.

6.9.3 Faucet Aerators and Waterless Urinals

Faucet aerators are small devices installed on the end of a faucet to reduce water flow while maintaining consistent pressure. They can typically reduce water consumption by up to 50%.[24] Waterless urinals use no water, relying on a trap and a liquid sealant to prevent odors and blockages. Installing waterless urinals can save up to 40,000 gallons per unit per year.[25]

6.9.4 Greywater Systems

Greywater systems allow for capturing and reusing non-potable water from sinks, showers, and other sources. This water can be treated and reused for irrigation, toilet flushing, or other non-potable uses. The water-efficient fixtures can help green buildings reduce water consumption and save money on utility bills. By incorporating these fixtures into building designs, architects and builders can create sustainable buildings that are both environmentally responsible and cost-effective.

6.10 ENERGY-EFFICIENT LIGHTING

LED (light-emitting diode) lighting is a popular choice for green buildings because it uses less energy than traditional lighting and lasts longer. Energy-efficient lighting is a crucial aspect of green buildings, as lighting can account for up to 25% of a building's energy consumption. Here are some examples of energy-efficient lighting that can be used in green buildings.[26]

6.10.1 LED and Task Lighting

LED lighting is significantly more energy-efficient than traditional incandescent bulbs, using up to 75% less energy. LED bulbs last much longer than traditional bulbs, reducing maintenance costs. Task lighting can provide focused lighting in areas where it is needed rather than lighting an entire room. It can help reduce energy consumption by only lighting the areas that are being used.

6.10.2 Occupancy and Daylight Sensors

Rooms can be equipped with occupancy sensors to determine when people are present and change the illumination accordingly. It can drastically save energy usage in places like restrooms and conference rooms, where lights are frequently kept on when unnecessary. Daylight sensors can be installed to measure the quantity of natural light in a space and modify the artificial lighting as necessary. When enough natural light is present, it can reduce energy use by automatically lowering or turning off lights.

6.10.3 Light Shelves

Installing light shelves can increase the depth of natural light in a space and lessen the demand for artificial lighting. Most of the time, light shelves are positioned next to windows, using reflective surfaces to bounce light deeper into the space.

Green buildings must have energy-efficient lighting as an essential element. Architects and builders may drastically lower energy usage and produce sustainable buildings that are economical and environmentally friendly by integrating LED lighting, occupancy sensors, daylight sensors, task lighting, and light shelves into building designs.

6.11 SOLAR PANELS

Solar electricity can be produced and help a structure use less energy from the grid. Solar energy is converted into electricity by these panels, which are put on the walls or roofs of buildings, reducing the demand for grid power. Solar panels are essential to green buildings since they provide renewable energy and significantly reduce a structure's carbon footprint. Here are a few strategies for incorporating solar panels into green buildings.

6.11.1 Rooftop Solar Panels

Rooftop solar panels are the most common type of installation in structures. Solar panels can be installed on a building's roof to generate electricity from sunshine. Electricity can be used to power the structure or feed back into the grid to earn credits. "Building-integrated photovoltaics" is a type of solar panel installation that integrates solar panels into the design of a building, such as solar shingles or solar glass. Building-integrated photovoltaics has the potential to be both aesthetically pleasing and more energy efficient because the solar panels can blend in seamlessly with the building's architecture.

6.11.2 Solar Water and Air Heaters

Solar air heaters use sunlight to warm the air for a building's heating system. They may be placed on the ground's surface or top of a building. Solar air heaters warm air that can be used for ventilation or space heating by harnessing the sun's power. The greenhouse effect, which describes how solar energy is absorbed by a transparent surface (typically glass) and converted into heat, forms the basis of solar air heaters. A flat panel or box mounted to a wall or roof oriented southward often makes up a solar air heater. The panel or box is constructed of a substance of a dark color that may absorb sunlight and convert it to heat. As it passes over the absorber surface, air travels via tubes or channels inside the panel or box, absorbing heat. The two main classifications are active and passive solar air heaters. Passive solar air heaters rely on natural convection to move air through the system, unlike active solar air heaters, which use fans or blowers to force air through the system. There are several benefits when comparing solar air heaters to conventional heating systems, including cheaper running costs, fewer greenhouse gas emissions, and more energy independence.[27] They work better in colder and cloudier climates. Therefore, they are not appropriate for all environments.

6.11.3 Solar Shading Systems

A building's facade can be equipped with solar shading devices, like solar blinds or awnings, to provide shade from the sun and lower cooling requirements. In order to produce electricity, these systems can also incorporate solar panels. Solar panels are an essential part of green structures as a source of renewable energy and a means of lowering a building's carbon footprint. Rooftop solar panels, building-integrated photovoltaics, solar water heaters, solar air warmers, and solar shading systems can all be incorporated into building designs to produce sustainable structures that are economical and considerate of the environment.[28]

6.12 SMART LIGHTING

Buildings with smart lighting systems have a network of connected lighting devices that can be automated and controlled by cutting-edge technology. These systems use a range of elements, including sensors, controllers, and communication protocols, to offer improved lighting control, energy efficiency, and convenience. The lighting levels in a building can be changed based on occupancy, natural light, and other variables using smart lighting systems that use sophisticated sensors and controllers. Building operating costs can be decreased, and energy can be saved. As it may drastically lower energy usage and increase a building's sustainability, smart lighting is a crucial component of structures. Here are some techniques for integrating smart lighting into structures.

6.12.1 Motion and Light Sensors

Motion sensors can be installed in rooms to detect human presence and control lighting accordingly. Lighting can be automatically turned off in vacant spaces to save electricity. By measuring how much natural light is there, light sensors installed in a location will enable us to manage artificial lighting. With enough natural light, lights can be set to dim or be turned off to help save electricity automatically.

6.12.2 Timer and Networked Lighting Controls

Timer controls can turn lights on/off at specific times, such as at the start and end of the workday. They can reduce energy consumption by ensuring that lights are not left on unnecessarily. Networked lighting controls can connect lighting systems and enable remote control and monitoring. They can also help build managers to optimize lighting usage and identify energy-saving opportunities.

6.12.3 Daylight Harvesting and Personalized Lighting

Sensors are used in daylight harvesting systems to gauge the amount of natural light in a space and adjust the artificial lighting accordingly. When there is enough natural light, this can cut energy use by lowering or shutting off lights. Individual control over lighting in a shared place can be provided by using customized lighting solutions. Enabling people to customize the lighting to their tastes can increase user comfort and lower energy use.

Smart lighting is a crucial part of green structures. Architects and builders can significantly lower energy consumption and produce sustainable buildings that are economical and environmentally friendly by integrating motion sensors, light sensors, timer controls, networked lighting controls, daylight harvesting systems, and personalized lighting into building designs.

6.13 GREEN ROOFING

Plants are grown atop green roofs, which offer insulation, lessen the impact of the urban heat island, and lessen stormwater runoff. Green roofing can lessen the energy required to heat and cool a building and its environmental impact. Green roofing is a viable and environmentally responsible choice for green structures, offering several advantages for the environment, such as better air quality, less energy use, and less stormwater runoff. Here are a few ways that green buildings can use green roofs.

6.13.1 Vegetative and Cool Roofs

Figure 6.7 shows that vegetative roofs, commonly called green roofs, are covered in vegetation like plants or grass. Numerous advantages of vegetative roofs include decreased energy use, enhanced air quality, and less stormwater runoff. By reflecting sunlight and lowering heat absorption, cool roofs can lessen the demand for air

FIGURE 6.7 The basic structure of a green roofing system.[29]

conditioning and reduce energy use. Reflective materials like white membranes or tiles can create cool roofs.

6.13.2 Solar and Blue Roofs

Solar rooftops harness the sun's energy to generate electricity. Solar energy can be produced for the building by adding rooftop solar panels. In order to lessen stormwater runoff, blue roofs are designed to collect rainfall and gradually release it into the environment. Blue roofs can be made of materials that absorb and release water, such as gravel, porous concrete, or green roofs.

6.13.3 Rooftop Gardens

Rooftop gardens are intended to give people and plants a place to congregate on a building's roof. Rooftop gardens have several advantages such as better air quality, less stormwater runoff, and enhanced mental and physical wellness.

The green roof is a crucial part of green structures and offers various environmental advantages, such as better air quality, less energy use, and less stormwater runoff. Architects and builders can develop cost-efficient, environmentally responsible, sustainable structures using vegetative roofs, cool roofs, solar roofs, blue roofs, and rooftop gardens.

6.14 SMART HVAC SYSTEMS

HVAC (heating, ventilation, and air conditioning) systems can be incorporated into green buildings. HVAC systems can be made smarter to improve performance and lower energy use by integrating cutting-edge sensors, controls, and algorithms. This can lower expenses, save energy, and enhance indoor air quality. Advanced buildings must have smart HVAC systems since they improve energy efficiency, lower carbon emissions, and improve indoor air quality, among other environmental advantages.[30] Here are some of the strategies for integrating smart HVAC systems into environmentally friendly structures are briefed.

6.14.1 Energy-Efficient Equipment and Smart Controls

Equipment for heating, ventilation, and energy-efficient air conditioning can dramatically lower energy use and improve the sustainability of a structure. This includes appliances like heat pumps, air conditioners, and high-efficiency boilers. HVAC systems may be optimized for optimal energy efficiency using smart controls, including devices like building automation systems (BASs), which employ sensors and algorithms to change the HVAC settings according to the number of people inside, the weather, and other variables.

6.14.2 Zoning and Demand-Controlled Ventilation

Zoning systems can provide individual temperature control in different areas of a building. Only heating or cooling occupied areas can improve comfort and energy efficiency. Demand-controlled ventilation systems use sensors to detect the number of occupants in a space and adjust ventilation accordingly. They can improve indoor air quality and reduce energy consumption by not ventilating unoccupied areas.

6.14.3 Heat Recovery Systems and Renewable Energy Integration

Utilizing heat recovery systems allows for heating incoming fresh air while also recovering heat from exhaust air. The energy required to heat or cool fresh air might be greatly decreased. HVAC systems can be linked with renewable energy sources like solar panels or geothermal systems to reduce reliance on fossil fuels and lower carbon emissions.

Smart HVAC systems are crucial to green buildings because they offer several environmental advantages, including increased energy efficiency, decreased carbon emissions, and better indoor air quality. Architects and builders may design sustainable buildings that are both economical and environmentally responsible by integrating renewable energy sources, smart controls, zoning, demand-controlled ventilation, heat recovery systems, and energy-efficient equipment.

6.15 RADIANT HEATING AND COOLING

These systems can be more energy efficient than conventional HVAC systems because they use thermal radiation to deliver heating and cooling. Radiant heating and cooling are an energy-efficient method of providing appropriate indoor temperatures in green buildings that use less energy than conventional HVAC systems. Several methods exist for including radiant heating and cooling in environmentally friendly structures.

6.15.1 Radiant Floor Heating and Ceiling Panels

Radiant floor heating systems use pipes or electric heating elements installed on the floor to warm the room from the ground up. They provide a uniform temperature throughout the space and can reduce energy consumption by heating only the occupied space. Radiant ceiling panels use water or electrical elements installed in the ceiling to provide cooling or heating. This technology can be used in spaces with high ceilings or where floor heating is impractical.

6.15.2 Chilled Beams and Radiant Walls

Chilled beams are passive heating and cooling devices that use convection to heat or cool a space. Chilled water is circulated through the beam, which cools the surrounding air and provides a comfortable indoor environment. Radiant walls use pipes or electric heating elements installed in the walls to provide heating. This technology can be used in spaces without practical floor heating or supplemental heating in colder climates.

6.15.3 Geothermal and Solar Radiant Systems

Geothermal radiant systems heat or cool buildings using the earth as a heat source or sink. By controlling indoor temperatures with the earth's constant temperature, this method can drastically save energy usage. Solar-powered radiant systems can be used to warm or cool a structure. By controlling the temperature of indoor spaces with renewable energy sources, this technology can drastically cut energy use.

An energy-efficient technology that can be employed in green buildings to deliver pleasant indoor temperatures with less energy usage than conventional HVAC systems is radiant heating and cooling.[31] Architects and builders can construct cost-effective, environmentally responsible, and sustainable buildings by adding radiant floor heating, ceiling panels, chilled beams, radiant walls, and geothermal and solar radiant systems.

6.16 SMART THERMOSTATS

Smart thermostats can learn a building's heating and cooling trends and change the temperature accordingly, saving energy and enhancing comfort. The smart thermostat is an innovative technology that can be used in green buildings to improve energy efficiency and reduce energy consumption. Here are several methods for incorporating intelligent thermostats into green buildings.

1. Adopting smart thermostats, which may modify the temperature settings in a building automatically, improves energy efficiency. Learning the occupants' preferences, planning, and regulating heating and cooling can result in significant energy savings.
2. Smart thermostats can reduce energy use, decreasing energy expenses and providing financial savings for building owners and tenants.
3. Smart thermostats can ensure people are always at ease by accurately controlling temperature. Additionally, this can result in increased productivity and a cozier indoor environment.
4. Smart thermostats may be remotely controlled with a smartphone app, allowing building owners and renters to adjust the temperature whenever and wherever they like. Adjusting when the building is vacant can help save money and use less energy.
5. Smart thermostats can be connected with other smart building systems, such as lighting and security, to provide a completely automated and ideal building environment.
6. By consuming less energy, smart thermostats can considerably improve the sustainability of green buildings and minimize greenhouse gas emissions.

Modern technology like smart thermostats can make green buildings more energy efficient, use less energy, and make residents more comfortable. By incorporating smart thermostats into their designs, architects, and builders may create affordable, environmentally responsible, sustainable structures that provide a comfortable and healthy indoor atmosphere.

6.17 VENTILATION SYSTEMS

Natural ventilation systems can save energy and enhance indoor air quality by reducing the need for mechanical ventilation. A vital component of the design of green buildings is natural ventilation. It refers to using natural air movement rather than artificial ventilation systems to maintain indoor air quality and thermal comfort in a structure. Natural ventilation systems use passive air movement to ventilate interior areas. It can be done by leaving windows, doors, or vents open, which lets fresh air enter the building and stale air leave. An energy-efficient alternative to mechanical systems or materials is natural ventilation. Fans or air-handling devices are used in mechanical ventilation systems to move air inside a structure. These systems may be built with energy or heat recovery components, which can cut back on energy use. Filters, fans, and air ducts are components of mechanical ventilation systems. Here are several strategies for including natural ventilation in green building designs.

6.17.1 Designing for Prevailing Winds

Ventilation is greatly influenced by a building's design, which considers the direction of the wind at a specific site. Designing structures to use natural airflow can be aided by knowing the direction of the prevailing winds. Buildings can benefit from cross-ventilation and natural air flow by being oriented to take advantage of prevailing breezes.

6.17.2 Using Operable Windows, Vents, and Thermal Mass

The controlled intake and exhaust of air are made possible by incorporating movable windows and vents. Louvers, movable skylights, and roof vents can all help with this. "Thermal mass" describes a substance's capacity to take in and hold heat. High-thermal-mass building materials, such as concrete, stone, or adobe, can control temperature swings and enhance indoor air quality.

6.17.3 Using Shading and Natural Vegetation

Overhangs, awnings, and vegetation can provide shade to block solar gain and keep the interior of a building warm. Different ventilation techniques may be needed when using natural ventilation in various building spaces. A bathroom or kitchen might need more ventilation than a bedroom or living room. Designing outdoor areas like gardens or parks can benefit from the color, texture, and interest that natural vegetation can bring. Visitors will feel more at ease in the area because of the shade trees and other plants can offer from the sun.

6.17.4 Considering Indoor Air Quality

By preventing the accumulation of contaminants and moisture, natural ventilation can enhance the quality of indoor air. However, considering potential outside pollution sources such as nearby traffic or industrial areas is essential. In order to save energy, improve indoor air quality, and create a cozier and healthier indoor environment, natural ventilation can be incorporated into green building designs. Green buildings rely on effective ventilation systems to preserve a healthy and cozy internal climate while using less energy. The following is a list of the many ventilation materials utilized in advanced buildings.

6.17.5 Living Walls

Green or living walls are vertical gardens that use plants to clean the air and control humidity. Mechanical ventilation systems may not be as necessary if living walls act as natural air filters. Planters, irrigation systems, and supporting structures are used in living walls. Numerous advantages of living walls include better air quality, lower energy costs, better acoustics, visual appeal, and better mental wellness. There are numerous varieties of living walls, including direct-to-wall, tray, and modular systems. The optimal option will rely on various elements, including the available space, the intended aesthetic, and the type of plants utilized. Each type has benefits and drawbacks. Using living walls is a sustainable and eye-catching technique to bring nature within and enhance the health and well-being of building occupants.

6.17.6 Earth Tubes and Solar Chimneys

Earth tubes use underground pipes to preheat or precool the air before it enters a building. The air is circulated through the pipes buried in the earth, where the temperature is stable. Earth tubes have the potential to save energy use and enhance indoor air quality. Filters, heat exchangers, and ventilation pipes are used in earth tubes. Solar chimneys generate a natural airflow that ventilates a structure by using the heat from the sun. Fresh air is drawn in through vents at the foot of a vertical chimney as warm air rises through it. Solar chimneys can increase energy efficiency and reduce mechanical ventilation systems' needs. The chimney, vents, and dampers are some materials utilized in solar chimneys. Advanced buildings must have efficient ventilation systems, and the chosen ventilation materials can significantly impact both energy efficiency and interior air quality. Green buildings can achieve optimal performance while reducing their environmental effect by using suitable materials for ventilation systems. Buildings' energy efficiency and sustainability can be increased through smart building materials, which can also lessen the adverse effects of construction and use on the environment while enhancing occupant comfort.

REFERENCES

1. Tabadkani A, Roetzel A, Li HX, Tsangrassoulis A. 2021. Design approaches and typologies of adaptive facades: A review. Autom Constr. 121:103450. doi:10.1016/J.AUTCON.2020.103450
2. Zhou XY, Zheng F, Li HG, Lu CL. 2010. An environment-friendly thermal insulation material from cotton stalk fibers. Energy Build. 42(7):1070–1074. doi:10.1016/J.ENBUILD.2010.01.020
3. Manohar K. 2012. Experimental investigation of building thermal insulation from agricultural by-products. Curr J Appl Sci Technol [Internet]. [accessed 2023 Mar 23] 2(3):227–239. doi:10.9734/BJAST/2012/1528
4. Gaujena B, Agapovs V, Borodinecs A, Strelets K. 2020. Analysis of thermal parameters of hemp fiber insulation. Energies. [accessed 2023 Mar 23] 13(23):6385. doi:10.3390/EN13236385
5. Korjenic A, Zach J, Hroudová J. 2016. The use of insulating materials based on natural fibers in combination with plant facades in building constructions. Energy Build. 116:45–58. doi:10.1016/J.ENBUILD.2015.12.037

6. Knapic S, Oliveira V, Machado JS, Pereira H. 2016. Cork as a building material: A review. European Journal of Wood and Wood Products [Internet]. [accessed 2023 Mar 23] 74(6):775–791. doi:10.1007/S00107-016-1076-4/FIGS/11
7. Paiva A, Pereira S, Sá A, Cruz D, Varum H, Pinto J. 2012. A contribution to the thermal insulation performance characterization of corn cob particleboards. Energy Build. 45:274–279. doi:10.1016/J.ENBUILD.2011.11.019
8. Ates M, Karadag S, Eker AA, Eker B. 2022. Polyurethane foam materials and their industrial applications. Polym Int [Internet]. [accessed 2023 Mar 23] 71(10):1157–1163. doi:10.1002/PI.6441
9. Peyrton J, Avérous L. 2021. Structure-properties relationships of cellular materials from biobased polyurethane foams. Mater Sci Engi: R: Rep. 145:100608. doi:10.1016/J.MSER.2021.100608
10. ABTG Wall Calculator | Applied Building Technology Group, LLC. [accessed 2023 Mar 17]. www.appliedbuildingtech.com/fsc/woodcalculator
11. Berardi U, Gallardo AA. 2019. Properties of concretes enhanced with phase change materials for building applications. Energy Build. 199:402–414. doi:10.1016/J.ENBUILD.2019.07.014
12. Sharma A, Shukla A, Chen CR, Dwivedi S. 2013. Development of phase change materials for building applications. Energy Build. 64:403–407. doi:10.1016/J.ENBUILD.2013.05.029
13. Hailu G. 2021. Energy systems in buildings. Energy Serv Fund Finan 181–209. doi:10.1016/B978-0-12-820592-1.00008-7
14. Sathre R, O'Connor J, FPInnovations (Institute). 2013. A synthesis of research on wood products and greenhouse gas impacts. desLibris.
15. Wang Y, Xu G, Yu H, Hu C, Yan X, Guo T, Li J. 2011. Comparison of anti-corrosion properties of polyurethane based composite coatings with low infrared emissivity. Appl Surf Sci. 257(10):4743–4748. doi:10.1016/J.APSUSC.2010.12.152
16. 3 Main Benefits of Low E Windows | Harvey Windows + Doors. [accessed 2023 Mar 23]. https://harveywindows.com/inspiration/ideas-advice/3-main-benefits-of-low-e-windows-harvey-windows-doors
17. De Luca, Pierantonio & Carbone, I. & Nagy, J.. (2017). Green building materials: A review of state of the art studies of innovative materials. Journal of Green Building. 12:141–161. 10.3992/1943-4618.12.4.141.
18. Wang W, Zmeureanu R, Rivard H. 2005. Applying multi-objective genetic algorithms in green building design optimization. Build Environ. 40(11):1512–1525. doi:10.1016/J.BUILDENV.2004.11.017
19. Gamayunova O, Gumerova E, Miloradova N. 2018. Smart glass as the method of improving the energy efficiency of high-rise buildings. E3S Web of Conferences [Internet]. [accessed 2023 Mar 17] 33:02046. doi:10.1051/E3SCONF/20183302046
20. Smart Glass – EDSL. [accessed 2023 Mar 23]. www.edsl.net/smart-glass/
21. Drück H, Pillai RG, Tharian MG, Majeed AZ, editors. 2019. Green Buildings and Sustainable Engineering [Internet]. [accessed 2023 Mar 24]. doi:10.1007/978-981-13-1202-1
22. Lai WY, Lai CM, Ke GR, Chung RS, Chen CT, Cheng CH, Pai CW, Chen SY, Chen CC. 2013. The effects of woodchip biochar application on crop yield, carbon sequestration and greenhouse gas emissions from soils planted with rice or leaf beet. J Taiwan Inst Chem Eng [Internet]. [accessed 2023 Mar 17] 44(6):1039–1044. doi:10.1016/j.jtice.2013.06.028
23. Reasons to avoid low-flow shower heads and pick the better alternative—Smart water saving shower. [accessed 2023 Mar 17]. www.oasense.com/post/low-flow-shower-heads
24. Silva LCC da, Filho DO, Silva IR, Pinto ACV e., Vaz PN. 2019. Water sustainability potential in a university building – Case study. Sustain Cities Soc. 47:101489. doi:10.1016/J.SCS.2019.101489

25. Sakthivel R. Waterless Urinals A Resource Book TY BOOK AU -Sakthivel, Ramesh AU - Chariar, Vijayaraghavan PY - 2009/06/01 SP - T1 - Waterless Urinals: A Resource BookVL -ER-
26. Muhamad WNW, Zain MYM, Wahab N, Aziz NHA, Kadir RA. 2010. Energy efficient lighting system design for building. ISMS 2010 — UKSim/AMSS 1st International Conference on Intelligent Systems, Modelling and Simulation:282–286. doi:10.1109/ISMS.2010.59
27. Zhai XQ, Wang RZ, Dai YJ, Wu JY, Xu YX, Ma Q. 2007. Solar integrated energy system for a green building. Energy Build. 39(8):985–993. doi:10.1016/J.ENBUILD.2006.11.010
28. Hammad M, Ebaid MSY, Al-Hyari L. 2014. Green building design solution for a kindergarten in Amman. Energy Build. 76:524–537. doi:10.1016/J.ENBUILD.2014.02.045
29. Green Roofs on Historic Buildings: What is a Green Roof? (U.S. National Park Service). [accessed 2023 Mar 22]. www.nps.gov/articles/000/green-roofs-on-historic-buildings-what-is-a-green-roof.htm
30. Gholamzadehmir M, Del Pero C, Buffa S, Fedrizzi R, Aste N. 2020. Adaptive-predictive control strategy for HVAC systems in smart buildings–A review. Sustain Cities Soc. 63:102480. doi:10.1016/J.SCS.2020.102480
31. Shen L, Tu Z, Hu Q, Tao C, Chen H. 2017. The optimization design and parametric study of thermoelectric radiant cooling and heating panel. Appl Therm Eng. 112:688–697. doi:10.1016/J.APPLTHERMALENG.2016.10.094

7 Energy Storage in Building Components

Vikash Kumar Chauhan and Tjokorda Gde Tirta Nindhia

7.1 INTRODUCTION

A building-integrated energy storage system (BESS) is a technology that allows buildings to store and use energy more efficiently and cost effectively—integrating energy storage systems into the building's infrastructure to provide backup power, peak shaving, load shifting, and other energy management functions. BESSs can be installed in different locations within a building, such as a basement, rooftop, or parking lot, depending on the available space and energy demands. These systems can use various energy storage technologies such as batteries, flywheels, capacitors, and thermal storage systems. The benefits of building-integrated energy storage systems are numerous. They can reduce energy bills by storing and using energy during off-peak hours when energy prices are lower. They can also provide backup power during power outages, which is critical for specific applications such as hospitals, data centers, and other mission-critical facilities.

In addition, BESSs can help reduce the strain on the electric grid during peak demand periods, improving grid stability and reliability. They can also help increase the use of renewable energy sources by storing excess energy generated by solar panels or wind turbines, allowing the building to use that energy during periods of low generation. Overall, building-integrated energy storage systems are a promising technology that can help buildings become more energy-efficient, reduce energy costs, and contribute to a more reliable and sustainable energy system.

Energy and water are the essential gradient of life for survival and development. From domestic purpose to industrial work, energy is required everywhere. Figure 7.1 represents the end user energy demand in different sectors including transport, construction, residential building, and other industries. The report indicates that total energy demand crosses over 580 million terajoules.[1] To fulfill the demand for energy, there is a need the explore alternative resources, among which nuclear power has shown a significant rise from last two decades, and hydropower generation has also increased at rapid rate.

 DOI: 10.1201/9781003415695-7

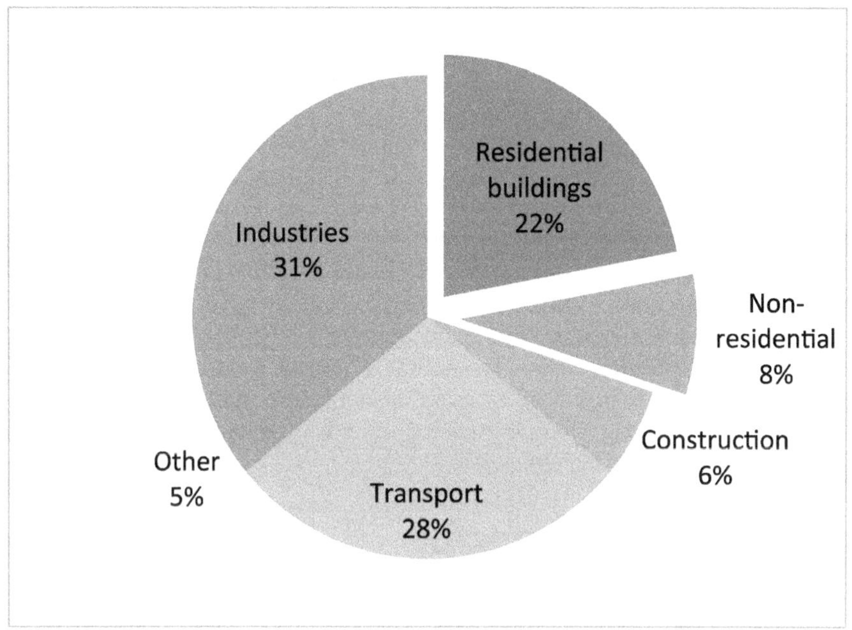

FIGURE 7.1 Share of final energy consumption sector-wise.

7.2 NEED FOR ENERGY STORAGE SYSTEMS

Excessive use of fossil fuel adversely affects climate change, depleting the ozone layer and polluting the atmosphere. Anthropogenic greenhouse gas is primarily CO_2 emitted during the fossil fuel burning process. Apart from this, a significant proportion comes from the building sector. Carbon dioxide (CO_2) is a greenhouse gas emitted into the atmosphere due to a range of operations in various industries. This includes using fossil fuels, deforestation, industrial operations, and transportation. Understanding how each sector contributes to global CO_2 emissions is critical in designing effective measures to reduce greenhouse gas emissions and alleviate the effects of climate change. The energy sector contributes the most to global CO_2 emissions, accounting for over 73% of total emissions in 2019.[2] This sector comprises activities such as the combustion of fossil fuels to generate power and heat and the production of oil and gas. Coal, oil, and gas combustion emit huge volumes of CO_2 into the atmosphere, and the energy industry accounts for approximately 41% of worldwide energy-related CO_2 emissions in 2021 c. Another significant contributor to global CO_2 emissions is the transportation sector, which accounted for around 23% of total emissions in 2019.[3] This sector includes transportation by road, air, and sea, and the use of fossil fuels for these activities emits considerable volumes of CO_2 into the atmosphere. The industrial sector also contributes significantly to global CO_2 emissions, accounting for roughly 11% of total emissions in 2019.[3] This industry comprises cement and steel production, and the use of fossil fuels in industrial operations emits considerable volumes of CO_2 into the environment. Another significant contributor to world CO_2 emissions is agriculture, which accounted for around 6% of total emissions in 2019.[4] This sector comprises animal husbandry and land-use changes, which emit considerable amounts of CO_2 into the atmosphere.

Finally, residential, and commercial sectors contribute to global CO_2 emissions, accounting for about 4% of total emissions in 2019.[3] These industries include heating and cooling buildings, as well as appliance use, and the use of fossil fuels emits considerable volumes of CO_2 into the atmosphere. To summarize, worldwide CO_2 emissions are a substantial contributor to climate change, and understanding the contributions of each sector is critical in devising successful emission-cutting strategies. The energy sector is the most significant contributor to worldwide CO_2 emissions, followed by transportation, industry, agriculture, and the residential and commercial sectors. Reduced emissions from these sectors will necessitate considerable efforts, such as developing and adopting low-carbon technology and introducing policies and laws to incentivize emission reductions.

7.3 THERMAL ENERGY STORAGE

Thermal energy storage (TES) is a technique that allows for the efficient storage of thermal energy for various purposes. When energy demand is low, TES systems store excess energy, such as heat, in a storage medium and subsequently release that energy when demand is high. This technology has grown in prominence in recent years as demand for renewable energy has increased. It needs energy storage solutions to assist in balancing the intermittent supply of renewable energy sources like solar and wind power. There are various varieties of TES systems, each with its own set of qualities and benefits. The magnetic field has been widely used in several applications, including industrial grade, robot drives, medical help, and so on, as a non-contact, real-time, precise, and adaptable control technology. On the other hand, nonmagnetic phase change materials (PCMs) are challenging to alter using magnetic fields. In a recent study,[5] magnetically tightened form-stable PCMs with the integrated multi-functions of being leakage-proof, possessing excellent thermal/electrical conductivities, robust mechanical strength, dynamic assembly, and shape reconfiguration have been discovered. Also, a self-contained temperature management and high-performance thermoelectric conversion systems using magnetically tightened form-stable PCMs was reported.[5] There are three techniques, as shown in Figure 7.2, to store thermal energy.

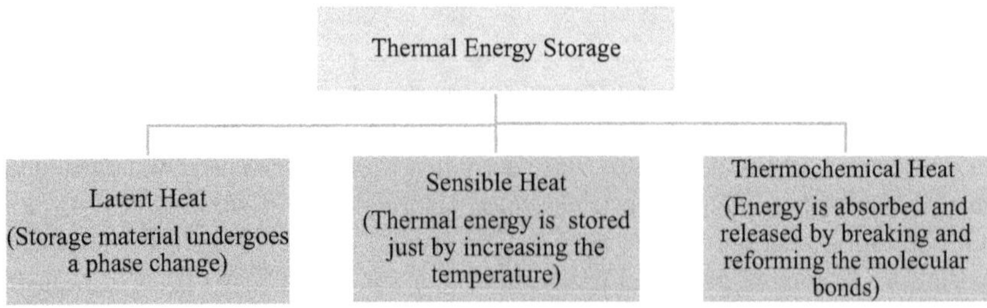

FIGURE 7.2 Thermal energy storage techniques.

7.3.1 Sensible Heat Storage

Sensible heat storage, which involves storing heat in a solid or liquid substance, is one of the most popular types of TES systems. Good heat storage can be accomplished using water, molten salt, and concrete. The stored heat is later used for space heating, hot water generation, and industrial processes. Sensible heat storage (SHS) takes place by fundamental mode of heat transfer, that is, conduction, convection, and radiation. During cool ambient temperatures, heat is liberated with a gradual decrease in temperature that causes unequal heat distribution with time. Underground hot aquifers are a natural example of the SHS principle. These materials have two lucrative benefits, namely availability and compatibility, as these materials are easily available in our surroundings, including domestic, industrial, and agricultural waste that can offer sensible heat storage material during sunshine periods. The most common sensible heat storage material is water, metal scrap, and rock-type material like pebbles, stones, brick, and gravel. By the use of these materials, raw water can be heated up to 40 to 80 °C.[1] Currently, utility-based solar thermal electric plants utilize liquid, powder, and metal, synthetic oil, and molten salts.[2] Table 7.1 represent the different sensible heat storage materials and their thermophysical properties.

The governing equation for sensible heat storage can be expressed as

$$Q = m\, C_p\, \Delta T$$

Where Q (J) is the heat,
m (kg) is the mass of the material,
C_p (J/kg.K) is the specific heat of the material, and
ΔT (K) is the change in the temperature of the material.

TABLE 7.1
Sensible Heat Storage Materials and their Properties[3–5]

S.N.	Material	Specific heat (J/kg.K)	Density × 10^3 (kg/m^3)	Thermal conductivity (W/m.K)	Thermal diffusivity × 10^6 (m^2/s)	Heat capacity (J/m^3.K)
1.	Marble	800	2.6–2.8	2.07–2.94	0.995–1.413	2.08
2.	Sandstone	710	2.1–2.4	1.83	1.172	1.56
3.	Sand (dry)	830	1.4–1.6	0.15–0.25	–	–
4.	Aluminum	8.96	2.70	204 @ 293 K	84.10	2.42
5.	Cast iron	837	7.90	29	4.43	6.61
6.	Copper	383	8.90	385 @ 293 K	112.3	3.42
7.	Potassium chloride	670	1.98	6.53 @ 322 K	–	1.32
8.	Therminol 55 oil	872	1.91	0.13 @ 293 K	–	2400
9.	Isobutanol	2303	0.80	0.13@293 K	–	3000
10.	Ethylene glycol	2433	1.11	0.256@293K	–	2382
11.	Jute cloth	324	1.50	0.427	–	–

7.3.2 Latent Heat Storage

Latent heat storage is another TES system that involves storing thermal energy in the form of a phase transition. During the charging process, energy converts a substance from a solid to a liquid or from a liquid to a gas. The stored energy is released when the material undergoes a reverse phase shift. In latent heat storage systems, phase change materials (PCMs) such as paraffin wax, salt hydrates, and eutectic mixes are widely utilized.[6] Another form of TES system is thermochemical energy storage (TCES), which uses reversible chemical reactions to store and release energy. These systems often employ metal oxides or carbonates that, when heated, emit oxygen or carbon dioxide, which can subsequently be used for a variety of applications such as power generation or fuel manufacturing.

Buildings, industrial processes, and power generation are just a few of the applications of TES systems. Excess heat from solar panels or district heating systems can be stored in buildings and used for space heating or hot water production. Industrial processes that require high-temperature heat can also benefit from TES systems, which can supply a consistent heat source even when energy demand is low. TES systems can be used in power generation to store excess energy from renewable sources and release it during high energy demand. One of the primary benefits of TES systems is their ability to balance the intermittent supply of renewable energy sources like solar and wind power. TES systems can help to provide a stable supply of energy during periods of high demand by storing extra energy during periods of low demand, decreasing the requirement for fossil fuel-based power generation. Another advantage of TES systems is their capacity to improve energy efficiency. TES systems can eliminate the need for additional energy generation and transmission infrastructure, which can be costly and have a significant environmental impact, by storing excess energy that would otherwise be wasted. While TES systems have many benefits, they suffer from several problems that must be addressed. The high cost of TES system installation and maintenance is one of the critical issues that can be a barrier to broader adoption. Another area for improvement is the need for suitable storage materials, especially for high-temperature applications.

TES is a promising technology with the potential to play a significant part in the transition to a more sustainable energy system. TES systems can help to balance energy supply and demand, eliminate the need for additional energy generation equipment, and increase energy efficiency by storing excess energy from renewable sources and releasing it when needed. While there are obstacles to overcome, the development and deployment of TES systems are expected to grow in the future years as demand for renewable energy and energy storage solutions grows. The most effective and efficient way to develop energy-efficient building envelopes is to store solar thermal energy using *building thermal mass* (building materials) and then utilizing it as per the requirement (hot and cold). TES enhances the efficiency of renewable energy sources.

TES offers better capability to store available heating and cooling energy in off-peak load conditions to effectively match the on-peak demand periods.

7.3.3 Phase Change Materials

The heat acquisition and discharge process when a storage medium undergoes a reversible phase shift from solid to liquid, liquid to gas, solid to solid, solid to solid, etc., provides the basis for TES using PCM. Figure 7.3 depicts the probable phase transition of materials. However, having a high phase transition Latent heat (LH) solid–gas or liquid–gas PCMs with substantial volume changes have limited use in TES systems.[7] Some other features that make PCMs good candidates for energy storage are,

- Volume changes are noticeable in solid–solid and solid–liquid conversions, frequently about 10% or less.
- They provide high-energy storage density.
- They store heat at constant temperature corresponding to the phase transition temperature of the PCM.
- They store 5 to 14 times more energy per unit volume than sensible heat materials like water, rock, etc.
- LHS can be accomplished through solid–liquid, liquid–gas, solid–gas, and solid–solid phase transformations.
- The solid–liquid system is the most studied and most commonly commercially available.

As a result, while having a lower heat of phase transition, they are still desirable materials for TES systems from an economical and practical standpoint.[8]

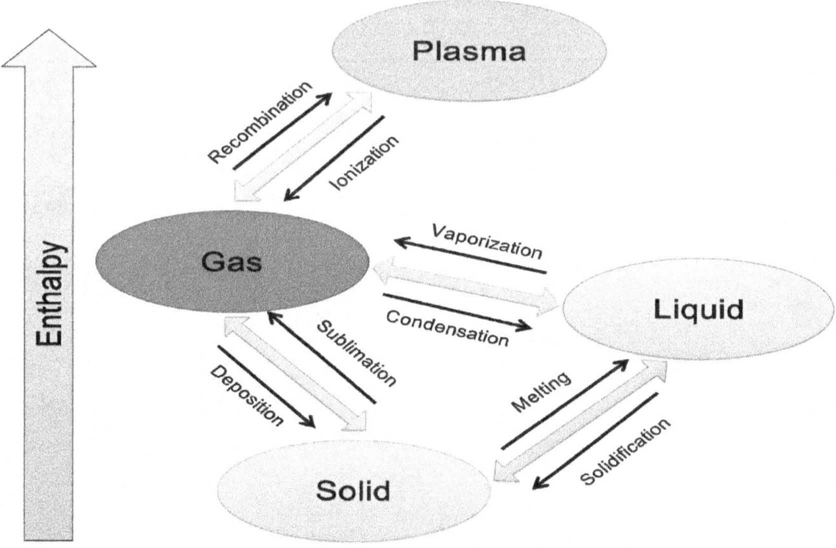

FIGURE 7.3 Phase change materials with respect to enthalpy.[10]

As an application, desalination is the most extensively employed approach to alleviate water constraints with solar energy in conjugation with heat storage materials.[9] Numerous studies showed that the solar desalination method steadily increases the amount of distilled water produced. Yet to boost production and efficiency, solar desalination systems have undergone numerous modifications. Solar heat storage technologies, for example, find a way to balance the sun's irregular activity with the demands for supply and demand.

The usage of LHS and SHS materials in conjunction with solar stills has showed great responsiveness in ensuring a consistent supply of fresh water. SHS materials are widely available and inexpensively priced due to their natural abundance. LHS material, on the other hand, increases production at night and during cloudy conditions.

PCMs can be classified according to the state of the substance before and after its application such as gas–liquid PCM, solid–gas PCM, solid–liquid PCM, and solid–solid PCM, as shown in Figure 7.4. In the current scenario, solid–liquid and solid–solid PCMs are commonly used due to low volume variation and high LH capacity. In solid–liquid PCMs, the temperature rises until the temperature of phase change (TPC) is reached, and then massive LH is stored in the material while the phase change happens from solid to liquid.[11] On a chemical nature basis, the solid–liquid PCMs can be subdivided into organic, inorganic, and eutectic. The organic PCMs are further classified into paraffins, alcohols, fatty acids, and esters, while on the other hand, the inorganic PCMs are classified into salts and salt hydrates.[12] In comparison

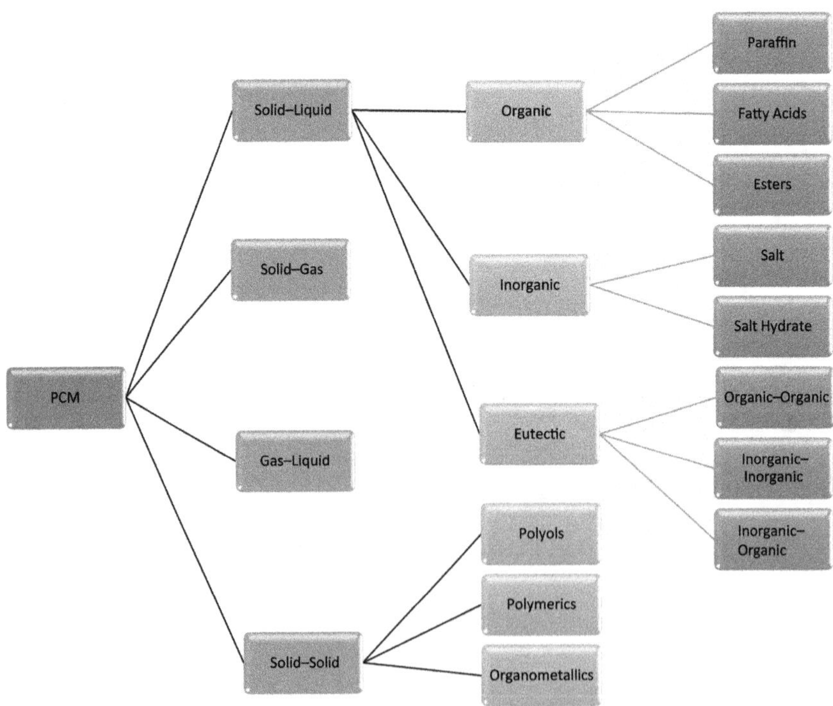

FIGURE 7.4 Types of phase change materials (PCMs).[13]

to organic PCMs, inorganic PCMs offer the advantages of having high LH per unit mass, non-flammability, and being less expensive. Eutectic PCMs are composed of two or more soluble components that are combined and have the property of simultaneously melting and solidifying without material separation. Solid–solid PCMs involve polyols, polymerics, and organometallics.

7.3.3.1 Organic PCMs

Most of the organic chemicals used as PCMs are hydrocarbons, which have a long chain of macromolecules primarily composed of carbon and hydrogen. This family of PCMs is the most commonly employed for TES because of its excellent melting or solidification temperature, stable chemical, and physical properties, high storage density, and abundance.[14] Non-paraffin and paraffin are two types of organic PCMs. Paraffin is further classified into two types: (1) saturated hydrocarbons with several carbons ranging from 12 to 40 and having the empirical formula C_nH_{2n+2} of the linear alkanes, and (2) "paraffin wax," which is a mixture of alkanes and other hydrocarbons with the formula $CH_3(CH_2)$ nCH_3.[15] Although pure alkanes are more expensive than blends, the $T_{melting}$ and heat of fusion increase in both cases as more carbons are present. Also, the price of paraffin rises with its purity.

For buildings, non-paraffin PCMs are more popular and include alcohols, esters, fatty acids, and glycols in general. Their thermophysical properties are nearly same as organic PCMs derived from paraffin. Their enhanced flammability, however, limits their suitability for applications at high temperatures. Group of alcohols have been explored for energy storage applications since the 1970s. However, more recent discoveries on these PCMs are connected to nanomaterials or unique composites coupled with specific properties, such as the inclusion of electrical conductivity. When used as conductive materials, they can endure some thermal shock, which is a significant property. For example, with 1-tetradecanol (TD)/nano-Ag, a composite material for energy storage, as *silver* nanoparticles increased, thermal conductivity improved.[16] TD/polyaniline (PANI) composites were also investigated, which are created using the polymerization approach, and it was discovered that the composites could conduct electricity while also storing heat energy.[17] They have properties that allow them to endure significant thermal stresses when used as conducting materials.

Because of their advantageous thermodynamic and kinetic properties for low-temperature LHS, the use of organic materials for energy storage, such as fatty acids ($CH_3(CH_2)_{2n}COOH$), has recently increased. Although they have not received as much attention as paraffin and salt hydrates, saturated fatty acids are quite beneficial for TES. Furthermore, they have a wide $T_{melting}$ range, ranging from 8 °C to 64 °C, with fusion enthalpy varying from 149 kJ/kg to 222 kJ/kg. Also, fatty acids have a low heat conductivity, which is usually undesirable. Organic phase change materials (PCMs) have various advantages in the construction industry:

- **Energy efficiency:** Organic PCMs have a high latent heat storage capacity, which means they can absorb and release significant thermal energy during the phase change process. Because of this, they are a good alternative for thermal energy storage and can help to lower a building's overall energy usage.

- **Indoor comfort:** Using organic PCMs in building construction can help to maintain a more comfortable indoor atmosphere. Organic PCMs can help to manage the temperature of a building by absorbing excess heat during the day and releasing it at night, making it more comfortable for residents.
- **Decreased HVAC requirements:** By regulating the temperature inside the building, using organic PCMs in building construction can help lessen the burden on the HVAC system. This can reduce the size and expense of the HVAC system, resulting in lower energy bills for the building owner.
- **Lifespan:** Organic PCMs have a longer lifespan as compared to other building materials. They do not deteriorate or lose effectiveness over time. Thus, they can give long-term energy savings to building owners.
- **Non-toxic:** Organic PCMs are non-toxic and safe for the environment. They do not emit toxic chemicals or pollutants into the environment, making them a safe and environmentally friendly solution for building construction.

Organic phase change materials (PCMs) have some restrictions that should be considered before using them in a specific application.

- **Low thermal conductivity:** Organic PCMs often have low thermal conductivity as compared to inorganic PCMs, which can limit their capacity to transmit heat effectively. To accomplish the necessary cooling or heating effect, higher PCM volumes or other cooling systems may be required.
- **Limited temperature range:** Organic PCMs typically have a restricted operating temperature range, limiting their applicability in situations requiring a more excellent temperature range.
- **Chemical stability:** Organic PCMs may degrade chemically over time, reducing their effectiveness and longevity. This can be reduced by using PCMs with improved chemical stability or encasing the PCM in a protective material.
- **Flammability:** Certain organic PCMs are flammable or combustible, which might present a safety risk in specific applications. The flammability of the PCM should be carefully considered, and suitable safety precautions should be followed.
- **Phase separation:** Certain organic PCMs may experience phase separation during repeated melting and solidification cycles, reducing their effectiveness over time. This can be reduced using PCMs with excellent phase stability or by including stabilizers in the PCM composition.

7.3.3.2 Organic Polyols

Among the polyalcohols that are frequently used as solid–solid PCMs are pentaglycerine (PG), pentaerythritol (PE), and, more recently, 2-amino-2methyl-1,3, propanediol (AMPL) and amino glycol. According to studies, polyols initially exhibit poor symmetry and a layered crystal structure at low temperatures. However, as the temperature is raised over the phase change temperature, they convert into a high-symmetry and face-centered cubic structure.[18]

Polyalcohol phase transition temperatures and enthalpies are determined by the degree of hydrogen bonding between the hydroxyl functional groups. Because PG and PE have relatively similar structural formulae, they can be consistently combined to generate solutions while keeping the original polyols' solid–solid phase transition qualities.[19]

7.3.3.3 Inorganic PCMs

Many inorganic materials, eutectics, and mixes have been investigated as feasible PCMs for low and high-temperature applications, suggesting that inorganic PCMs have a broad operating temperature range. Inorganic PCMs have the same LH per unit mass as organic PCMs but have a greater LH per unit volume due to their higher density. Salt hydrates are a type of inorganic PCM composed of inorganic salts (AB) and one or more water molecules (H_2O), resulting in a crystalline solid form[20] (AB_xH_2O). Several salt hydrates with melting points ranging from 5 °C to 130 °C are suitable for a wide range of applications. Metallic materials include rocks, concrete, stones, and other inorganic materials with a high melting point.

PCMs are substances that can collect and release a considerable quantity of thermal energy during the transition from solid to liquid and vice versa. When it comes to construction applications, inorganic PCMs such as salt hydrates, metal alloys, and eutectics have various advantages over biological PCMs. Following are some of the benefits of using inorganic PCMs in construction.

- **High thermal conductivity:** Inorganic PCMs have a better thermal conductivity than organic PCMs, which means they can transmit heat more quickly and efficiently. This makes them excellent for use in building applications requiring rapid and efficient thermal energy transmission.
- **High melting and freezing points:** Inorganic PCMs have higher melting and freezing temperatures than organic PCMs, which allows them to store and release thermal energy across a larger temperature range. As a result, they are suited for use in construction applications where temperature control is critical.
- **Non-flammable and non-toxic:** Inorganic PCMs are normally non-flammable and non-toxic, making them suitable for use in construction. This is especially critical in buildings where fire safety is a priority.
- **Long lifespan:** Because inorganic PCMs have a longer lifespan than organic PCMs, they can be utilized for many years before needing to be replaced. As a result, they are a cost-effective choice for building applications.
- **Availability:** Inorganic PCMs are readily available and can be obtained locally, making them more accessible and inexpensive than organic PCMs. This is critical for building applications when cost and availability are critical.

Overall, the benefits of inorganic PCM make them an appealing solution for construction applications requiring efficient and reliable thermal energy storage and release.

TABLE 7.2
Comparison of Characteristic Features of Phase Change Materials

S.N.	Organic	Inorganic	Eutectic
1.	Chemically stable	High latent heat value	Wide range of freezing and melting temperatures
2.	Melt congruently	High heat of fusion	Undergoes phase segregation
3.	High heat of fusion	High thermal conductivity	High heat storage density
4.	Low thermal conductivity	Non-flammable	High cost
5.	Low volumetric latent heat storage capacity	Corrosiveness	–
6.	Moderately flammable	Incongruent melting	–
7.	High initial cost	Instability	–

Two main disadvantages with Inorganic PCMs are:

- Corrosive nature
- High volume change during phase transition

7.3.3.4 Polymerics

Polymeric PCMs are composed of a solid–liquid ("soft") macromolecule chemically linked to a stiffer polymer. A solid–solid phase transition occurs when the softer component melts and is restricted by the more challenging component.

These solid–liquid PCMs were discovered as form-stable PCMs that undergo a solid–solid phase transition, yielding a range of polymeric SS-PCMs. One such strategy is physically restricting the PCM in a polymer-supporting matrix to prevent leakage.[21] Another method involves attaching the S-L PCM to a supporting high-melting-point polymer utilizing chemical techniques such as grafting, blocking, and crosslinking copolymerization. The latter was preferred since it creates a homogeneous polymer. When a phase transition occurs, the active component is a phase change material such as a fatty acid or PEG; however, the supporting polymer stabilizes and keeps it from melting.[22]

7.3.3.5 Eutectic Materials

Certain combinations of two or more chemicals that have a lower melting point than any of the constituent parts are known as eutectic mixes. The thermal energy storage uses of these mixes are common, especially with phase change materials (PCMs). Some examples of eutectic mixes that are frequently employed to store thermal energy are as follows.

- **Salt hydrates** like mixture of potassium nitrate (KNO_3) and sodium sulphate decahydrate ($Na_2SO_4 \cdot 10H_2O$).
- **Eutectic temperature:** about 30 °C
- **Use:** Compared to either component alone, these salt hydrates form a eutectic mixture with a lower melting point. They are employed as medium-temperature thermal energy storage phase transition materials.

- **Eutectic organic mixtures** like a combination of fatty acids, such as stearic acid ($C_{18}H_{36}O_2$) and myristic acid ($C_{14}H_{28}O_2$).
- **Eutectic temperature:** Changes based on the particular fatty acids utilized
- **Use:** Phase change materials for low-to-moderate-temperature thermal energy storage in buildings and other applications are frequently made from organic eutectic mixtures.
- **Metal alloys**, for instance, alloys made of metals like bismuth (Bi) and indium (In).
- **Eutectic Temperature:** 60 °C or below.

Researchers continue to explore and develop new eutectic mixtures with improved properties for various temperature ranges and applications. Table 7.3 highlights some organic, inorganic and eutectic phase change materials used for different applications.

7.3.3.6 Time Lag

Heat transmission in buildings happens via conduction, convection, and radiation. The time lag of heat transmission refers to the amount of time that passes between applying a temperature differential across a building envelope (such as a wall or roof) and the consequent temperature change on the other side.

Heat transfer time lag is determined by various factors, including the building materials' thermal characteristics, the building envelope's thickness, and the ambient conditions within and outside the structure. For example, a structure with a thick brick wall may have a more significant time lag than a building with a thin wood-framed wall. Heat transfer time lag is generally advantageous for maintaining a pleasant interior temperature, especially in areas with considerable temperature changes between day and night. A building with a longer time lag will tend to collect heat during the day and slowly release it at night, which can assist in balancing out temperature swings.

On the other hand, the time lag in heat transmission might be a disadvantage when quick temperature changes are sought, such as in reaction to changes in occupancy or equipment use. Building design and HVAC systems must be carefully calibrated in such instances to guarantee that the facility responds swiftly and effectively to changing conditions, as shown in Figures 7.5 and 7.6.

- Buildings and construction account for more than 35% of global final energy use and nearly 40% of energy-related CO_2 emissions.
- Energy-efficient and low-carbon heating and cooling technology investments would reduce final energy demand in buildings by 25% over current levels.
- Nearly two-thirds of the global buildings sector energy consumption is supplied by fossil fuels for direct use or for upstream power generation.

Energy-efficient building envelopes will help in following:

- Shifting the heating and cooling load from peak load to the off-peak load
- Creating an envelope to improve indoor thermal comfort
- Reducing the heat gains caused due to solar radiation

TABLE 7.3
Thermophysical Properties of Some PCMs.[23–25]

Class of PCM	PCM name	Melting point (°C)	Latent heat of fusion (KJ/kg)	Thermal conductivity (KJ/kg K)		Density (kg/m³)	
				Solid	*Liquid*	*Solid*	*Liquid*
Organic	Paraffin wax	32	251	0.514	0.224	830	–
	RT-10	10	190	0.2	0.2	880	770
	Polyethylene	110–135	200			910	870
	RT-9 HC	–9	260	0.2	0.2	880	770
	Naphthalene	80	147.7	0.341	0.132	1145	976
	Erythritol	118	339.8	0.733	0.326	1480	1300
	Caprylic acid	16	149	–	0.149	981	901
	RT-28 HC	28	245	0.2	0.2	880	770
	Capric acid	32	153	–	0.153	1004	886
	n-Octadecane	27.7	243.5	0.19	0.148	865	785
	Vinyl stearate	27–29	122	–	–	–	–
	Palmitic acid	61	185	–	0.162	989	850
	Stearic acid	69	202.5	–	0.172	989	965
	Lauric acid	42–44	178		0.147	1007	862
Inorganic	Calcium chloride hexahydrate ($CaCl_2 . 6H_2O$)	29.8	190.8	1.088	0.54	1802	1562
	$Ba(OH)_2 \cdot 8H_2O$	78	265–280	1.225	0.653	2070	1937
	$Mg(NO_3)_2 \cdot 6H_2O$	89	162.8	0.611	0.49	1636	1550

(Continued)

TABLE 7.3 (Continued)
Thermophysical Properties of Some PCMs.[23–25]

Class of PCM	PCM name	Melting point (°C)	Latent heat of fusion (KJ/kg)	Thermal conductivity (KJ/kg K)		Density (kg/m³)	
	Glauber's salt ($Na_2SO_4{\cdot}10H_2O$)	32	254	0.554	–	1485	1458
	$Zn(NO_3)_2{\cdot}6H_2O$	36	146.9	–	0.464	1937	1828
	$Ba(OH)_2{\cdot}8H_2O$	78	265.7	1.255	0.653	2180	1937
	$Mg(NO_3)_2{\cdot}6H_2O$	89	162.5	0.611	0490	1636	1550
	$MgCl_2{\cdot}6H_2O$	117	168.6	0.694	0.570	1569	1442
	$Na_2HPO{\cdot}12H_2O$	35–45	279.6	0.514	0.476	–	1520
Eutectics	61.5% $Mg(NO_3)_2{\cdot}6H_2O$ + 38.5% NH_4NO_3	52	125.5	0.552	0.494	1596	1515
	66.6% urea + 33.4% NH_4Br	76	161	0682	0.331	1548	1440
	58.7% $Mg(NO_3){\cdot}6H_2O$ + 41.3% $MgCl_2{\cdot}6H_2O$	59	132.2	0.678	0.510	1630	1550
	45.8% LiF + 54.2% MgF_2	746	–	–	–	2880	2305
	35.1% LiF + 38.4% NaF + 26:5% CaF_2	615	–	–	–	2820	2225
	48.1% LiF + 51:9% NaF	652	–	–	–	2720	2090

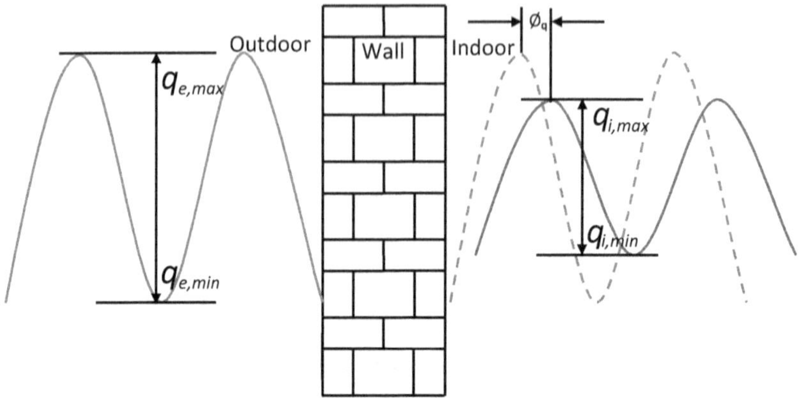

FIGURE 7.5 Thermal lag due to a building wall causes a reduction in heat flux.[26]

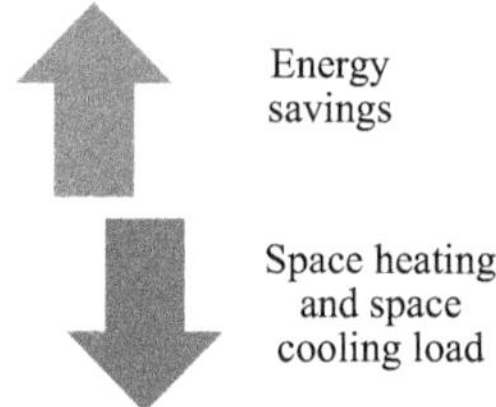

FIGURE 7.6 Advantage of phase change materials in buildings.

7.4 BUILDING-INTEGRATED ENERGY STORAGE

Building energy conservation has benefited from PCM energy storage technology. The application of PCM to buildings, including both roofs and walls, increases the building's ability to store heat, potentially increasing energy efficiency and lowering the amount of electricity required for heating and cooling. When a PCM is implemented into a structural component, the sun and high temperatures create a heat wave that infiltrates the building walls during the day. By absorbing extra heat through melting, the PCM delays and even reduces the peak of the heat wave inside the building. The room's average temperature remains soothing most of the day, and the cooling system uses less electricity. At night, when temperatures are lower, phase change material releases the heat stored to both the interior and external environment, maintaining a proper room temperature and solidifying itself.

The critical difference between PCMs and more traditional thermal mass is that, unlike masonry, which absorbs and releases heat slower, PCMs do it quickly. As a result, they are instrumental in buildings with lightweight construction, where traditional passive design concepts are challenging to apply. A PCM, however, will only be effective if it is effectively incorporated into a structure.

PCMs (phase change materials) are widely used in building construction and design. The following are some examples of how PCM might be used in building applications:

- **Thermal energy storage:** PCMs can be used to store thermal energy in buildings, such as heat or cold. This is especially beneficial for lowering energy usage by eliminating the need for air conditioning and heating systems. PCM can be integrated into building materials such as walls, ceilings, and floors to absorb surplus heat during the day and release it at night.

- **Temperature control:** PCMs can be used to adjust the temperature of building spaces, especially in places where temperatures fluctuate a lot. In hot and humid areas, for example, PCMs can absorb excess heat during the day and release it at night, keeping the indoor atmosphere cool.
- **Passive cooling:** PCMs can be used to cool buildings passively. When exposed to sunlight, PCMs can absorb and store heat, which can then be released to cool the building space. This is especially useful in hot, sunny climates.
- **Thermal mass:** PCMs can also be utilized to boost the thermal mass of building materials. This is especially valuable in locations with high-temperature swings, where it can aid in temperature stabilization and reduce the need for heating and cooling.

Using PCMs in building construction can help minimize energy consumption, improve thermal comfort, and improve building system efficiency. And these can be achieved by proper selection of building materials that make buildings energy efficient and comfortable for human beings. Different types of material such as insulating materials (cellulose insulation, recycled denim insulation, wool insulation, hemp insulation, and cork insulation), sustainable wood products (reclaimed wood, engineered wood, bamboo, and cork), glass-based material (highly polished smart glazing material, low-emissivity material, energy-efficient windows, daylight harvesting, aesthetics), self-healing concrete, and light emitting cement are used in modern building construction to make self-sustainable. Moreover, there are different technologies to use PCMs inside building walls, as mentioned in Figure 7.7.

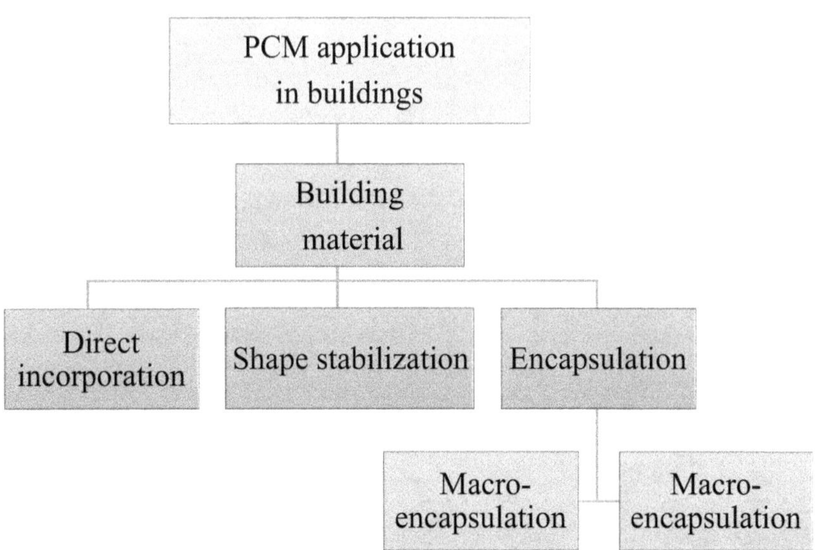

FIGURE 7.7 Applications of PCMs as building materials.

7.5 DIRECT INCORPORATION

The direct incorporation method applies the PCM in powder or liquid form directly to the construction material, such as gypsum mortar, cement mortar, or a concrete mixture.[27] This strategy is the simplest and most cost effective because it requires no experience and is simple.[28] On the contrary, the main disadvantage of this approach is PCM leakage during the melting process. This leaking causes material incompatibility and increases the fire risk (for flammable PCMs). Furthermore, because the PCM is introduced to the mixture in a liquid condition, the water content ratio is reduced, which degrades the mechanical properties of built elements at high temperatures.

In the ***immersion method***, a porous building material is immersed in liquid PCM and absorbed via capillary action. The main disadvantages of this technology include leaking, incompatibility with the building, and corrosion of reinforced steel when merged with concrete elements, which reduces its service life.

7.6 SHAPE STABILIZATION

The PCM is contained within a carrier matrix in the shape-stabilized technique. This approach is promising since it offers improved thermal conductivity, high specific heat, and the ability to keep the shape across several phase change cycles. To optimize thermal performance in shape-stabilized phase change materials, numerous vital aspects should be examined, including shell thickness in encapsulation, dimension, particle size and pore size in nano- and porous materials, chemical composition, and synthesis technique.[29] Shape stabilization can be conceived as an amalgamation of PCMs, porous materials, and nanoparticles, as illustrated in Figure 7.8. With the advancement of nanoscience, several novel techniques for fabricating shape-stabilized phase change materials have emerged due to their benefits such as

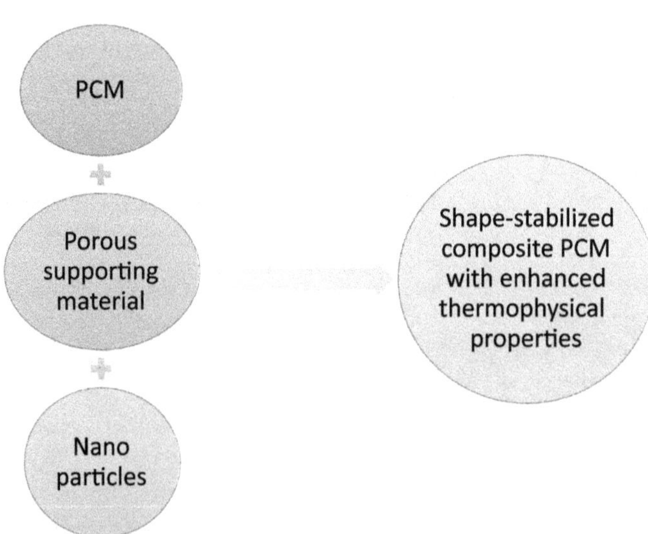

FIGURE 7.8 Shape-stabilized composted PCM.

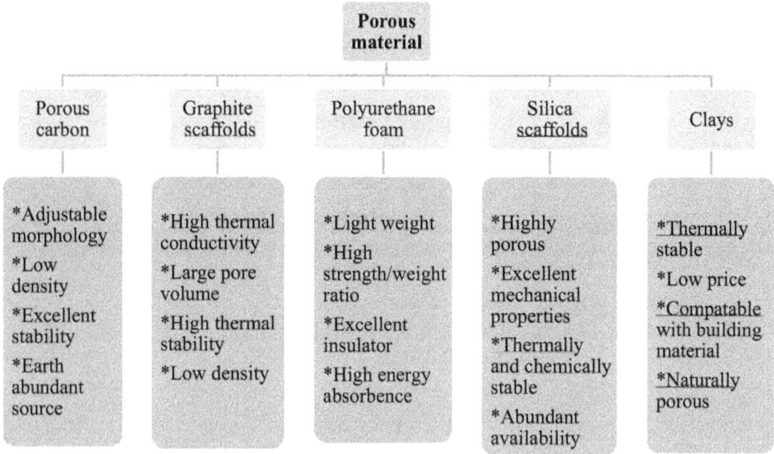

FIGURE 7.9 Porous material for shape stabilization.

cost effectiveness, shape stability, and ease of preparation with desirable dimensions, such as using nanomaterials and encapsulation through core–shell and nanostructured foams. Nanomaterials can be applied from one to three dimensions (1-D to 3-D) using spinning methods and carbon derivatives. Moreover, nanoencapsulation and nanofoams are created by encasing phase transition materials (core) in a protective shell or media (organic/inorganic).

Form-stabilized PCM is another sophisticated incorporation approach. It is a specific form of composite material that retains the most significant quantity of one or more types of PCM while exhibiting no leakage at melting temperatures. Although the two latter strategies are the most expensive to adopt, they are also the most reliable. Reliability analysis shows that the PCM cycles (melting/solidification) are repeated with high performance without degradation, which is critical for long-term applications like buildings. As shown earlier, porous materials possess critical a role in shape stabilization by providing accommodation space for the PCM and nanoparticle. Hence, various types of porous materials have been investigated such as porous carbon, graphite scaffolds, polyurethane foam, silica scaffolds, and clays, as shown in Figure 7.9, which highlights the properties also.

Type of shape stabilization based on dimension of shape-stabilized material made via electrospinning.

1. One dimension in relation with electrospinning
2. Two dimensions in relation with carbon derivatives
3. Three dimensions in relation with carbon derivatives

7.7 ENCAPSULATION

Encapsulation is an effective way to prevent PCM leakage and improve its compatibility with building structures. Encapsulation involves wrapping the PCM with a shell to protect it from the outside environment and prevent leakage. This approach

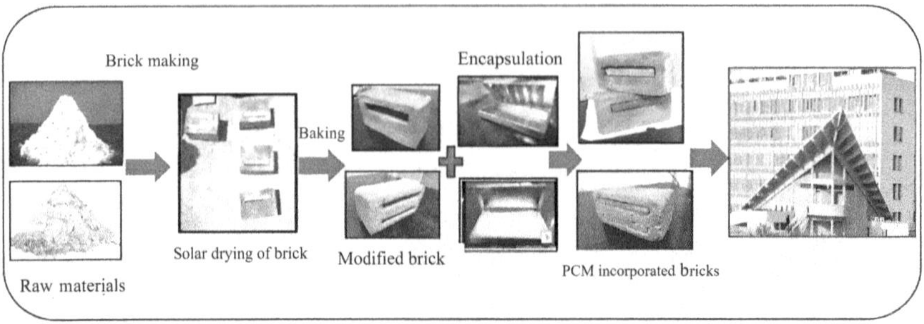

FIGURE 7.10 Steps of fabrication of macroencapsulated bricks in buildings.

is also required to improve the heat transfer area and, as a result, the thermal conductivity of a PCM in order to maximize its storage capacity.[30] The PCM can be macroencapsulated using shells, tubes, channels, or thin plates or microencapsulated with a unique polymeric substance covering the micro-sized PCM.

The encapsulation material in both ways should have distinct properties, such as preventing leakage, keeping all thermal properties of the PCM, not reacting with the PCM, being compatible with the PCM and its application, giving structural stability, and securing handling. Additionally, it should regulate any volumetric change in the PCM during phase shifts and provide enough protection for the PCM against environmental degradation, as well as strong thermal conductivity and mechanical strength across the PCM life cycle. Pipes, panels, and foils constructed of aluminum, copper, and stainless steel are often used for macroencapsulation because they provide excellent thermal conductivity, compatibility, and mechanical strength support to construction materials. The encapsulation of PCM can be done in bricks, as shown in Figure 7.10, and used to make the building.

7.7.1 Macroencapsulation

PCMs have weak heat conductivity in general. Because the PCM is covered with low-thermal-conductivity polymeric materials, microencapsulation raises this matter. The thermal conductivity of a PCM can be enhanced by employing metal pipes made of copper, brass, and aluminum and their corrosion resistance over time. PCM macroencapsulation pipes have greater potential than microencapsulation pipes since they can be made quickly and cheaply, have a broader space that enables more PCM quantity to be incorporated, and preserve volumetric changes during cycles. The passive inclusion of PCMs into the building envelope is not always sufficient due to the failure of melting or solidification processes due to intense solar radiation, which overheats the PCM, or a shortage of the solidifying medium. Active strategies are indicated in such instances.

The research explored the possibility of PCM-integrated building envelopes, a developing technology for improving building performance. The PCM technology demonstrated a significant increase in building thermal energy, either by reducing unwanted thermal loads or regulating thermal demand, consequently positively influencing thermal comfort and building energy savings. A general review[6] of recent studies was conducted considering the major PCM types, encapsulation methods, influential parameters, and incorporation strategies with building envelope materials, primarily for roofs and external walls. The following are some of the findings that can be taken from the current studies.

The appropriate location is very crucial in installing PCMs into the building shell. However, the most appropriate assertion is that the PCM layer should be placed closer to the heat source. For example, the PCM should be near the outside envelope layers in warmer areas. The primary reason for this is that the PCM acts as a heat barrier (insulation) in these conditions. Hence, the stored heat should be as far away from the indoors as feasible to avoid any unwanted heat emittance and to make use of the night cooling effect throughout the evening. Under cold conditions, the PCM layer acts as a heat provider (that is, it prevents/restricts heat from escaping from the interior to the exterior, stores the heat, and then releases it back to the interior due to temperature difference); consequently, it should be as near to the interior as feasible. For example, $MgCl_2{\cdot}6H_2O$ and $CaCl_2{\cdot}6H_2O$ filled in aluminum compound foil in cuboid, cylindrical, plate-shaped, and spherical macroencapsulations, as shown in Figure 7.11. The results showed a cuboid block having 1.5 kg PCM melting slowly as compared to a spherical shape. In another example, steel container can be used as a container (Figure 7.12a), copper tubes as a container (Figure 7.12b), and aluminum panels (Figure 7.12c) for PCM encapsulation, as shown[31,32] in Figure 7.12.

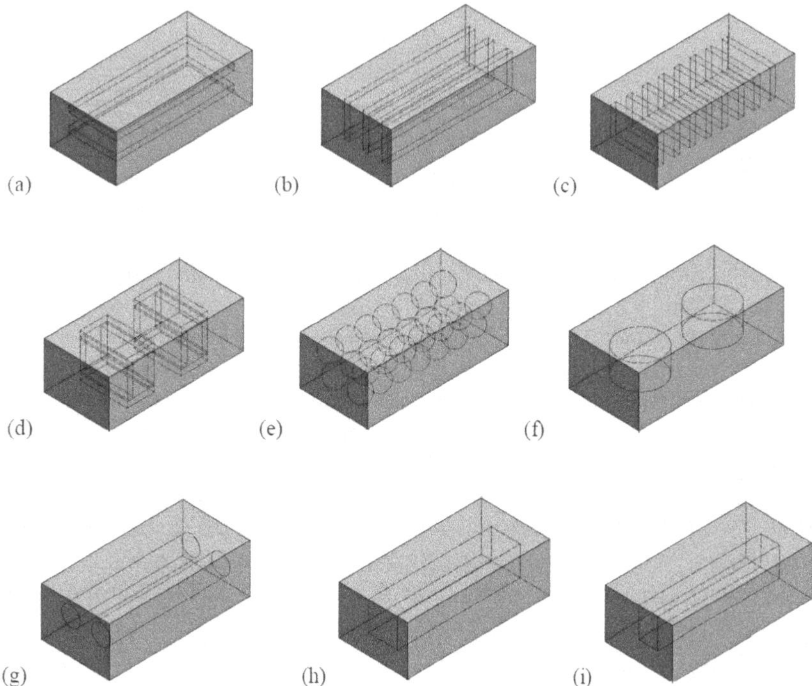

FIGURE 7.11 Encapsulation of PCM in plate-shaped (a,b,c,d), spherical (e), cylindrical (f,g), and cuboid (h,i) aluminum containers.[33]

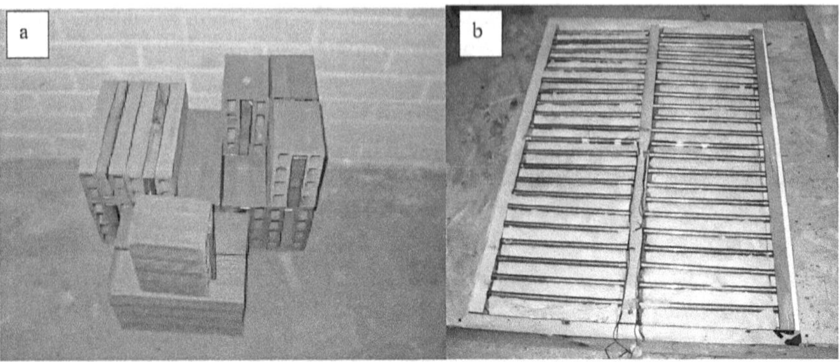

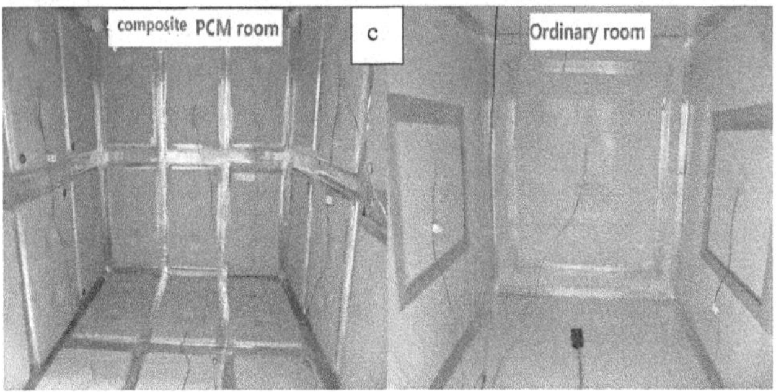

FIGURE 7.12 Macroencapsulation in (a) wooden containers, (b) copper tubes, and (c) aluminum panels.[31,32]

7.7.2 Microencapsulation

At the advanced level, microencapsulation is tested for the building envelope. The various microencapsulation types, technologies, applications, and release mechanisms have been covered with additives in building construction materials. As a result of microencapsulated additives in building materials, the following advancements have been mentioned in patents: increased fireproofing; improved freeze and freeze–thaw resistance; reduced expansion and degradation of concrete and mortar; improved hydration of concrete and mortar mixes in the compression-molding production of building elements; reduction in thermal cracking due to heat release by cement hydration; reduction in water absorption of hydraulic cement sheets; insulation or noise absorption; protection of building materials against mildew, bacteria, insects, and rodents. Insulating materials based on microencapsulated phase change materials (PCMs) for active heat collection and release are the fastest-expanding category. Microcapsules with high mechanical resistance are required to allow reversible liquid–solid–liquid phase transitions and to safeguard the PCM during the entire product life cycle. Microencapsulated light walls can be conceived for green buildings, as shown in Figure 7.13.

The disadvantages of the passive integration approach are mostly connected with the inability to fully use PCMs' storage capacity, unguaranteed phase transition (charging/discharging), and uncontrolled direction of stored heat. Also, little

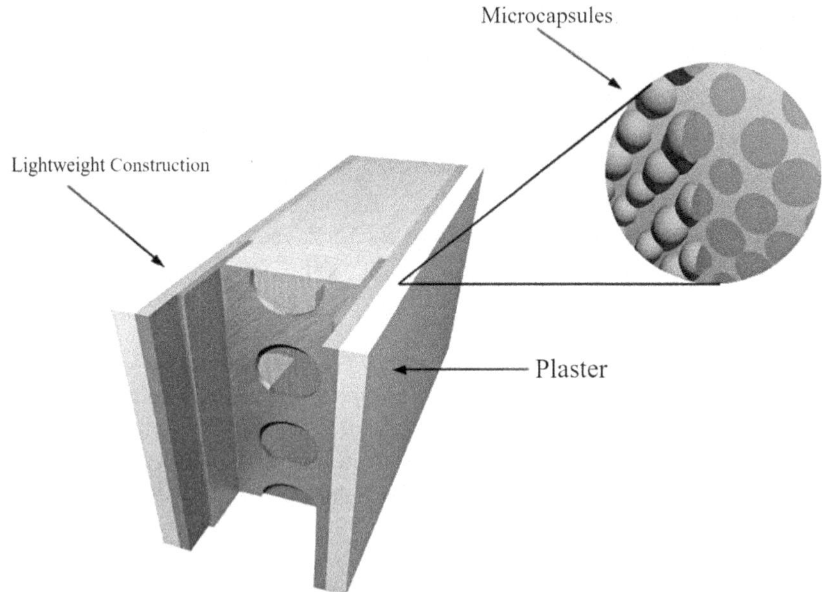

FIGURE 7.13 Lightweight microencapsulated PCM wall.[34]

emphasis has been placed on the thermal behavior and positive characteristics of PCM-incorporated structures in adverse weather conditions. With extreme heat, the PCM melts entirely in the early morning hours. To avoid malevolent behavior, the stored heat must be discharged immediately. The passive approach is insufficient in this circumstance, and the nocturnal ventilation strategy is worthless. As a result, an alternate discharge medium, such as geothermal energy, is required to prepare the PCM for the next day's cycle. Under chilly conditions, the sun-exposed building element passively stores heat in the PCM throughout the day and then releases it when the temperature drops. Nevertheless, solar radiation is often minimal in frigid places, making it unable to heat the envelope layers, including the PCM. As a result, active solar energy use, such as solar collectors, is necessary. The main disadvantage cited in building applications is PCMs' low heat conductivity. This problem results in partial charging and discharging during the phase transition, impacting the PCM storage capacity in the subsequent cycle. Some researchers have investigated this issue experimentally using various methods such as high-thermal-conductivity encapsulation materials, nanoparticle immersion, fins, copper fumes, metal matrices, and carbon fibers to accelerate the melting and freezing processes. Further experimental research on the use of such approaches is necessary.

7.8 GEOTHERMAL ENERGY STORAGE

A great way to incorporate renewable and sustainable energy sources into green building design is through geothermal energy. The following are some ways that geothermal energy might support green construction initiatives:

1. **Renewable energy resource:** Geothermal energy is a practically limitless source of renewable energy. It is dependent on the heat that the Earth naturally retains, which the planet's internal energy sources constantly renew. Geothermal energy is used by green buildings to lessen their dependency on fossil fuels, making their energy mix more sustainable.
2. **Reduced carbon footprint:** Green structures work to reduce their carbon footprint and environmental impact. This objective is aided by geothermal energy because it lessens the need for fossil fuel-based heating and cooling systems. Green buildings can dramatically reduce their carbon dioxide (CO_2) and other greenhouse gas emissions related to building operations by installing geothermal heat pumps.
3. **Ambient quality:** Geothermal systems provide effective and reliable heating and cooling, resulting in improved indoor comfort and air quality. They can keep indoor temperatures constant all year round, reducing temperature swings and enhancing occupant comfort. Geothermal systems also don't need combustion processes, which lowers the danger of indoor air pollution and raises indoor air quality.
4. **Robust construction:** Geothermal systems are renowned for their longevity and robust construction. The subsurface parts, like the plumbing system, can last for many years with little upkeep. The concepts of green buildings, which emphasize the use of robust materials and systems to limit waste and the need for frequent replacements, are in line with this lifespan.
5. **LEED certification:** LEED certification is a well-known green building rating system. LEED stands for Leadership in Energy and Environmental Design. Energy and atmosphere and indoor environmental quality are two categories where geothermal systems might help achieve LEED points. Green buildings can raise their total LEED certification level by using geothermal energy.
6. **Resilience and energy independence:** When compared to traditional heating and cooling systems, geothermal energy systems are less sensitive to changes in energy pricing and supply. Green buildings can increase their resilience and lessen reliance on outside energy sources by utilizing a renewable energy source that is already present on the property.

When planning and installing geothermal systems for green buildings, it's crucial to collaborate with knowledgeable architects, engineers, and geothermal specialists. They can assist in ensuring proper system sizing, effective fusion with other architectural elements, and compliance with regional laws and standards.

7.8.1 Geothermal Heat Pumps

Geothermal heat pumps (GHPs) employed in green structures are incredibly energy efficient. When compared to conventional heating and cooling systems, they can achieve significant energy savings. GHPs may provide space heating, cooling, and hot water with less energy use by utilizing the Earth's constant temperature, which lowers greenhouse gas emissions and lowers overall energy demand. Heat pump-based geothermal energy utilization is the current trend for energy-efficient and long-term housing complexes provided the availability of this energy source. Building-integrated geothermal energy systems have the potential to minimize the built environment's energy consumption. Every attempt is made within the green building concept to reduce energy use. Using electricity, heat pumps transfer heat from one location to another. Refrigerators and air conditioners are two typical heat pump examples. Buildings can be heated and cooled using heat pumps. Year round, the temperature at about 30 feet below the surface ranges between roughly 50 °F (10 °C) and 59 °F (15 °C), as shown in Figure 7.14. This means that soil temperatures are typically warmer in the winter and cooler in the summer than the surrounding air at

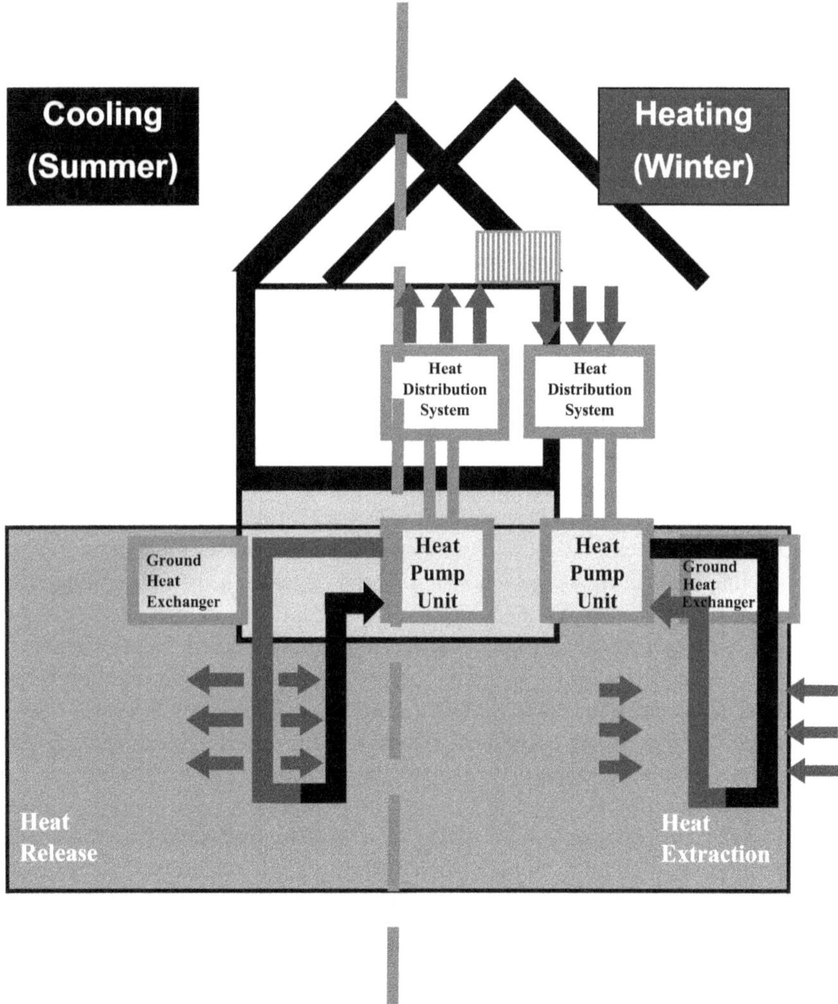

FIGURE 7.14 Geothermal heat pump circuit diagram for a building.[35]

some specific places. The steady underground temperatures are used by geothermal heat pumps (GHPs) to effectively exchange heat, cooling homes in the summer and heating them in the winter. Many of the passive dwellings rely heavily on ventilation and GHP.

7.9 CHEMICAL ENERGY STORAGE

Chemical energy storage in buildings is the method of storing energy in the form of chemical compounds inside the infrastructure of a building. This energy storage may be utilized to power the building's numerous electrical and mechanical systems. The usage of batteries is a frequent example of chemical energy storage in structures. Batteries are chemical compounds that store energy and may be used to power various pieces of equipment within a structure, including emergency lights, security systems, and backup generators. Hydrogen fuel cells are another example of chemical energy storage in buildings. By mixing hydrogen and oxygen, hydrogen fuel cells generate energy that may be used to operate a building's heating and cooling systems and other electrical equipment.

Supercapacitors and compressed air energy storage devices are examples of other types of chemical energy storage in buildings. These technologies store energy in various chemical compounds and can power various building systems and gadgets. Overall, chemical energy storage in buildings is a critical technology for increasing energy efficiency and lowering the environmental effect of building operations. As renewable energy sources, chemical energy storage in buildings is anticipated to become more significant in guaranteeing a dependable and sustainable energy supply.

7.9.1 Battery Storage

Battery storage devices are becoming increasingly significant in the construction of green buildings. Instead of relying only on the electrical grid, these systems store energy from renewable sources such as solar and wind. This minimizes the carbon footprint of the building and contributes to a more sustainable future. The green building design has numerous ways to take advantage of battery storage. For starters, battery storage systems provide greater energy independence. With the increased usage of renewable energy sources such as solar and wind power, battery storage systems can store energy generated during peak hours and utilize it during off-peak hours. This lessens reliance on the electrical grid, which is frequently fueled by fossil fuels. Buildings may lower their carbon impact and save money on electricity by storing energy locally.

Second, during power outages, battery storage systems provide energy resiliency. Power outages are becoming more widespread as extreme weather events become more regular due to climate change. During severe outages, battery storage devices can supply backup power, ensuring that essential systems like lights and refrigeration continue to work. This is especially vital for hospitals, data centers, and other critical infrastructure that requires continuous electricity.

Third, battery storage devices can aid in the reduction of peak demand on the electrical grid. The electrical grid in many locations cannot meet the heavy demand during peak hours, resulting in power outages or blackouts. Battery storage devices may store energy during off-peak hours and discharge it during peak hours, minimizing grid pressure. This can assist in averting power outages and blackouts and reducing the need for costly grid improvements. In another way, battery storage devices can aid in the balance of energy supply and demand. Renewable energy sources such as solar and wind are intermittent, producing electricity only when the sun shines or the wind blows. Excess energy produced during peak hours can be stored in battery storage systems and released during off-peak hours when demand is high. This aids in the balance of energy supply and demand, ensuring that energy is accessible when it is required.

Lastly, battery storage devices can assist buildings in lowering their carbon impact. Buildings may lessen their dependency on fossil fuels and the electrical grid by storing energy locally. This lowers greenhouse gas emissions and contributes to a more sustainable future. Furthermore, by alleviating strain on the electrical grid, battery storage devices can assist in lessening the demand for new power plants frequently fueled by fossil fuels.

7.10 MECHANICAL ENERGY STORAGE

Building-integrated flywheel energy storage is a technique that stores energy in buildings using flywheels. A flywheel is a mechanical device that rotates and stores energy as rotational kinetic energy. In this technique, a flywheel is coupled to an electrical generator, and energy is stored when the flywheel is spun up to a high speed. When energy is required, the flywheel slows down and converts the stored kinetic energy back into electrical energy.

The advantages of building-integrated flywheel energy storage are its high efficiency, quick response times, and capacity to supply a consistent and stable power source. It also has a long lifespan and requires minimal upkeep. Nevertheless, the technology is still in its early phases of development, and significant hurdles remain, such as the requirement for accurate flywheel balance and safety considerations connected to high-speed spinning.

Notwithstanding these obstacles, building-integrated flywheel energy storage has the potential to become a vital element of the future energy mix, especially for structures that require a high level of reliability and resilience. It may also be used with renewable energy sources like wind and solar, assisting in the transition to a cleaner and more sustainable energy system.

REFERENCES

1. Global Energy Consumption. [accessed 2024 May 2]. https://www.theworldcounts.com/challenges/energy/global-energy-consumption
2. Grant D, Zelinka D, Mitova S. 2021. Reducing CO_2 emissions by targeting the world's hyper-polluting power plants *. Environ Res Lett [Internet]. [accessed 2024 May 2] 16(9):094022. doi:10.1088/1748-9326/AC13F1
3. Transport. 2023. Clim Chang 2022 - Mitig Clim Chang [Internet]. [accessed 2024 May 2]:1049–1160. doi:10.1017/9781009157926.012
4. Global Emissions - Center for Climate and Energy SolutionsCenter for Climate and Energy Solutions. [accessed 2024 May 2]. https://www.c2es.org/content/international-emissions/

5. Lu Y, Yu D, Dong H, Lv J, Wang L, Zhou H, Li Z, Liu J, He Z. 2022. Magnetically tightened form-stable phase change materials with modular assembly and geometric conformality features. Nat Commun 2022 131 [Internet]. [accessed 2024 May 2] 13(1):1–12. doi:10.1038/s41467-022-29090-1
6. Yang T, Ding Y, Li B, Athienitis AK. 2023. A review of climate adaptation of phase change material incorporated in building envelopes for passive energy conservation. Build Environ. 244:110711. doi:10.1016/J.BUILDENV.2023.110711
7. Xu J, Wang RZ, Li Y. 2014. A review of available technologies for seasonal thermal energy storage. Sol Energy. 103:610–638. doi:10.1016/j.solener.2013.06.006
8. Kuravi S, Trahan J, Goswami DY, Rahman MM, Stefanakos EK. 2013. Thermal energy storage technologies and systems for concentrating solar power plants. Prog Energy Combust Sci. 39(4):285–319. doi:10.1016/j.pecs.2013.02.001
9. Gautam A, Saini RP. 2020. A review on sensible heat based packed bed solar thermal energy storage system for low temperature applications. Sol Energy. 207:937–956.
10. Potassium Chloride (KCl) Optical Material. [accessed 2021 Jun 25]. www.crystran.co.uk/optical-materials/potassium-chloride-kcl
11. Bandyopadhyay SK, Ghose PK, Bose SK, Mukhopadhyay U. 1987. The thermal resistance of jute and jute-blend fabrics. J Text Inst [Internet]. [accessed 2021 Jun 25] 78(4):255–260. doi:10.1080/00405008708658249
12. Chauhan VK, Shukla SK, Rathore PKS. 2022. A systematic review for performance augmentation of solar still with heat storage materials: A state of art. J Energy Storage. 47:103578. doi:10.1016/J.EST.2021.103578
13. Abhat A. 1983. Low temperature latent heat thermal energy storage: Heat storage materials. Sol Energy. 30(4):313–332. doi:10.1016/0038-092X(83)90186-X
14. Sharma A, Tyagi V V., Chen CR, Buddhi D. 2009. Review on thermal energy storage with phase change materials and applications. Renew Sustain Energy Rev. 13(2):318–345. doi:10.1016/J.RSER.2007.10.005
15. Chauhan VK, Shukla SK. 2023. Performance analysis of Prism shaped solar still using Black phosphorus quantum dot material and Lauric acid in composite climate: An experimental investigation. Sol Energy [Internet]. [accessed 2023 Feb 24] 253:85–99. doi:10.1016/J.SOLENER.2023.02.017
16. Kant K, Biwole PH, Shamseddine I, Tlaiji G, Pennec F, Fardoun F. 2021. Recent advances in thermophysical properties enhancement of phase change materials for thermal energy storage. Sol Energy Mater Sol Cells. 231:111309. doi:10.1016/J.SOLMAT.2021.111309
17. Lin Y, Jia Y, Alva G, Fang G. 2018. Review on thermal conductivity enhancement, thermal properties and applications of phase change materials in thermal energy storage. Renew Sustain Energy Rev. 82:2730–2742. doi:10.1016/J.RSER.2017.10.002
18. Milián YE, Gutiérrez A, Grágeda M, Ushak S. 2017. A review on encapsulation techniques for inorganic phase change materials and the influence on their thermophysical properties. Renew Sustain Energy Rev. 73:983–999. doi:10.1016/J.RSER.2017.01.159
19. Chauhan VK, Shukla SK, Tirkey JV, Singh Rathore PK. 2021. A comprehensive review of direct solar desalination techniques and its advancements. J Clean Prod. 284:124719. doi:10.1016/J.JCLEPRO.2020.124719
20. Lamrani B, Kuznik F, Draoui A. 2020. Thermal performance of a coupled solar parabolic trough collector latent heat storage unit for solar water heating in large buildings. Renew Energy. 162:411–426. doi:10.1016/J.RENENE.2020.08.038
21. Thermophysical comparison of five commercial paraffin waxes as latent heat storage materials. [accessed 2023 Apr 7]. www.researchgate.net/publication/44858087_Thermophysical_Comparison_of_Five_Commercial_Paraffin_Waxes_as_Latent_Heat_Storage_Materials

22. Zeng JL, Sun LX, Xu F, Tan ZC, Zhang ZH, Zhang J, Zhang T. 2007. Study of a PCM based energy storage system containing Ag nanoparticles. J Therm Anal Calorim [Internet]. [accessed 2023 Apr 7] 87(2):371–375. doi:10.1007/S10973-006-7783-Z/METRICS
23. Zeng JL, Zhang J, Liu YY, Cao ZX, Zhang ZH, Xu F, Sun LX. 2008. Polyaniline/1-tetradecanol composites: Form-stable PCMS and electrical conductive materials. J Therm Anal Calorim [Internet]. [accessed 2023 Apr 7] 91(2):455–461. doi:10.1007/S10973-007-8495-8/METRICS
24. Murrill E, Breed L. 1970. Solid—solid phase transitions determined by differential scanning calorimetry: Part I. Tetrahedral substances. Thermochim Acta. 1(3):239–246. doi:10.1016/0040-6031(70)80027-2
25. Benson DK, Burrows RW, Webb JD. 1986. Solid state phase transitions in pentaerythritol and related polyhydric alcohols. Sol Energy Mater. 13(2):133–152. doi:10.1016/0165-1633(86)90040-7
26. Luo L, Le Pierrès N. 2015. Innovative systems for storage of thermal solar energy in buildings. Sol Energy Storage:27–62. doi:10.1016/B978-0-12-409540-3.00003-7
27. Benson DK, Webb JD, Burrows RW, Mcfadden JDO, Christensen C. 1985. Materials Research for Passive Solar Systems: Solid-State Phase-Change Materials [Internet]. [place unknown]; [accessed 2024 May 3]. https://www.nrel.gov/docs/legosti/old/1828.pdf.
28. Peng S, Fuchs A, science RW-J of applied polymer, 2004 undefined. 2004. Polymeric phase change composites for thermal energy storage. Wiley Online Libr [Internet]. [accessed 2023 Apr 7] 93(3):1240–1251. doi:10.1002/app.20578
29. Zalba B, Marín JM, Cabeza LF, Mehling H. 2003. Review on thermal energy storage with phase change: materials, heat transfer analysis and applications. Appl Therm Eng. 23(3):251–283. doi:10.1016/S1359-4311(02)00192-8
30. Rathod MK, Banerjee J. 2013. Thermal stability of phase change materials used in latent heat energy storage systems: A review. Renew Sustain Energy Rev. 18:246–258. doi:10.1016/J.RSER.2012.10.022
31. Khan Z, Khan Z, Ghafoor A. 2016. A review of performance enhancement of PCM based latent heat storage system within the context of materials, thermal stability and compatibility. Energy Convers Manag. 115:132–158. doi:10.1016/J.ENCONMAN.2016.02.045
32. Jin X, Zhang X, Cao Y, Wang G. 2012. Thermal performance evaluation of the wall using heat flux time lag and decrement factor. Energy Build. 47:369–374. doi:10.1016/J.ENBUILD.2011.12.010
33. Al-Yasiri Q, Szabó M. 2021. Incorporation of phase change materials into building envelope for thermal comfort and energy saving: A comprehensive analysis. J Build Eng. 36. doi:10.1016/J.JOBE.2020.102122
34. Lu S, Li Y, Kong X, Pang B, Chen Y, Zheng S, Sun L. 2017. A review of PCM energy storage technology used in buildings for the global warming solution. Lect Notes Energy [Internet]. [accessed 2023 Apr 8] 33:611–644. doi:10.1007/978-3-319-26950-4_31/COVER
35. Jia S, Yu D, Zhu Y, Wang Z, Chen L, Fu L. 2017. Morphology, crystallization and thermal behaviors of PLA-based composites: Wonderful effects of hybrid GO/PEG via dynamic impregnating. Polymers (Basel). 9(10). doi:10.3390/POLYM9100528
36. Chandel SS, Agarwal T. 2017. Review of current state of research on energy storage, toxicity, health hazards and commercialization of phase changing materials. Renew Sustain Energy Rev. 67:581–596. doi:10.1016/J.RSER.2016.09.070
37. Silva T, Vicente R, Soares N, Ferreira V. 2012. Experimental testing and numerical modelling of masonry wall solution with PCM incorporation: A passive construction solution. Energy Build. 49:235–245. doi:10.1016/J.ENBUILD.2012.02.010

38. Lee KO, Medina MA, Raith E, Sun X. 2015. Assessing the integration of a thin phase change material (PCM) layer in a residential building wall for heat transfer reduction and management. Appl Energy. 137:699–706. doi:10.1016/J.APENERGY.2014.09.003
39. Erlbeck L, Schreiner P, Schlachter K, Dörnhofer P, Fasel F, Methner FJ, Rädle M. 2018. Adjustment of thermal behavior by changing the shape of PCM inclusions in concrete blocks. Energy Convers Manag. 158:256–265. doi:10.1016/J.ENCONMAN.2017.12.073
40. Singh Rathore PK, Shukla SK, Gupta NK. 2020. Potential of microencapsulated PCM for energy savings in buildings: A critical review. Sustain Cities Soc. 53:101884. doi:10.1016/J.SCS.2019.101884
41. Cui Y, Zhu J, Twaha S, Chu J, Bai H, Huang K, Chen X, Zoras S, Soleimani Z. 2019. Techno-economic assessment of the horizontal geothermal heat pump systems: A comprehensive review. Energy Convers Manag. 191:208–236. doi:10.1016/J.ENCONMAN.2019.04.018

8 Passive and Active Exploitation of Renewable Energy

Vikash Kumar Chauhan and Ali A. F. Al-Hamadani

8.1 INTRODUCTION

Energy consumption is increasing due to rising living standards and digitalization. However, the proportion of renewable energy sources in total energy consumption is increasing. Global warming and environmental issues are still rising due to the use of renewable energy. Fossil-fuel basedenergy sources pollute the environment and deplete resources more than others. According to environmental pollution studies,[1] the most significant polluters are fossil-fuel based energies. The most common type of energy is still fossil-fuel based energy. Furthermore, the consumption of renewable energy sources is very low. As a result, it is critical to increase the use of renewable energy, which produces cleaner and lower-emissions energy. Clean energy use incentives for buildings should be appropriate. Renewable energy can meet needs such as heating, refrigeration, and lighting.

The construction industry rapidly expanding and investing 30–40% of essential global resources.[2] After industry and agriculture, modern buildings have become the third largest consumer of fossil energy.[3] The Asia-Link program is an initiative by the European Commission to promote and spread knowledge on sustainable built environments with a nearly zero-energy approach. Various application such as water heating, heating/cooling and electricity generation are mature technologies integrated with green building technology. As a result, building sustainability assessment is becoming increasingly crucial for long-term development, particularly in the building sector worldwide. Reducing the loss of vital resources like energy, water, and raw materials; halting environmental deterioration brought on by buildings and infrastructure throughout their lives; and designing constructed environments that are secure, effective, and economical for water and solar energy use were the main objectives of sustainable design.

Rainwater harvesting, solar heating and cooling are passive solutions that make buildings self-sustainable. Passive heating designs are integral to passive buildings in cold countries, such as Sunspace, Trombe walls, air handling units with air-air heat exchangers, and air tightness with the required air change per hour. Similarly, passive cooling, such as water evaporation, roof cooling, earth-water heat exchange, roof texture design, and downdraft space cooling, provides comfort in hot and dry regions.

DOI: 10.1201/9781003415695-8

Because buildings are the world's largest energy-consuming sectors, causing energy inefficiency, they can serve as a promising target with the most significant potential to achieve the common goal of sustainable development. Nonetheless, excessive building energy consumption has negative environmental consequences such as air pollution, the greenhouse effect, the urban heat island effect, and others, which can harm human health and social economy development.[1] The most significant source of environmental pollutants is derived from fossil fuels. However, fossil-fuel based energy accounts for 84.7% of total global energy consumption. Renewable and nuclear energies account for 5% and 4% of total primary energy consumption, respectively.[2]

The International Energy Agency (IEA) suggests that buildings are a significant energy consumer, accounting for half of total consumed electricity and one-third of total consumed natural gas. It also highlights that construction activities are responsible for one-third of global greenhouse gas emissions.[3] Using alternative energy like solar energy, wind energy, geothermal energy, and biomass energy instead of fossil fuels or limited natural resources is one of the most effective ways to save energy in buildings. This way, both preserving and protecting our resources for future generations and environmental values are not jeopardized.

8.2 RENEWABLE ENERGIES FOR BUILDINGS

To make buildings energy efficient and cost effective for residents, various renewable energy sources can be utilized. Also, some sustainable practices will lead to better management of natural resources. Solar energy is the most prominent energy source for buildings; other than this, wind energy, biomass energy, hydrogen energy, and geothermal energy are used. These energy sources can be used directly or indirectly for environmentally friendly buildings. *Renewable energy sources* are derived from the existing energy flow in continuous natural processes. Figure 8.1 shows the different renewable energy sources that can be used as passive and active method.

The importance of energy consumption gradually crept into engineering considerations during the early stages of design. Furthermore, sustainability is regarded as the most critical factor in system design in the first decade of the twenty-first century. Following the industrial revolution is unquestionably the best time for another revolution. This transition confronts the consequences of the previous century's resource exploitation. Throughout the life of a building, energy is utilized for various purposes. Heating/ventilation/air conditioning (HVAC) systems that provide comfort conditions during the usage phase consume 94.4% of the total energy. To lessen this rate, passive strategies and renewable energy sources should be used instead of mechanical systems to provide occupant comfort. More appropriate physical situations for human health can thus be developed within the buildings.

8.3 SOLAR ENERGY

A limitless source of light and heat energy is the sun. The fundamental idea behind using solar energy in buildings is to utilize thermal radiation falling on the earth's surface in the 300 to 3000 nm wavelength range. Outside the earth's atmosphere, solar

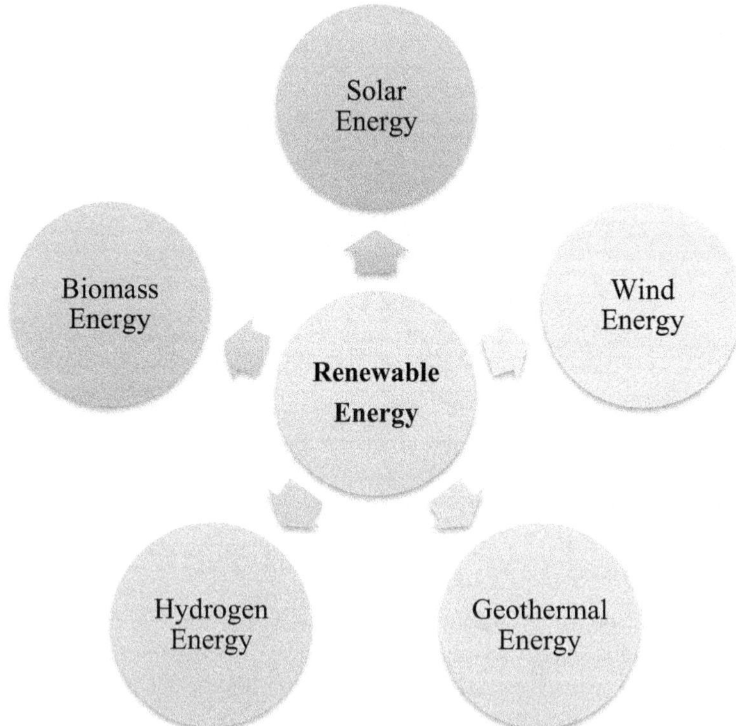

FIGURE 8.1 Renewable energy available for building usage.

flux reaches up to 1360 W/m^2; however, a significant proportion of it gets reflected and lost into the atmosphere while traveling toward the earth's surface. The sun's thermal energy flows through conduction, convection, and radiation. Building design controls these natural processes and facilitates warming and cooling of the structure. Figure 8.2 illustrates the solar radiation projection on a building that highlights the proportion of solar radiation that gets reflected and absorbed. Throughout the day, the apparent motion of the sun varies the solar flux intensity. Photovoltaics, solar heating and cooling, and concentrating solar power are the three primary approaches for utilizing solar energy. To power anything from small devices like calculators and traffic signs up to homes and substantial commercial enterprises, photovoltaics directly convert sunshine into electricity via an electrical process. Both solar heating and cooling (SHC) and concentrating solar power (CSP) applications employ the heat produced by the sun to run conventional electricity-generating turbines in the case of CSP power plants or to offer space or water heating in the case of SHC systems. Solar energy is a very adaptable energy source that may be produced by utility-scale, central-station solar power plants or distributed generating systems that are placed at or close to the point of use (like traditional power plants). Both approaches utilize cutting-edge solar + storage technologies to store the energy they generate for distribution when the sun sets.

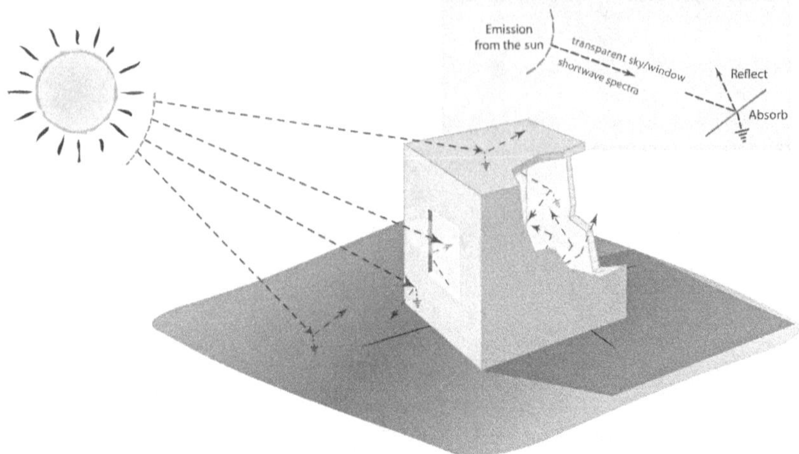

FIGURE 8.2 Interaction of solar irradiation with building.[4]

8.3.1 Passive Use

Solar energy can be directly used for buildings as per the comfort of occupants. Passive uses of solar energy consist of the direct use of beam radiation for air conditioning of inside spaces with the help of construction design and architecture of the building without including any auxiliary active equipment for the same. The passive solar design uses a building form and shell to accept, store, and distribute energy from renewable sources appropriate for buildings. Passive systems primarily use solar energy and fresh air to heat, cool, and light spaces without mechanical or electrical equipment.

8.3.1.1 Passive Heating

The idea behind passive heating is to utilize the solar energy directly as thermal energy for heating, storage, or distribution in building. In a passive system, building orientation like south-facing windows, dark flooring materials, and designs like facades and louvers are selected in a way to optimize the parameters that influence the building performance. Worldwide, on average, solar irradiance is around 5 kWh/m^2 on the earth surface. India receives 4–7 kWh/m^2 on an average day. The amount of energy entered is based on the size of the opening of a window, roof, or door. The building envelope's thermal insulation and sealants are essential for conserving the energy acquired. The placement of the building's components and their thermal efficiency affect how much energy can be stored. The residence of the philosopher Socrates, who lived between 470 and 399 BC, is the most straightforward illustration of passive heating, as shown in Figure 8.3. This home is designed with a compact construction and a trapezoidal plan, with the long side facing the sun and the northern side minimized to provide maximum productivity. When the sun's orbit is in the summer, the overhang on the roof's south side offers protection; in winter, it lets the sun below illuminate the structure. To protect against winter winds, the roof slopes down in the back.[5]

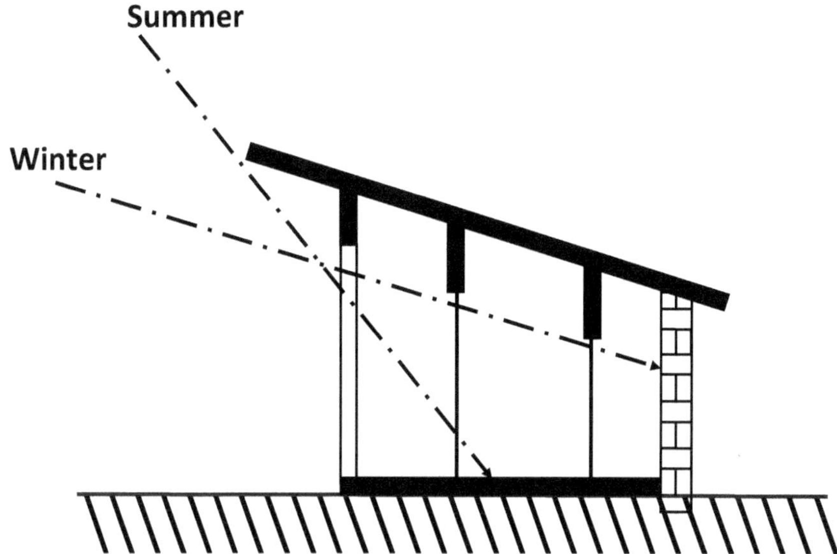

FIGURE 8.3 Design of a building for passive heating.[6]

8.3.1.2 Natural Lighting

Natural lighting, also referred to as daylighting, is a method that effectively brings natural light into your home using exterior glazing (windows, skylights, etc.), which lowers the need for artificial lighting and saves energy. Lighting-related energy use accounts for 70% of global energy consumption. Seventy percent of the lighting requirements can be met by the sun with the appropriate design. The rate in typical buildings is 25%. The need for artificial lighting can be decreased by using daylight as much as possible to illuminate buildings' spaces following visual comfort requirements. Using the ample apertures still present in the building envelope is the simplest method for obtaining natural lighting. These are examples of traditional homes with window configurations that let in enough natural light. With roof windows or light tubes, natural illumination can be given in locations lacking ideal facades for direct sunlight. Opening skylights on roofs when the roof element covers the inner space can bring in natural lighting in Figure 8.4.

8.3.1.3 Natural Ventilation

The energy needed for mechanical ventilation of buildings could be significantly reduced with natural ventilation. Compared to mechanical ventilation systems, these natural ventilation systems may result in lower initial and ongoing expenditures while maintaining ventilation rates consistent with acceptable indoor air quality. Also, those living in apartments with natural ventilation reported fewer symptoms than those living in buildings with ventilators.[4] Natural ventilation can enhance indoor environmental conditions, which enhances occupant productivity by lowering absenteeism, healthcare expenses, and worker output. Natural ventilation uses the wind and the "chimney effect" to circulate air and keep a house cool. Natural ventilation

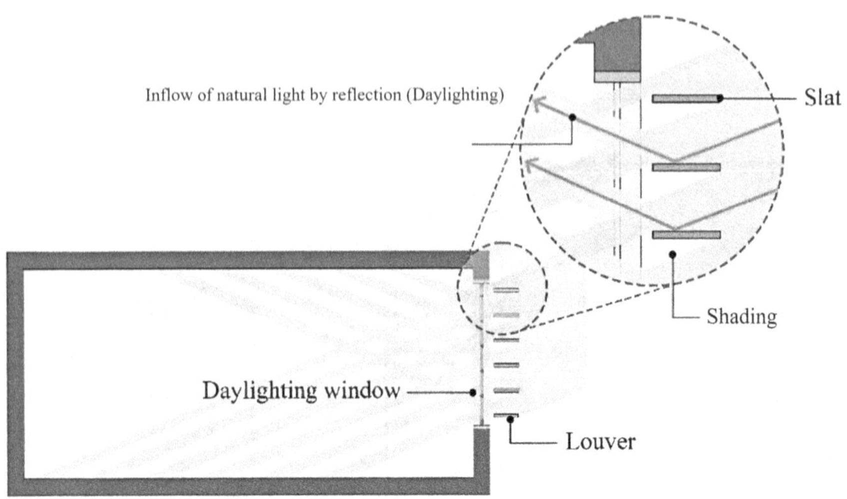

FIGURE 8.4 Natural lighting in a building.[7]

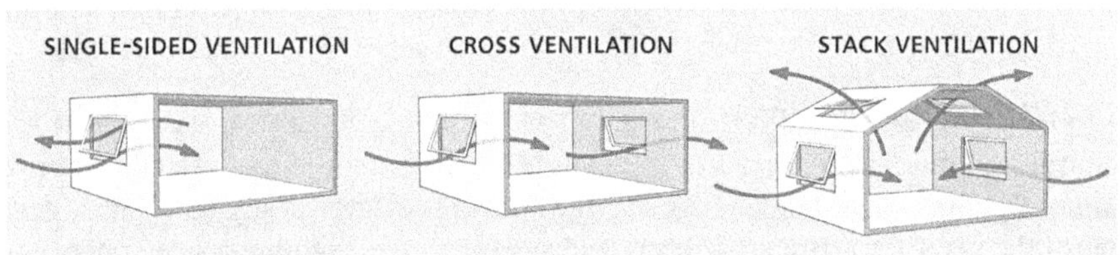

FIGURE 8.5 Configuration of opening in a building for natural ventilation.[8]

performs best in dry climates and in other climates during moderate weather when evenings are cool.

Depending on how our windows face the wind, it will air our home naturally by blowing through or leaving them open. When the wind blows against your house, it forces air into the windows on the side that faces the wind, while the leeward (downwind) side of the home experiences a natural vacuum effect that tends to pull air out of the windows. To illustrate this, Figure 8.5 shows the configurations of openings for natural ventilation. Broadly, three types of arrangement can be done: single-side ventilation, cross ventilation, and stack ventilation. To use the cooling sea breezes in coastal areas, many seaside buildings feature large, openable windows facing the ocean. Natural ventilation for drier areas entails limiting heat accumulation during the day and venting at night.

- **Benefits of natural ventilation**
  - **Direct cooling:** Outdoor air replaces the warm air generated inside the building due to internal work output. Fresh outdoor air creates a pleasant and comfortable ambiance for the occupants.
  - **Air quality:** Natural ventilation sweeps the inside low-oxygenated air and replaces it with fresh air that improves the air quality for the

occupants. It became very significant in a heart patient with a natural breathing problem.[5]

- **Nighttime cooling:** Cold nighttime exterior air is indirectly used to pre-cool thermally significant building fabric components or a thermal storage system to cool interior building spaces.

8.3.1.4 Desalination

Solar energy can be used to desalt brackish or saline water by evaporation and condensation to obtain fresh water. For this purpose, a solar still device is used that was conceived at an early stage. The passive mode of desalination achieved by conventional solar stills does not support any additional support to enhance productivity. Its performance depends mainly on solar flux and ambient conditions (temperature and wind). However, with some significant design modifications and material selection, the performance of solar stills can be improved. Solar stills can be installed on the rooftop of the building just like a solar panel to utilize solar energy. Figure 8.6 describes the overall heat transfer phenomenon inside and outside of a solar still. Inside the solar still, mainly convective, radiative, and evaporative heat transfer takes place. Upon solar heating, basin water starts radiative (Q_{rw}) and convective (Q_{cg}) heat transfer along with evaporative mass transfer (Q_{ew}). Simultaneously, convective heat transfer ($Q_{conv.b\text{-}w}$) takes place from the boundary wall to the basin water. From the condenser surface, ambient, radiative (Q_{rg}), and convective (Q_{cg}) heat loss takes place under the solar intensity "I". Additionally, bottom loss from the solar still takes place via conduction heat transfer.

Solar stills use the sun's energy to evaporate water and then condense the vapor to produce clean drinking water. The process involves several heat transfer mechanisms.

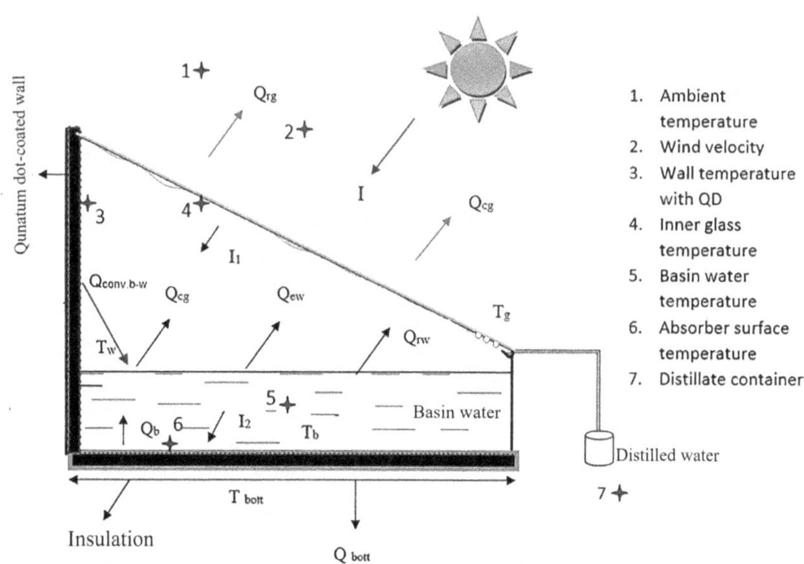

FIGURE 8.6 Heat transfer involved in a solar still.[9]

The first mechanism is solar radiation. Solar stills are designed to absorb and trap as much solar radiation as possible. The absorbed solar radiation heats up the still's interior, which in turn heats up the water inside the still.

The second mechanism is conduction. The heat generated by the absorbed solar radiation is transferred to the still's walls through conduction. The walls of the still act as a barrier between the hot interior and the cooler external environment, reducing heat loss.

The third mechanism is convection. As the water inside the still is heated, it becomes less dense and rises to the top of the still. Cooler water from the top of the still then flows down to replace it, creating a natural convection current. This helps to distribute heat evenly throughout the still.

The fourth mechanism is evaporation. The heated water inside the still evaporates and turns into water vapor. The vapor rises to the top of the still and comes into contact with the still's cool surface. As the vapor cools, it condenses back into liquid water, which is collected and used as drinking water. Overall, solar stills use a combination of solar radiation, conduction, convection, and evaporation to transfer heat and produce clean drinking water.

The use of solar desalination for buildings is crucial because it encourages sustainable methods for managing water resources and lessens reliance on traditional water sources. Here are some of the main arguments in favor of solar desalination in green construction:

- **Water scarcity:** Due to the world's growing population and the effects of climate change, water resources are becoming increasingly scarce. Green buildings may help by preserving and managing water resources. Saline or brackish water can be converted into fresh water utilizing renewable energy sources through solar desalination, relieving demand for freshwater supplies.
- **Energy effectiveness:** Solar desalination systems use a lot less energy than conventional desalination methods, which use a lot of electricity. Green buildings can lessen their carbon footprint and aid in the fight against climate change by utilizing solar energy.
- **Savings:** Compared to conventional desalination technologies, solar desalination systems can be significantly less expensive, especially in regions with expensive power. Solar desalination can also be applied on a smaller scale, which makes it more accessible and economical for towns or specific structures.
- **Resilience:** Green buildings equipped with solar desalination systems can withstand water shortages or interruptions in the water supply. Buildings can become independent and less reliant on outside forces by having a decentralized source of water.
- **Health and safety:** Finding access to clean, safe drinking water is a serious problem in many places of the world. Solar desalination can enhance public health by supplying a reliable source of fresh water and lowering the threat of waterborne illnesses.

In conclusion, solar desalination is a crucial technological advancement for encouraging sustainable water management strategies and minimizing the environmental impact of construction. Solar desalination systems in green buildings can help create a more sustainable future by preserving resources, consuming less energy, and fostering resilience.

8.3.2 Active Use

In terms of gathering solar energy, storing collected heat, and distributing heat to spaces, active solar systems are distinct from passive techniques, which rely on the structure of the building. The active systems use devices for storing, collecting, and distributing solar energy for residents' comfort. Active solar energy systems include all of the mechanical and electronic elements that convert the solar radiation absorbed by purpose-built collectors into the desired form of energy and allow it to be used in the structure. These systems can convert solar radiation into heat and electrical power.[10] These systems convert solar radiation into energy and are classified into two types based on the energy they produce: solar thermal systems, which generate heat energy, and photovoltaic (PV) systems, which generate electrical power. These systems are described briefly herein.

- ***Solar Space Heating***

Solar heating systems may directly heat water, air, etc. by converting solar radiation into thermal energy through collectors; alternatively, all mechanical and electronic systems that are used in a storage unit for evaluation and usage are referred to as "solar heating systems." Structures employ solar active heating systems for space heating, pool water heating, air conditioning, and air preheating. The basic idea behind heating systems is to collect heat via collectors, store it, and then distribute it to the appropriate locations for later consumption.[10] Solar energy can also be used for water heating and air heating for different purposes, as follows.

1. Active solar water heating
2. Active solar air heating

In active solar water heating, water is circulated by a pump; it is heated by solar radiation. To achieve this, copper tubes are used through which water runs, and to increase the solar absorption, area augmentation is done by providing long fins. The maximum temperature reaches the desired limit by altering the solar concentration. In a natural solar water or passive water heater, a greenhouse effect comes into play that helps to achieve a maximum 80°C temperature. The commercial name is evacuated tube collectors (ETCs). Figure 8.7 shows a prototype model at CERD Lab, Mechanical Engineering Department, Indian Institute of Technology Banaras Hindu University, IIT (BHU) that has wide applications in food processing and in chemical, pharmaceutical, textile, and heavy metal industries for pre-processing.

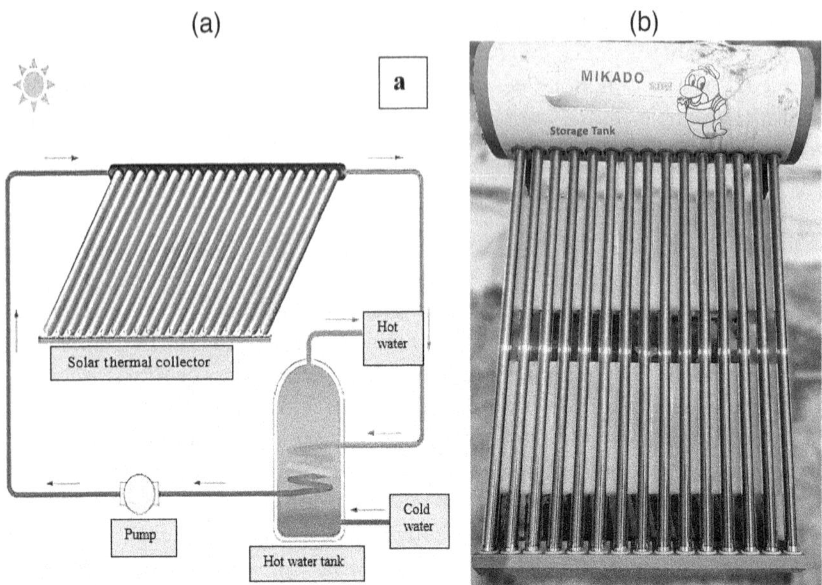

FIGURE 8.7 Solar water heaters (a). Schematic diagram of an active water heater.[11] (b). Prototype model of a passive water heater.

- ***Solar Air Heater***

The primary goal of the air heater is to utilize solar energy to heat the air and to enhance heat extraction from the absorber while minimizing heat losses from the collector's front cover. This can be accomplished by forcing air to pass through the front glass cover before entering the absorber (preheating the air). To illustrate a conventional design of solar air dryer, Figure 8.8 highlights the dryer section, solar absorption panel and blower section available at the CERD lab, Mechanical Engineering department, IIT (BHU). The design requires an additional cover (double pass) to create a counter-flow heat exchanger. A large area for heat transfer is created by porous media, where the volumetric heat transfer coefficient is relatively high. A porous absorber should be used to improve heat transfer from the absorber to the airstream. The pressure drop should be kept to a minimum when designing this sort of collector, which combines a double air route with a porous material.

Insolation, the solar energy from the sun, is absorbed by an absorbing media and used in solar air heating, a solar thermal technique, to heat the air. Solar air heating is a renewable energy heating method to heat or condition the air in buildings or for heat process applications.

- ***Photovoltaic Panels***

Photovoltaics (often abbreviated as PV) is a technology that converts sunlight directly into electricity. PV cells are made of semiconductor materials, usually silicon, that absorb photons from sunlight and release electrons, which are captured by

FIGURE 8.8 Solar air dryer.

an electric circuit to produce an electric current. This technology is widely used in solar panels to generate electricity for residential, commercial, and industrial applications. The solar power industry is a growing industry in India. In 2015, the Indian government set a target to achieve 100 GW of solar capacity, including 40 GW from rooftop solar.

PV technology has rapidly advanced in recent years, improving efficiency, cost, and durability. Solar panels can now be found in various sizes and configurations, from small portable chargers to large-scale power plants. PV technology is a vital component of the global transition to renewable energy, as it provides a clean and sustainable source of electricity. These panels are conveniently used for building power supplies to cut down the fossil-fuel basedpower to make green buildings without any carbon footprint. Although this system is not worth it in case of cloudy and nighttime conditions, to counter this, battery storage can be used with an inverter to continue the supply. Solar photovoltaic (PV) materials can convert sunlight directly into electricity. There are several solar PV materials, but the most commonly used ones are silicon-based. Silicon is the second most abundant element on Earth, and it is widely used in solar cells due to its ability to convert sunlight into electricity efficiently. Silicon solar cells are made by doping two silicon layers with impurities to create a p–n junction. When sunlight hits the cell, it creates an electric field at the p–n junction, which separates the charges and generates electricity.

Other types of solar PV materials include:

- **Thin-film materials:** These are made by depositing thin layers of semiconductor material on a substrate. Examples include cadmium telluride (CdTe) and copper indium gallium selenide (CIGS).
- **Plastic solar cells:** These are made from organic molecules that can absorb sunlight and generate electricity. They belong to organic electronics in which conductive organic polymers are studied that absorb sunlight and transport charge for electricity generation. Examples include polymers and small molecules like fullerene.
- **Perovskite materials:** These are a class of materials that have recently emerged as promising candidates for solar cells. They have high efficiency and can be easily fabricated. Perovskite solar cells have shown promising improvement in their efficiency from 3% in 2009 to 25% in 2022. Apart from their stability and durability, which indicates their natural degradation issue in operational life, these materials have shown their capability to capture the market soon and replace silicon-based photovoltaic cells.

Buildings can produce electricity in a number of ways using photovoltaic (PV) panels. Here are a few typical ways that PV panels are used in architecture:

- PV panels are frequently installed on rooftops to produce electricity for the building. Rooftop solar panels are a well-liked alternative because they don't require any extra room and can lessen the building's reliance on the power grid.
- Building-integrated photovoltaics (BIPV) is a design strategy that inserts PV panels onto the roof outside of buildings, such as at walls, facades, or windows. This strategy can generate electricity while also assisting in lessening the PV panels' visual impact on the building. As shown in Figure 8.9, a grid-connected solar panel provides the power to run the house, and excess power is supplied to other places via grid connections.
- Solar carports are structures that are made to shade parked cars and produce power from PV panels that are put on the top. This is a common choice for business structures with sizable parking lots, as shown in Figure 8.10. There are several advantages to using solar carports. Solar carports provide straightforward and affordable alternatives to sophisticated and pricey roof-mounted systems. Solar-powered carports, instead of giving up precious real estate, make use of already-existing parking lots to generate electricity. Solar carports have the potential to significantly boost the overall energy production of your solar project by using parking lots in addition to the roofs of existing buildings. Solar trellis and carport structures that are permanently installed require little to no maintenance and provide simple access to the panels for servicing and maintenance. Autos can park under cover or in the shade thanks to solar carports.
- PV systems can be ground mounted. PV panels may also be positioned close to a structure on the ground. This strategy is frequently applied to larger structures or structures with a small amount of roof space.

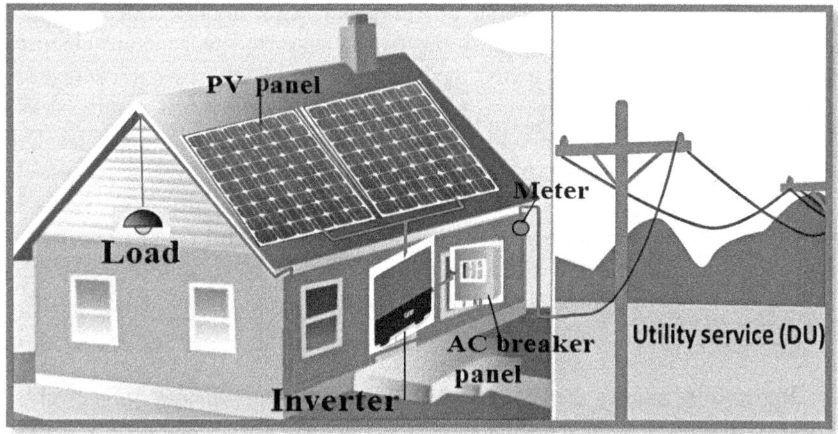

FIGURE 8.9 Photovoltaic panels on the rooftop of a building.[12]

FIGURE 8.10 Carports near business structures.[13]

- Mobile offices and construction trailers can both be powered by portable solar panels, which are compact, lightweight constructions. These panels are simple to move and set up anywhere is necessary.

In a community solar project, several residences or businesses share the electricity produced by a bigger PV system that has been built nearby. This strategy is gaining popularity as a means of supplying solar energy to structures that are unable to install PV panels on their own.

PV panels are designed to capture sunlight and convert it into electricity. They are made up of multiple solar cells that are connected together and enclosed in a protective casing. When sunlight strikes the solar cells, it causes electrons to be released, which are then captured by the panels and used to generate electricity. There are several types of PV panels available, each with their own advantages and disadvantages. The most common types of PV panels include monocrystalline, polycrystalline, and thin-film panels. Monocrystalline panels are the most efficient and have the highest power output, but they are also the most expensive. Polycrystalline panels are less expensive than monocrystalline panels, but they are also less efficient. Thin-film panels are the least expensive and have the lowest power output, but they are also the most flexible and can be used in a variety of applications.

There are numerous advantages to using PV panels in building design. Here are some of the most significant benefits.

- **Renewable energy source:** PV panels use sunlight, which is a renewable energy source, to generate electricity. This means that they have a lower environmental impact than fossil fuel-based energy sources, which contribute to climate change.
- **Energy efficiency:** PV panels can be used to generate electricity on-site, reducing the need for energy from the grid. This can result in lower energy bills and a more energy-efficient building.
- **Reduced carbon footprint:** By generating electricity from a renewable energy source, PV panels can help to reduce the carbon footprint of a building. This can help to reduce greenhouse gas emissions and mitigate climate change.
- **Increased property value:** Buildings with PV panels installed are often considered more valuable than those without. This is because they are seen as being more energy efficient and environmentally friendly.
- **Energy independence:** PV panels can provide energy independence, allowing buildings to generate their own electricity. This can be particularly useful in off-grid locations or during power outages.
- **Long lifespan:** PV panels have a long lifespan, with many panels lasting for 25 years or more. This means that they require little maintenance and can provide a reliable source of electricity for many years.

- ***Concentrated Solar Power***

Green buildings with the use of concentrated solar power (CSP) technology have the potential to dramatically reduce a structure's carbon footprint and lessen the effects of climate change. In this section, we will go over the advantages and drawbacks of employing CSP technology for green construction and a few real-world instances of CSP-powered green structures.

For green buildings, CSP technology can be used to generate both thermal and electrical energy. Through solar thermal air conditioning systems, CSP technology for green buildings is most frequently used. This device uses focused sunlight to heat a fluid that can be utilized to power a thermally activated cooling system, like an absorption chiller. This technology can cool the building without using fossil fuels or grid electricity.

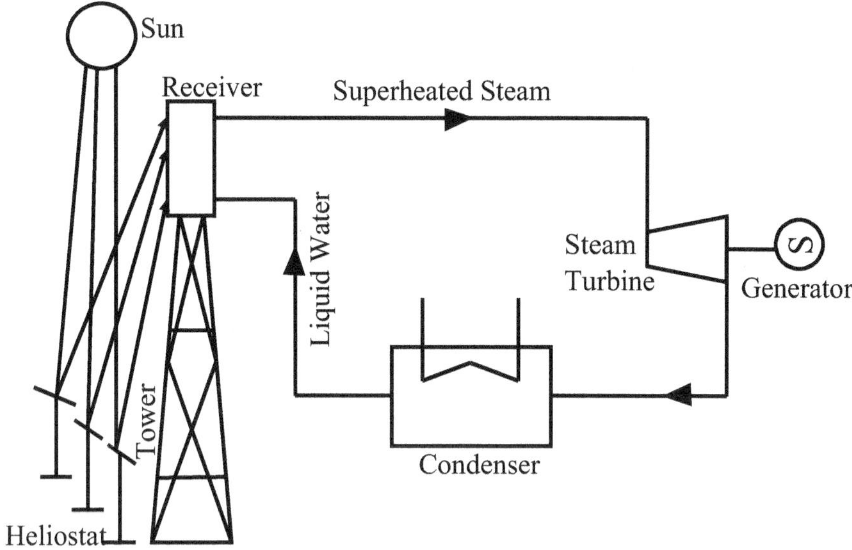

FIGURE 8.11 Schematic diagram of solar power systems using water/steam working fluid.[14]

Combination heat and power (CHP) systems are yet another way CSP technology is used in green construction. A fluid in a CSP–CHP system is heated by concentrated sunlight, which creates steam that powers a turbine and produces energy. The excess heat from the turbine can subsequently be used for residential hot water or space heating. Figure 8.11 shows different types of arrangements from various investigations that have been tested to convert solar energy into thermal energy such as a parabolic trough arrangement with absorber tubes as line concentration, heliostats with a receiver tower as point concentration, and a parabolic dish collector with receiver engine (from left to right).

- **CSP's advantages for green buildings**

The use of CSP technology in green construction has various advantages. First, CSP is an environmentally favorable substitute for fossil fuels because it is a renewable energy source that does not release greenhouse gases or other air pollutants while it is in use. As a result, buildings powered by CSP can dramatically lower their carbon footprint and help lessen climate change's effects.

Second, CSP technology is very effective and can deliver consistent energy for green structures. CSP systems are a dependable energy source for buildings because they can be made to produce constant energy output even under cloudy or partially shaded conditions. Thirdly, CSP technology is adaptable and may be created to satisfy the energy requirements of structures of all sizes. CSP systems are a flexible and adaptable choice for green construction because they can be tailored to meet the unique energy requirements of different structures.

- **CSP for green building challenges**

Although CSP technology has numerous advantages for green building, some obstacles must be overcome to realize its potential fully. The price of CSP technology is one of its biggest problems. CSP systems can be costly to install and operate, which may prevent smaller building owners or those with tighter budgets from using them.

The land use constraints of CSP technology present another difficulty. CSP systems need a lot of space to be placed, which could be a problem in cities or other highly inhabited places. Additionally, local wildlife habitats may be impacted by CSP systems, which requires careful consideration and mitigation.

Finally, there may be some technical difficulties with CSP technology that need to be resolved, such as storing thermal energy for use during times of low sunlight or integrating it with current building systems.

- **Examples of green structures powered by CSP**

Despite these obstacles, several instances of CSP-powered green buildings worldwide show the technology's promise for eco-friendly building design. Here are a few examples.

- A renowned research institution in Saudi Arabia, King Abdullah University of Science and Technology (KAUST), runs entirely on renewable energy, including CSP technology. The university's CSP system powers the campus with electrical and thermal energy while making a significant dent in the institution's carbon footprint.[6]
- A CSP–CHP system supplies electricity and thermal energy to the Arizona State University Polytechnic Campus in the United States. The system is made to satisfy the campus's energy requirements while lowering the university's greenhouse gas emissions.[7]
- The De Aar Solar Power Plant in South Africa is a CSP facility that fuels the nation's grid with power. The facility concentrates sunlight onto a receiver using parabolic troughs.[8]

8.4 WIND ENERGY

Wind energy can be a great source of renewable energy for buildings. There are a few different ways that wind energy can be harnessed for buildings:

1. **Wind turbines:** Small-scale wind turbines can be installed on the roofs or walls of buildings to generate electricity. These turbines typically have a capacity of 1–10 kW and can provide power for a portion of a building's electricity needs.
2. **Windcatchers:** Windcatchers are passive ventilation systems that use wind energy to cool and ventilate buildings. They capture and direct the wind into the building, where it is used to cool and circulate air.

3. **Wind scoops:** Wind scoops are similar to windcatchers but are designed to capture and direct current into specific building parts, such as a courtyard or an atrium. This can help to create natural ventilation and reduce the need for mechanical cooling.
4. **Wind-assisted ventilation:** In some cases, wind energy can power mechanical ventilation systems, such as fans or air handlers. This can help to reduce energy consumption and costs.

Overall, wind energy can be a valuable addition to a building's energy mix, helping to reduce reliance on fossil fuels and lower energy costs. The wind has long been used as an energy source, an essential source of environmentally friendly energy that has grown in importance in recent years. It can be used to benefit both passive and active systems. These methods are discussed further herein.

8.4.1 Passive Use

In warm, arid areas like in North Africa and the Middle East, passive cooling systems are an excellent method to create a completely clean, perpetual, and cost-free air conditioning system for buildings or business places.

8.4.1.1 Passive Cooling

Passive wind cooling is a method of cooling a building or other structure using natural wind movement without mechanical or electrical devices. The technique involves creating openings in the structure that allow for air circulation, taking advantage of natural convection currents to cool the interior. Using passive systems makes it feasible to provide the building with the necessary levels of comfort for worker productivity and human health without consuming any energy. In particular, the impact of natural ventilation methods is significant in establishing thermal comfort and indoor air quality. Preventing heat gain in buildings is the fundamental tenet of passive cooling. Planning for this objective should be incorporated into the house's design process. High thermal mass, thick sectional structural elements like mudbrick or stone, and shade components may be used to reduce building heat gain. Different passive cooling methods, such as the ones listed in the next paragraph, have been developed for various climates.

Shading, solar heat reflection, building element insulation, wind cooling, ground cooling, water cooling, humidification, evaporative cooling, night radiant cooling, night cooling of thermal mass in buildings, exotic passive cooling methods, and seasonal cold storage are all examples of passive cooling strategies. *Windcatchers* are the most basic example of using natural ventilation to cool a building as shown in Figure 8.12. Thermal chimneys are collectors, drawing in fresh air from outside the building. The entrance of air from the outside environment into the building is accelerated by using a hot or warm region with an outflow. The basic principle behind windcatchers is the buoyancy effect, in which hot air lifts up in the presence of cold air that occupies a lower level.

In essence, wind catchers are towers built on building rooftops, often with four directional ports that face each of the four directions. Air flows down the shaft and

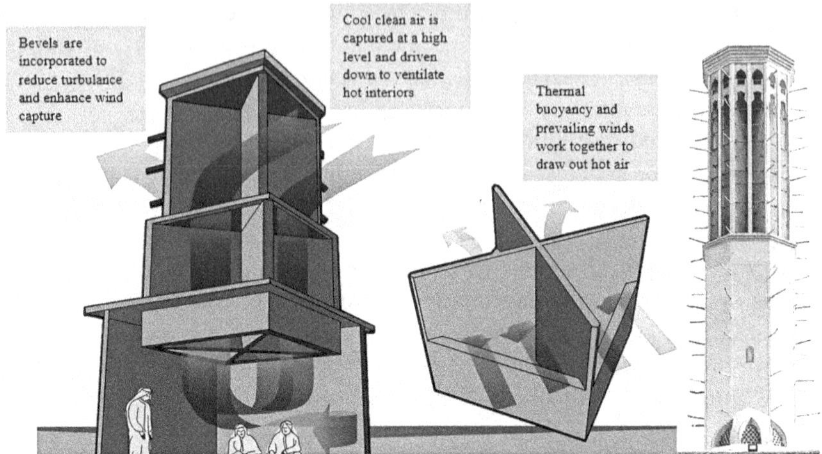

FIGURE 8.12 Three-dimensional illustration of windcatchers in buildings.[15]

into the building when an open port faces the direction of the dominant wind. Once inside, a pool of water fueled by aqueducts is passed over by the air (called a qanat). The warm air cools by evaporative cooling, bringing down the temperature within. At night, the windcatchers automatically draw the colder outside air within.

Passive wind cooling takes advantage of the natural tendency of hot air to rise and cool air to sink. This creates convection currents that can be used to cool a building. By positioning openings in the structure in such a way as to take advantage of these currents, it is possible to create a natural cooling effect.

Some examples of passive wind cooling strategies include:

1. **Ventilation:** By creating openings in the structure, such as windows or vents, hot air can escape, and cool air can be drawn in.
2. **Cross ventilation:** By positioning openings on opposite sides of the structure, air can be drawn through the building, creating a cooling effect.
3. **Stack ventilation:** This involves creating a vertical shaft that allows hot air to rise and escape, drawing in cooler air from below.
4. **Courtyard design:** By creating a central courtyard, air can be drawn in and circulated through the building, providing natural cooling.

Passive wind cooling can be an effective way to cool a building without the use of energy-intensive mechanical or electrical systems. However, it is essential to consider the design carefully.

8.4.2 Active Use

Turning the kinetic energy of moving air into mechanical energy is known as wind energy. Wind energy is a quick-developing energy source because it is renewable, inexhaustible, doesn't produce trash while being used, doesn't have a radioactive

FIGURE 8.13 Wind turbines in buildings. (a) Strata Tower,[16] London. (b) Pearl River Tower,[17] China.

effect, and doesn't harm the environment or people's health. On the surface of the Earth, wind can generate power. According to predictions, 40% of the world's energy will come from wind by 2040 [8]. Wind turbines are systems that actively use wind energy. The buildings make use of both medium- and small-scale wind turbines. These turbines can be positioned in an appropriate location in the garden or on the roofs. Figure 8.13 represents the examples of wind turbines built inside multistory high-rise buildings. In Asia, the Pearl River Tower in Guangzhou, China has a vertical axis turbine. It is a 309 m high, 71-story building that consists of an 8 m high vertical axis wind turbine paper. Furthermore, the Pearl River Tower is aerodynamically formed with a concave wall on the southern surface and a convex wall on the northern surface that enables better intensification and funneling of the wind through the openings.

Macro wind turbines, which are installed for large-scale energy generation such as wind farms, and micro wind turbines, which are used for local electricity production, are the two main types of wind energy technology. Building-integrated wind turbines are small wind turbines appropriate for use in buildings. A wind turbine's rotor, gearbox, generator, and blades are essential. Tiny wind turbines are called horizontal axis wind turbines (HAWTs). Vertical axis wind turbines (VAWTs) are becoming more and more common for integrated building applications to lessen the requirement for a tall tower and for aesthetic reasons. Moreover, VAWTs operate more quietly than HAWTs throughout the operation, causing less noise disturbance.

Wind turbines can operate on or off the grid. Off-grid systems require battery storage to store excess electricity, resulting in a more stable power supply. Their application is best suited for rural and remote areas where grid power is unavailable, such as remote villages and small isolated islands. Grid-connected systems traditionally require power converters to convert the generated DC electricity to AC electricity to be compatible with the power grid and AC electricity-based appliances. Modern wind turbines can directly generate alternating current (AC) as technology advances.

8.5 GEOTHERMAL ENERGY

Geothermal energy is a type of renewable energy derived from the heat generated by the Earth's core. This energy is found in the form of heat and steam deep within the Earth's crust, and it can be harnessed through geothermal power plants. Geothermal power plants work by drilling deep into the Earth's crust to access hot water and steam. The steam is then used to drive turbines, generating electricity. The hot water can also be used directly for heating and other purposes, such as industrial processes and greenhouse heating.

Geothermal energy can be a significant renewable energy source for building heating and cooling systems. Geothermal systems use the earth's stable temperature to transfer heat, making them more efficient and cost effective than traditional HVAC systems.

Here are some ways geothermal energy can be used for building:

1. **Geothermal heat pumps:** These systems use underground pipes filled with a fluid to transfer heat between the building and the earth. They can be used for both heating and cooling, and they are very efficient, reducing energy costs by up to 70%.
2. **Direct-use geothermal:** In some cases, geothermal energy can be used directly for heating and cooling buildings. This involves drilling a well into a geothermal reservoir and using the hot water or steam to heat the building.
3. **Geothermal energy storage:** Geothermal energy can also be used to store energy, which can be used to heat or cool the building when needed. This involves using a geothermal reservoir as a heat sink to keep excess energy during low demand.

Overall, geothermal energy can significantly reduce a building's carbon footprint and energy costs while providing reliable and sustainable heating and cooling, as shown in Figure 8.14. It shows the two cycles that work in winter and summer and provide the hot and cold water as per requirement. In the winter season, cold water is sent through the pipes under the earth surface and upon heat interaction, hot water is obtained from other side. Similarly, in the summer season, hot water is send at partial depth where hot water gets cooled and then sucked through pump at the other end.

The surface and subsoil temperature difference can also be used to heat and cool buildings. The system consists of installing underground pipes next to the building

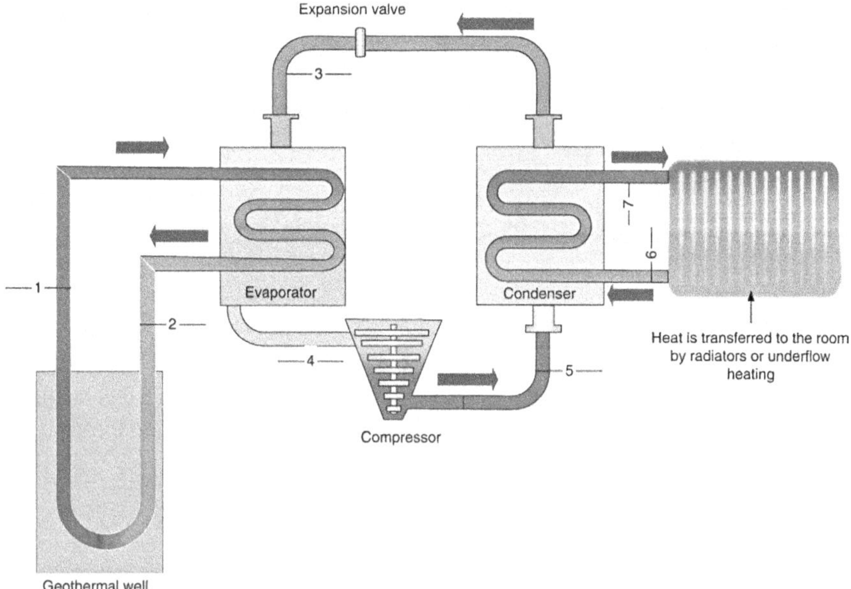

FIGURE 8.14 Geothermal energy utilization for buildings.

that are filled with water or another fluid, as well as a heat pump. This will result in permanent liquid flow in pipes from underground to the surface, exchanging heat with the ground. This device can then be fed into the building's air conditioning system via ducts or used to heat water. If the ground temperature is higher than the surrounding air temperature, the heat pump will transfer heat from the ground to the building. It can also work in reverse, transferring heat from a building's ambient air to the ground and thus cooling it.

8.5.1 Passive Use

Passive use of geothermal energy refers to utilizing the natural heat from the Earth's crust without the use of mechanical systems or pumps. Passive use of geothermal energy can be a cost-effective and environmentally friendly way to heat buildings, particularly in areas with a high geothermal gradient. It is also a renewable energy source, as the Earth's heat is constantly replenished. However, the availability of this resource may be limited to specific regions, and building designs may need to be tailored to take full advantage of geothermal energy. This type of geothermal energy is generally used for heating purposes, and it can be employed in a variety of ways.

- Balneology (hot spring and spa bathing)
- Agriculture (greenhouse and soil warming)
- Aquaculture (fish, prawn, and alligator farming)
- Industrial uses (product drying and warming)
- Residential and district heating

- Heating/cooling
- Technologies for direct uses like district heating, geothermal heat pumps, greenhouses, and for other applications are widely used.
- The geothermal energy can be utilized in summer as well as in the winter season.
- Hot water from one or more geothermal wells is piped through a heat exchanger plant to heat city water in separate pipes. Hot city water is piped to heat exchangers in buildings to warm the air.

8.5.2 Active Use

Geothermal power plants are another way that geothermal energy is utilized. These power plants use the heat from the Earth to generate electricity, which can be distributed to homes and businesses. The largest geothermal power plants are located in the United States, the Philippines, and Indonesia.

Different configurations to harness the geothermal energy

- Dry steam power stations
- Flash steam power stations
- Binary cycle power stations

Dry steam power plants are one of the oldest and most straightforward designs that can utilize geothermal energy. In this method, geothermal energy is used directly to make steam with a temperature and pressure around 165°C and 7 atm, respectively, to turn turbines. The dry steam technology allows the steam from a geothermal production well to be fed directly into a steam turbine without a secondary heat exchanger. The turbine then converts the change in steam pressure to mechanical rotational energy, which a generator converts into electrical energy. The famous dry steam field are the Geysers region in California, which may be the largest, and Larderello in Italy.

In *flash steam power stations*, the boiling point of a fluid increases with an increase in pressure. In a high-temperature, liquid-dominated reservoir, the water temperature is above 175°C and present in a liquid state due to high pressure (say 40 atm). Superheated water is pumped from the ground at 175°C or higher temperatures. When this superheated liquid comes onto the Earth's surface, it instantly flashes out into steam and water due to low pressure on the Earth's surface. The arrangement is made in such a manner that this process occurs in the flash tank, from which steam is fed to the turbine directly and hot water is used either for a heating application or reinjected into the ground. Such a facility is available in the Wairakei field in New Zealand. These types of plants are more common in the modern age.

Binary cycle power plants are the most recent development and can accept fluid temperatures as low as 57°C. A secondary fluid passes the moderately hot geothermal water with a much lower boiling point than water. In liquid-dominated low-temperature systems, flash steam plants are not suitable. This causes the introduction of a secondary fluid (organic fluid) that flashes out with the interaction of hot water, which then drives the turbines. Both organic and Kalina cycles are used in this perspective. The thermal efficiency of this type of plant is typically about 10–13%.

8.6 BIOMASS ENERGY

Biomass energy refers to using organic materials such as wood, agricultural residues, and other plant-based materials to generate power. Biomass energy can be used in buildings to provide heat and electricity. One common way to use biomass energy in buildings is through a biomass boiler, which burns wood chips or pellets to hot water or air, which can then be distributed throughout the building. This can be a renewable and sustainable alternative to traditional fossil fuel-based heating systems.

Another way to use biomass energy in buildings is by installing a biomass stove or fireplace. These can burn wood or other biomass to provide heat and ambiance in a single room or space. It's important to note that while biomass energy can be a renewable and sustainable alternative to fossil fuels, it also has its limitations and challenges. For example, sourcing biomass materials needs to be carefully managed to avoid deforestation or other negative environmental impacts, and biomass combustion can produce emissions that need to be carefully monitored and controlled.[18]

Physical procedures (size reduction—crushing and grinding, drying, filtering, extraction, and briquette) and conversion processes (biochemical and thermochemical processes) are used to produce fuel from biomass [26]. Biogas obtained by airless digestion from biomass sources in dwellings is used in energy generation. Ethanol obtained through the pyrolysis process is utilized for heating, and hydrogen acquired through the direct burning method is used for heating. Household waste management is a challenging task because of the increasing amount of waste globally and the large quantity of different materials in this waste stream. Sorting the waste at the source, the place where it is generated, is a crucial task to promote recycling and circular economy. Domestic waste has immense potential to be converted into energy. Biomass-to-biogas conversion is one of the methods to obtain methane gas.

8.6.1 Passive Use

Biomass can be converted into biogas through a process called anaerobic digestion. Anaerobic digestion is a natural process in which microorganisms break down organic material without oxygen, producing biogas as a byproduct. This gas has enough calorific heat capacity to cook the food at a domestic level and to keep the building warm at a human comfort level. For this, biomass is first collected and loaded into an anaerobic digester to convert it to biogas. The digester is a sealed container that provides the ideal conditions for the microorganisms to thrive and break down the organic material. During anaerobic digestion, the microorganisms consume the biomass and produce a mixture of gases, including methane and carbon dioxide. This mixture is known as biogas and can be used as a renewable energy source. The biogas can be purified and compressed as fuel for vehicles, heating, or electricity generation. The remaining material, digestate, can be used as a fertilizer. Converting biomass to biogas is an environmentally friendly way to produce renewable energy while reducing greenhouse gas emissions and waste.

8.6.2 Active Use

The combustion process in a limited oxygen supply is known as *gasification*.[19] By the combustion of biomass in a limited supply of oxygen, producer gas is obtained, a mixture of H_2, CO, and CO_2. In this process, a large quantity of biomass is collected from the local society and combusted in the gasifier. In the agricultural domain, the most common biomass is wood pellets and bagasse. In this way, net-zero emissions of CO_2 can be achieved to comply with the international greenhouse reduction norms. So, its location plays a crucial role in utilizing biomass for sustainable building. To achieve green or sustainable energy, building design, the material of the building, and the location of the building have a crucial role and must comply with the national sustainability level standards. Figure 8.15 shows the schematic diagram of a *downdraft gasifier*, in which biomass is fed from the top and gasifying medium is inserted from the throat, and the resulting producer gas can be obtained from bottom of the grate.

Gasifiers are machinery that create a gas called syngas from carbon-based resources like wood, agricultural waste, and municipal solid waste. Syngas is a fuel that can be used for a number of purposes, such as transportation, power generation, and heating. Following are a few ways gasifiers can be included into green construction ideas:

- **Heating:** Gasifiers can be used to heat green building-based structures. Syngas generated by gasifiers can be utilized in boilers or furnaces to generate heat for residential hot water or space heating. This lessens the need for fossil fuels, which have higher carbon emissions, such oil and natural gas.

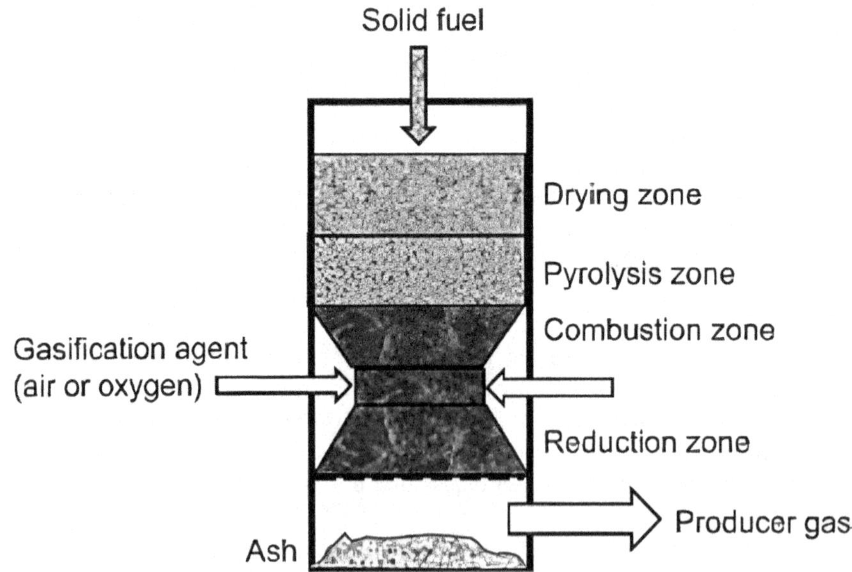

FIGURE 8.15 Different zones in a downdraft gasifier.[20]

- **Power generation:** Gasifiers can be used to produce electricity for environmentally friendly structures. An internal combustion engine or a gas turbine can produce energy using the syngas that gasifiers produce. In off-grid or isolated areas with limited access to the grid, this can be especially helpful.
- **Trash management:** Gasifiers can also be employed to handle the waste produced by green structures. By converting organic waste, such as food scraps and yard trash, into syngas by gasification, less garbage is dumped in landfills. This may lessen the impact of trash disposal on the environment and greenhouse gas emissions.
- **Transportation:** Gasifier-produced syngas can also be used as fuel for vehicles. Syngas-powered vehicles can be converted, eliminating the requirement for petrol or diesel. This can be especially helpful in places with limited access to public transport.
- **Carbon capture:** Gasifiers can also be utilized to capture and store carbon dioxide. Gasification produces carbon dioxide, which can be trapped and stored underground to reduce greenhouse gas emissions.

Overall, gasifiers can be a significant part of green building ideas because they offer a renewable energy source and can lower greenhouse gas emissions. They are versatile and essential tools for sustainable building practices because they may be utilized for heating, power generation, waste management, transportation, and carbon capture.

8.7 HYDROGEN ENERGY

Hydrogen energy can be used to heat homes, provide hot water, cook, and meet electricity needs. It must be first produced and stored safely, and then it can be transported before it is finally used. Renewable energy sources such as solar, hydroelectric, wind, and geothermal can be used to produce hydrogen. Hydrogen is a colorless, odorless, highly flammable gas that makes it compulsory to handle with care, although it is non-toxic and non-carcinogenic. Hence, to utilize it in the building, a robust and leak-proof connection is needed. The "solar–hydrogen hybrid" system is currently the most efficient among renewable energy sources. Some other methods to produce hydrogen gas include:

1. **Thermochemical processes**
   - Natural gas reforming (also called steam methane reforming or SMR)
   - Biomass gasification
   - Biomass-derived liquid reforming
   - Solar thermochemical hydrogen (STCH)
2. **Electrolytic processes**
   - Electrolysis
3. **Direct solar water splitting processes**
   - Photoelectrochemical (PEC)
   - Photobiological

4. **Biological processes**
   - Microbial biomass conversion
   - Photobiological

The natural gas reforming method is broadly implemented method worldwide. However, domestic hydrogen gas can be generated from diverse resources, including biomass, water electrolysis with electricity, and fossil fuels. Additionally, renewable energy fuels like *ethanol* will react with high-temperature steam to form hydrogen gas. From biomass fermentation, hydrogen gas can be produced at the domestic level for building use.

8.7.1 Passive Use

In passive modes, hydrogen energy can be used in buildings as a clean and renewable energy source. There are several ways hydrogen can be used to power buildings

1. **Heating:** Hydrogen can also be used for heating buildings. It can be burned in a hydrogen boiler, which produces heat that can be used to warm up buildings. Hydrogen boilers can be used as a direct replacement for natural gas boilers, with the only byproduct being water. However, it would be a costlier deal than an efficient heat pump.
2. **Cooking:** Hydrogen can be used as a cooking fuel in buildings. It can be burned in a hydrogen stove, which produces heat that can be used for cooking. Hydrogen stoves can be used as a direct replacement for natural gas stoves, with the only byproduct being water.
3. **Hydrogen vehicles:** Hydrogen gas can be directly used in an internal combustion engine to convert the chemical energy of hydrogen into mechanical power. It will be a more efficient vehicle as compared to fossil fuel-based vehicles by virtue of its calorific value. Some important physical properties of hydrogen are listed in Table 8.1 and compared to conventional fuels.

TABLE 8.1
Physical Properties of Different Fuels and their Comparison[21]

S.N.	Parameters	Hydrogen	Natural gas	Petrol-diesel	LPG
1.	Calorific value (MJ/kg)	120–142	49–54	41–44	46–50
2.	Density (kg/cm^3)	0.08	0.6	720–780	510
3.	Auto-ignition temperature (°C)	500–540	580	247–280	410–580
4.	Diffusion coefficient (cm^2/s)	0.61	0.16	0.05	0.11

8.7.2 Active Use

Hydrogen gas can be stored and utilized at multiple levels in different industries, like in the petroleum industry, fertilizer industry, heavy metal industry, and food processing industry. Apart from this it can be used in various ways.

1. **Fuel cells:** Fuel cells can be used to produce electricity. Sir William Grove invented the first fuel cells in 1838. The first practical usage of fuel cells emerged more than a century later, with Francis Thomas Bacon's discovery of the hydrogen–oxygen fuel cell in 1932. A fuel cell is an electrochemical cell that uses redox reactions to transform the chemical energy of a fuel (typically hydrogen) and an oxidizing agent (commonly oxygen) into electricity. This electricity can then be used to power buildings. Fuel cells are efficient and produce no harmful emissions, making them popular for powering buildings. The working principles of fuel cells and the electrolysis process are shown in Figure 8.16, which describes the both processes schematically.

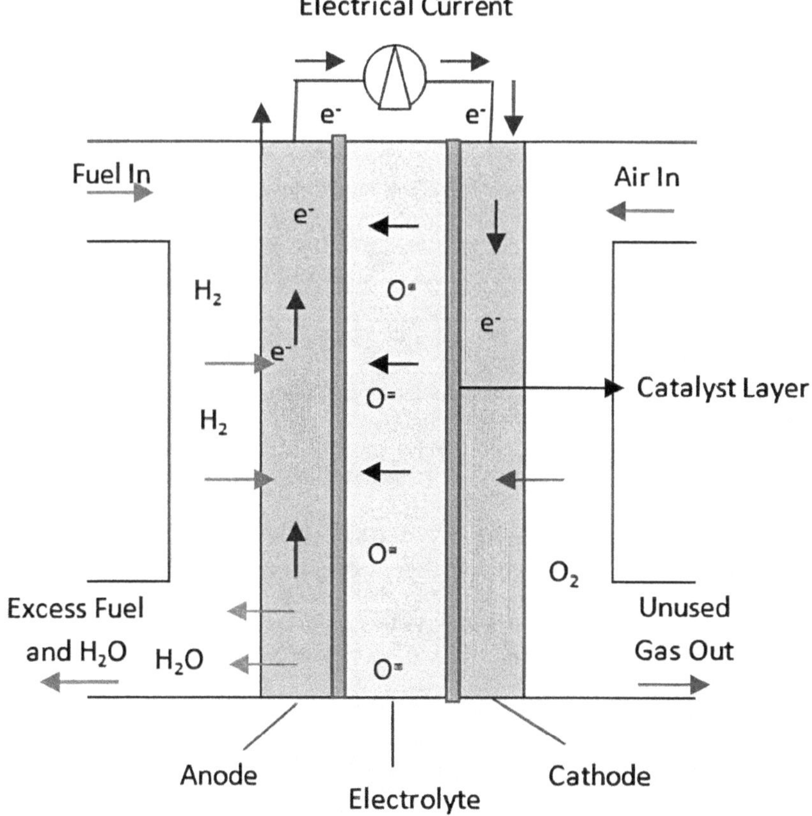

FIGURE 8.16 Schematic working diagram of a fuel cell.[22]

2. **Energy storage:** Hydrogen can be used as an energy storage medium. Excess energy from renewable sources, such as wind and solar, can produce hydrogen, which can then be stored and used when needed to power buildings.
3. **Hydrogen fuel cell vehicles:** Hydrogen-based vehicles run either directly by chemical energy of hydrogen gas or indirectly by electricity. Fuel cell-integrated vehicles convert hydrogen gas into electricity. These vehicles are different from electric vehicles that need charging of the batteries.

However, it is essential to note that some challenges are still associated with using hydrogen in buildings, such as the high cost of hydrogen production and the need for specialized infrastructure to transport and store hydrogen. With the increasing demand for clean and renewable energy sources, hydrogen is becoming a more viable option for powering buildings.

Some studies also reveal that building heating via hydrogen gas can cause air pollution to some extent compared to methane gas. Hydrogen gas requires more renewable electricity than a highly efficient heat pump. Blending hydrogen and methane gas could be another low-cost solution for building heating.

REFERENCES

1. Pan Y, Zhang L. 2020. Data-driven estimation of building energy consumption with multi-source heterogeneous data. Appl Energy. 268. doi:10.1016/J.APENERGY.2020.114965
2. Listing A, Content S. Oil revenue-economic growth nexus and greenhouse gas emissions in Africa. diasporaconnex.com [Internet]. [accessed 2023 Mar 18]. https://diasporaconnex.com/oil-revenue-economic-growth-nexus-and-greenhouse-gas-emissions-in-africa/
3. Energy efficiency–Topics—IEA. [accessed 2023 Mar 18]. www.iea.org/topics/energy-efficiency
4. Brownson JRS. 2013. Framing the sun and buildings as commons. Buildings. 3(4):659–673. doi:10.3390/BUILDINGS3040659
5. Schittich C, editor. 2003. Solar Architecture [Internet]. [accessed 2023 Mar 20]. doi:10.11129/DETAIL.9783034615198
6. Yüksek İ, Karadağ İ, Yüksek İ, Karadağ İ. 2021. Use of renewable energy in buildings. Renew Energy—Technol Appl [Internet]. [accessed 2023 Mar 24]. doi:10.5772/INTECHOPEN.93571
7. Jung S, Han S, Lee M-G, Lee H. 2022. Development of a solar tracking-based movable louver system to save lighting energy and create a comfortable light environment. Build [Internet]. [accessed 2023 Mar 21] 12(11):2017. doi:10.3390/BUILDINGS12112017
8. Back to basics: Natural ventilation and its use in different contexts. ArchDaily. [accessed 2023 Mar 21]. www.archdaily.com/963706/back-to-basics-natural-ventilation-and-its-use-in-different-contexts
9. Kumar Chauhan V, Kumar Shukla S. 2022. Analytical and experimental study of performance of Pyrex glass Q-dot based passive solar still glass evaporator. Therm Sci Eng Prog. 34:101387. doi:10.1016/J.TSEP.2022.101387
10. Sakınç E. An Approach to the evaluation of solar. Google Scholar. [accessed 2023 Mar 20]. https://scholar.google.com/scholar?hl=en&safe=off&as_q=Sakınç E. An Approach to the Evaluation of Solar Powered Factor Systems in Architecture in the Context of Sustainability [thesis]. İstanbul Yıldız Technical University; 2006

11. Numerical and experimental study of solar water heater in heating a space. [accessed 2023 Mar 21]. www.researchgate.net/publication/339029483_Numerical_and_Experimental_study_of_solar_Water_Heater_in_Heating_a_Space/figures?lo=1
12. Dondariya C, Porwal D, Awasthi A, Shukla AK, Sudhakar K, Murali MM, Bhimte A. 2018. Performance simulation of grid-connected rooftop solar PV system for small households: A case study of Ujjain, India. Energy Reports. 4:546–553. doi:10.1016/J.EGYR.2018.08.002
13. Chandra Mouli GR, Bauer P, Zeman M. 2016. System design for a solar powered electric vehicle charging station for workplaces. Appl Energy. 168:434–443. doi:10.1016/J.APENERGY.2016.01.110
14. Wang Z. 2019. Chapter 1_Introduction_design of solar thermal power plants. Des Sol Therm Power Plants [Internet]. [accessed 2023 May 22]:1–46. doi:10.1016/B978-0-12-815613-1.00001-8
15. Wind catchers—looking to the ancient past for new clean energy sources. Green Business Bureau. [accessed 2023 Mar 21]. https://greenbusinessbureau.com/green-practices/energy/wind-catchers-looking-to-the-ancient-past-for-new-clean-energy-sources/
16. Brodowicz D, Pospieszny P, Grzymala Z. 2015. Eco-cities: Challenges, trends and solutions. CeDeWu [Internet]. [accessed 2024 Jan 1]. doi:ISBN: 978-83-7941-216-7
17. Baker W, Besjak C, McElhatten B, Li X. 2014. Pearl river tower: Design integration towards sustainability. Struct Congr 2014 [Internet]. [accessed 2024 Jan 1]:747–757. doi:10.1061/9780784413357.067
18. Vassilev SV., Vassileva CG, Vassilev VS. 2015. Advantages and disadvantages of composition and properties of biomass in comparison with coal: An overview. Fuel. 158:330–350. doi:10.1016/J.FUEL.2015.05.050
19. Kumar Chauhan V, Kumar Shukla S, Sharma D, Sharma S, Meena G, Lal Jangir G. 2023. Study on effect of moisture content on the composition of producer gas. Mater Today Proc. 76:138–145. doi:10.1016/J.MATPR.2022.10.139
20. Pang S. 2016. Fuel flexible gas production: Biomass, coal and bio-solid wastes. Fuel Flex Energy Gener Solid, Liq Gaseous Fuels.:241–269. doi:10.1016/B978-1-78242-378-2.00009-2
21. Alternative fuels data center: Alternative fuels and advanced vehicles. [accessed 2023 Jun 22]. https://afdc.energy.gov/fuels/
22. Dharmalingam S, Kugarajah V, Sugumar M. 2019. Membranes for microbial fuel cells. Biomass, Biofuels, Biochem Microb Electrochem Technol Sustain Platf Fuels, Chem Remediat:143–194. doi:10.1016/B978-0-444-64052-9.00007-8
23. Gaur N, Sharma S, Yadav N. 2023. Environmental pollution. Green Chem Approaches to Environ Sustain Status, Challenges Prospect:23–41. doi:10.1016/B978-0-443-18959-3.00010-0
24. Chel A, Kaushik G. 2018. Renewable energy technologies for sustainable development of energy efficient building. Alexandria Eng J: 57(2):655–669. doi:10.1016/J.AEJ.2017.02.027
25. Buildings are the foundation of our energy-efficient future | World Economic Forum. [accessed 2024 May 3]. https://www.weforum.org/agenda/2021/02/why-the-buildings-of-the-future-are-key-to-an-efficient-energy-ecosystem/
26. Emmerich SJ, Dols WS, Axley JW. Natural ventilation review and plan for design and analysis tools. doi:10.6028/NIST.IR.6781
27. Abdul-Rahman T, Roy P, Bliss ZSB, Mohammad A, Corriero AC, Patel NT, Wireko AA, Shaikh R, Faith OE, Arevalo-Rios ECE, et al. 2024. The impact of air quality on cardiovascular health: A state of the art review. Curr Probl Cardiol [Internet]. [accessed 2024 May 3] 49(2). doi:10.1016/J.CPCARDIOL.2023.102174

28. KAUST Solar Center officially inaugurated. [accessed 2024 May 3]. https://www.kaust.edu.sa/en/news/solar-center-officially-inaugurated
29. Combined Heat and Power Facility | Arizona State University. [accessed 2024 May 3]. https://cfo.asu.edu/combined-heat-power-facility
30. De Aar Solar Park - Mainstream Renewable Power. [accessed 2024 May 3]. https://www.mainstreamrp.com/markets-projects/africa/south-africa/de-aar-solar-farm/

9 Emerging Technologies for HVAC System Efficiency

D. B. Jani, Manish Mishra, and P. K. Sahoo

9.1 INTRODUCTION

In areas especially with typical hot and humid climatic conditions, efficient air conditioning systems are essential for preserving optimum indoor thermal comfort. One of the primary criteria for indoor air conditioning is to maintain indoor humidity at an acceptable level since it leads to better thermal comfort, especially in times of extreme weather with very high moisture content for typical transient environmental changes. In continental places with exceptionally high humidity in parallel with severe temperatures, making indoor cooling system design is essential to handle humidity effectively in addition to good control over temperature. Air conditioning systems that are designed poorly can frequently lead to high internal humidity, health issues, decreased thermal comfort and greater energy use. Office job performance may suffer by up to 8–10% due to poor indoor air quality, which is typically insufficient to provide the necessary ventilation for better hygienic indoor conditions. Furthermore, airborne pollutants that are still present in the indoor re-circulating air are dispersed widely and have led to unhygienic indoor conditions. This explains why traditional cooling methods are often not preferred by building designers and HVAC professionals. In vapour compression-based traditional cooling systems, the same single coil action in the same unit simultaneously controls both the temperature and moisture level needed to produce the requisite indoor thermal comfort, which leads to inadequate control of either parameter. This built-in restriction in the system design achieves the objective of lowering the dry bulb temperature, but the relative humidity is too high to be pleasant [1–3]. Mould can grow if humidity is not controlled to the acceptable level, and so employing vapour compression air conditioning systems to achieve the finest indoor humidity level by use of single cooling coil having dual functions of temperature and humidity control resultsinpoor overall system performance. Desiccant cooling systems are especially helpful when the latent load is much greater than the sensible load (either larger occupancy or area). Desiccants can also assist in removing contaminants from indoor air streams during air conditioning to improve the indoor air quality [4–8]. Low-grade thermal energy available from renewable solar energy or waste industrial heat input may be used to replenish the desiccant. Regeneration energy is equal to the heat needed to raise the

DOI: 10.1201/9781003415695-9

desiccant's temperature to a point where its surface vapour pressure is higher than the surrounding air, as well as the heat needed to evaporate the adsorbed moisture that is present in the desiccant. Energy is also needed for the desiccant's water to be desorbed from it and exhausted to the outdoor atmosphere [9, 10].

Due to climate change and growing environmental pollution over the past few decades, air conditioning systems have become much more prevalent, especially in commercial buildings. Yet research suggests that theyhave detrimental effects on the rising electricity needs for cooling the built environment and creating cooling in buildings to maintain the necessary indoor thermal comfort. Several academics around the world have been working hard for many years to find alternatives to standard cooling systems, which require a lot of electrical energy to operate and run. Desiccant cooling is one such promising alternativeto the conventional way for producing a built indoor environment, although research into its application is still in its infancy and has to be examined more carefully [11–14].

One of the main problems facing building cooling in the modern period is the rising energy consumption of cooling applications. The vast majority of air conditioners in use today are vapour compression refrigeration-based versions, which operate using a tremendous amount of conventional energy. The use of this technology, which increases Chlorofluorocarbon (CFC) levels, is destroying the ozone layer. These applications have exacerbated environmental issues like global warming since they require greater electrical power for operation. As the need for space cooling applications increases, scientists are looking at different AC technologies to deal with the aforementioned issues [15, 16]. The fundamental principle of working air conditioners, such as the conventional vapour compression refrigeration technology, consumes a lot of electrical power for running the compressor because of simultaneous operation for cooling and dehumidification. So, from the perspective of energy conservation and significantly reducing major greenhouse gas emissions caused by the cooling of buildings, clean, efficient, sustainable technologies are chosen to ensure adequate conditions for users' comfort. One of these methods, desiccant cooling, may be able to resolve the many problems mentioned earlier [17–21]. Innovative dehumidification and cooling methods based on desiccants can be used to effectively control indoor air moisture while using comparatively less energy than the conventional HVAC systems in buildings. The desiccant material used in dehumidifiers of desiccant-based cooling systems can produce hot, dry air that can be used for the drying process in industrial applications like pharmaceutical coating, packaging, and dyeing, as well as in both commercial and residential buildings for airconditioning to reduce the moisture content in built environments and provide the required human comfort [22–24].

The best way to use desiccant-based dehumidification and cooling systems is for humid environments for effective moisture control, especially in areas having low sensible heat ratios. It saves the high cost of electrical energy by use of primary thermal energy sources like industrial waste heat, renewable solar energy, and compressed natural gas. By creating a dry environment, it avoids microbial growth in supply and return ducts. Desiccant-based dehumidification and cooling systems are successfully installed in buildings witheither large area for cooling or either large human occupancy like supermarket, restaurants, swimming pools, motels, hospitals, health clubs, hospitals and nursing homes, office buildings, etc. In supermarkets, the use of desiccant-based cooling systems saves about 38% on cooling energy due to large floor space area as compared to vapour-compression-based conventional cooling. In hotels and motels, the use of desiccant-based dehumidification and cooling can eliminate mould and mildew problems as a result of a moist environment. This is expected to be a major destination of this innovative cooling technology.

The performance, price, and dependability of desiccant-based dehumidification and cooling systems have significantly improved over the course of more than four decades of constant research and innovative development in the HVAC sector. For a few specialised uses as mentioned earlier, including in supermarkets, they have become more competitive among HVAC market. Desiccant cooling systems have been studied extensively over the past few decades, and it has become clear that these systems have a great deal of potential to compete with and complement traditional, electrically powered vapour compression systems.

As compared to solid desiccant cooling systems, liquid desiccant-based dehumidification and cooling systems can possibly offergreater design benefits and may be deployed in structures having distinct sites for the supplement ducts. The bigger size and lower capacity of liquid desiccant systems area main drawback forinstallingthemin limited space. Depending on the type of liquid desiccant used in the dehumidification and cooling system, corrosion or carry-over may occur, and this is the other limitation of this system. So, proper design should be incorporated while designing liquid desiccant-based dehumidification and cooling systems. The most recently designed and developed liquid desiccant-based dehumidification and cooling systems can have superior performance as compared with other conventional cooling systems. Commercial development of desiccant-integrated solid or liquid systems is underway, and several packaged systems should be available in commercial HVAC market in the coming days.

The HVAC industry has some promising options with desiccant-integrated hybrid cooling systems that can make integration of commercially viable desiccants for dehumidification and traditional vapour compression for cooling. Buildings with commercial and institutional uses have successfully implemented a number of integrated hybrid desiccant cooling systems. The HVAC industry has successfully used integrated cooling systems in supermarkets as well. The desiccant-based dehumidification and cooling systems have high capital costs that need to be reduced for them to seriously enter the HVAC industry. A higher rate of market acceptance and penetration will also be facilitated by increasing performance and dependability while lowering size.

The employment of solid-based desiccant-based rotary dehumidifiers is primarily focused on in areview of desiccant-based dehumidification and cooling systems. The investigations included a study of various combinations of integrating various cooling systems coupled with rotating desiccant dehumidifiers in order to be employed in hot and humid environments. The investigation shows that solid-based desiccant-based dehumidification and cooling devices have a great deal of potential for use in hot, humid areas. In terms of output temperature, solar-powered desiccant cooling systems are more efficient than traditional cooling types, but any advancement above the typical configuration, such as the inclusion of open/closed types, leads to still greater innovation. To provide the conditions for indoor comfort, the air conditioner must effectively manage the sensible and latent loads of the building. By choosing the right air flow rate and reactivation heater to effectively and efficiently desorb the desiccant while maintaining the right cooling rate, the standard mechanical vapour compression system normally manages the interior moisture level [25–27].

The main advantages of desiccant cooling are as follows:

1. Air and water are the only required operating fluids. The ozone layer is not impacted since fluorocarbons are not required (in the case of evaporative sensible post coolers).
2. Much potential exists for energy savings and reduced reliance on fossil fuels. Electrical systems require less than 20–28% of the energy required for conventional refrigeration systems.
3. A number of sources, including the sun, leftover heat such asindustrial waste heat (which is otherwise exhausted to open atmosphere), and natural gas, can produce thermal energy.
4. The desiccants enhance the quality of the indoor air by helping to maintain comfort criteria by removing particulates and bioaerosols from the air, and removing chemical pollutants from the air.
5. Desiccant systems operate at pressures that are close to air pressure, which makes leakage-proof system design, building, and maintenance easier.
6. Desiccant cooling systems can eliminate the need for a separate furnace orany other auxiliary heating source to heat rooms in the winter as it neglects any post re-heating of conditioned air due to over cooling in comparison to conventional VCR system.

The heat coefficient of performance(COP) is a common metric for comparing system performance. Comparisons need to account for the parasitic electrical energy requirements, which could negate gains in the thermal COP. Also, depending on the systems addressed in theresearch, the definition of the thermal COP varies. Some studies looked into how to employ the right kind of desiccant for a particular application in order to supply the heat energy required to fully reactivate the desiccant. Others [28–34] supply the thermal energy required to reactivate the dehumidifier from the condensate waste heat or the reactivation air side of the supplied room air.

Desiccant material-assisted cutting-edge cooling systems helpin discovering a clever, affordable, and eco-friendly replacement for standard, traditional air conditioning systems to reduce wasteful gas emissions and primary energy consumption [35–37]. This chapterdiscusses the significance and applications of desiccant cooling technology. The key environmental problems and main obstacles related to conventional energy requirements, as well as environmental problems related to poor indoor air quality, have also been covered in the second section of this chapter. Additionally, this covers numerous cooling and dehumidification systems that use cutting-edge solid and liquid desiccant cooling in a variety of residential and commercial applications to operate in typical hot, humid situations more effectively. How theperformance of the system obtained can be improved further when cheaper options like freely available solar energy or industrial waste heat areused for desiccant regeneration are also discussed [38–40]. Potential of this hybrid cooling can be demonstrated by this chapter, to implement this green cooling air conditioning successfully in typical transient atmospheric conditions for indoor comfort cooling.

9.2 SYSTEM DESCRIPTION

In a liquid desiccant cooling system, lithium or calcium chloride is used as a desiccant to eliminate humidity from the process supply air during dehumidification of the moist air [41, 42]. In the construction of a liquid desiccant cooling system, two towers, namely an absorber tower and regenerator tower, are provided to carry out the dehumidification of moist room air and the desorption of liquid desiccant, respectively (Figure 9.1). In the absorber tower, a downwards flow of strong desiccant solution absorbs moisture from the humid air travelling upward during contact among them. After absorbing moisture, weak solution collected at the bottom of the absorber tower is pumped back to the regenerator tower. Meanwhile, dried air is conveyed for further processing by transferring it out of the absorber tower at the top. Packed-bed construction may be used to increase the contact between moist air and the desiccant materials. In the regenerator tower, the weak desiccant solution received from the absorber tower is heated first in the heater to achieve the desired reactivation temperature according to the type of desiccant material used in the system to flow from the top of the tower andcontact the upward flow of hot reactivation air. Hot reactivation air collects the moisture from the weak desiccant material to convert into strong desiccant solution, which is collected at the bottom of the regeneration tower. Exhaust regeneration air collected at the top of the regeneration tower to expel it to the open outdoor atmosphere in form of the exhaust. This strong desiccant solution is again made to flow towards the absorber tower to complete next cycle this way. An intermediate cooler is provided prior to the absorber column to convey the strong solution to the top of the tower to pre-cool the strong desiccant solution for making better dehumidification of the moist process sir. Strong and weak desiccant solutions pass via a heat exchanger to pre-cool the strong solution and pre-heat the weak solution by making necessary heat exchange to ameliorate the performance of the system [43, 44]. In this way, liquid desiccant materials like LiCl, $CaCl_2$, LiBr, etc.

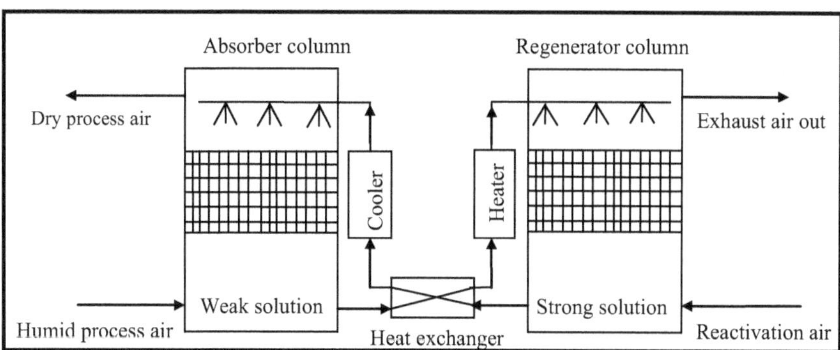

FIGURE 9.1 Schematic arrangement of a liquid desiccant cooling system.

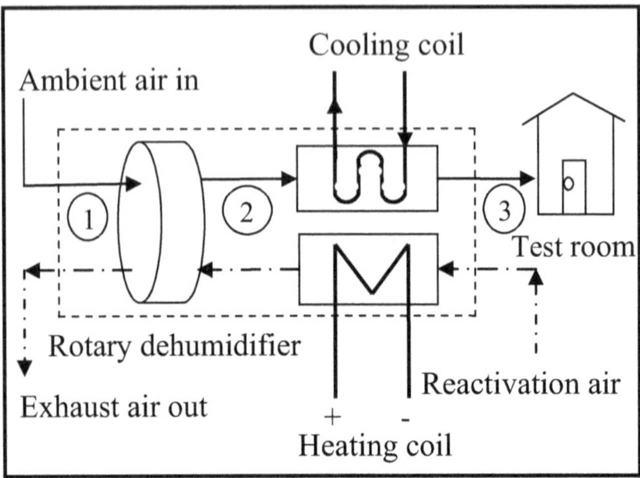

FIGURE 9.2 Schematic arrangement of a solid desiccant dehumidification and cooling system.

continuously circulate between the absorber tower and regenerator tower during the cycle. The equilibrium temperature of liquid desiccants is determined by the partial pressure of the water vapour in the humid moist air.

The traditional vapour compression-based air conditioning systems can be replaced potentially by novel desiccant-based cooling systems, which are primarily heat-driven dehumidification and air conditioning systems. A desiccant cooling system works by using a rotating dehumidifier commonly known as desiccant wheel, which is primarily used to remove moisture from the air. A sensible heat exchanger partially cools this generated dry air before supplying it to the indoor space by further cooling it as per design conditions. A room is filled with the resulting cool air. The system can be run in a closed cycle for ventilation or recirculation, or more often in an open cycle. To replenish the desiccant, a heat source is required. Regeneration only requires low-grade heat at a temperature of 60°C to 95°C, thus using renewable energies like freely available solar energy or industrial waste heat. A regenerator is required to reactivate the dehumidifier to make it continually work in the cycle without a loss of effectiveness. This system is simple in construction, and its thermal coefficient of performance is usually acceptable. The schematic arrangement for a solid desiccant cooling system is shown in Figure 9.2 [45, 46].

9.3 EMERGING TECHNOLOGIES TO ENHANCE COOLING POTENTIAL

The schematic arrangement of solar-powered desiccant cooling with a ground-source heat pump looped system is shown in Figure 9.3. The process air section and the regeneration air section are the two primary working sections or divisions of the cooling system. The humidity of the moist process air decreases while passing through the desiccant dehumidifier before it enters the cooling apparatus in the process

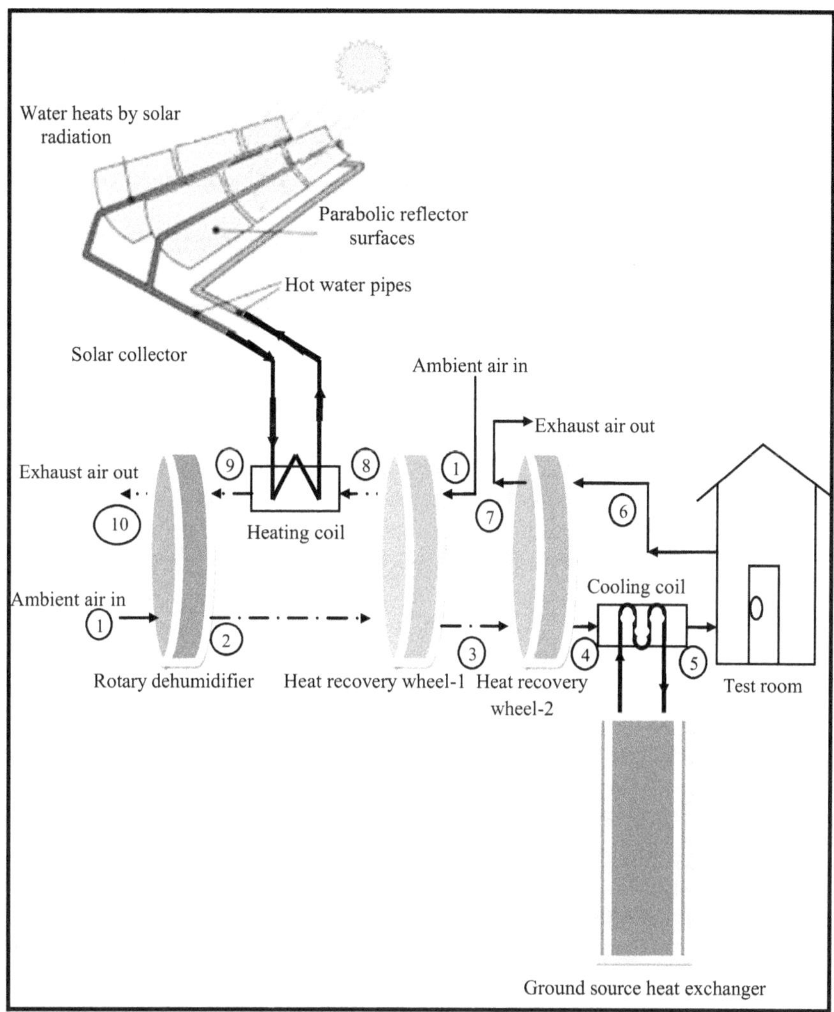

FIGURE 9.3 Solar-powered desiccant cooling with ground-source heat pump looped arrangement.

section. Ground-source heat pump heat exchangers, sensible cooling coils, and air-to-air heat recovery wheels are examples of equipment provided for the cooling section of the system. The ambient air is warmed in the regeneration portion to reach the temperature needed to regenerate the desiccant dehumidifier. The energy needed for the regeneration process is provided by a solar loop and an auxiliary heater working together. The ambient air enters the inlet section of the desiccant dehumidifier in the process air section (state point 1) to lower the humidity ratio and raise the temperature. After entering the first heat recovery wheel (HRW), the process air (state point 2) loses heat. The heat recovery wheel takes into account just sensible heat transfer; hence, the humidity ratio of air streams is unaffected by this component. State point 3 is where the process air enters the second heat recovery wheel, when its temperature drops further. The ground cooling coil, which is fed by the ground-source heat exchanger with cold water, cools the process air (state point 4) as it travels through it. This conditioned air (state point 5) enters the conditioned area. When ambient air (state point 1) enters the first heat recovery wheel in the regeneration area, it warms up due to a continuous heat recovery process. The solar loop heat exchanger is the next stop for the regeneration air (state point 8), where its temperature rises. The source and load sides of the solar loop are shown in the illustration. Water flows through collectors on the source side, soaking up any solar energy that is present [47, 48]. To raise the temperature of the regeneration air on the load side, hot water is transferred from the storage tank towards the heat exchanger. The auxiliary heater is then filled with the regeneration air (point 9), which has now reached the necessary regeneration temperature. State point 9 is where regeneration air enters the reactivation side of the desiccant dehumidifier; after that, it lowers the temperature and raises the humidity ratio by desorption. The regeneration air is discarded into the surrounding atmosphere as the process comes to a close (state point 10).

The system's cooling capacity has beenassessed in hot and humid climates anddemonstrates high feasibility. Earlier [15,17,19] findings show that this system can successfully provide comfort conditions in warm, humid climates, making it a fascinating and useful replacement for the traditional cooling systems used in these areas. The effect of the return air ratio on the amount of thermal comfort offered, system performance, energy delivered by the auxiliary heater, and SF has been researched. In simulations, the desiccant wheel is simultaneously employed before and after the ground-source heat exchanger. The effect of employing the ground-source heat exchanger before and after the desiccant wheelon the system's hourly behaviour, system energy use, COP, and solar fractionwas evaluated. In order to demonstrate the introduced configurations' viability with regard to three reference systems, an economic study was also conducted and validated. Aproposed desiccant system with no solar loop and no ground-source heat pump system, a proposed solar-powered desiccant system with no ground-source heat pump system, and a mechanical vapour compression system are the respective reference systems.

The primary findings on the system's potential for use in hot, humid climates show that even with low-grade regeneration temperatures, this arrangement may successfully provide thermal comfort in those areas. The value of the return air ratio from the conditioned zone has a big impact on the system's hourly performance. In particular, raising the return air ratio raises the system's performance and solar fraction and lowers the amount of energy needed for regeneration. Using the ground-source heat pump system before the dehumidifier and sensible cooler reduces the maximum regeneration temperature needed, raising the system's performance and solar fraction and lowering the amount of energy supplied by the auxiliary heater. The cost payback period varies between 5 and 7 years for the three scenarios in which the ground-source heat pump system is not used, is used after the desiccant dehumidifier, and is utilised both before and after the desiccant dehumidifier, using the mechanical system as the reference. The desiccant system's long-term performance with the ground-source heat pump integrated (just as a pre-cooler or as a simultaneous pre- and post-cooler) demonstrates the viability of this strategy.

9.4 COMPARISON BETWEEN DESICCANT COOLING AND TRADITIONAL COOLING

Efficient control can be obtained over the amount of water vapour or moisture in conditioned air by either condensing the water vapour or utilising suitable absorbents, like those found in desiccant cooling systems. Unlike desiccant systems, which first only dehumidify the air before cooling it, conventional air conditioners can cool and dehumidify the air at the same time [49]. Moreover, because it has efficient control over indoor temperature and humidity, a desiccant system can be utilised in conjunction with an evaporative cooling system. Desiccant cooling systems have previously been employed in the agricultural and industrial sectors, such as textile mills and better drying and storage of harvested agricultural crop, as well as for efficient humidity management and to carry out different drying operations for fruits and vegetables. Yet the development of desiccant cooling systems as an efficient approach to control humidity has been made possible by the energy crisis and the need to design more environmentally friendly solutions. The main distinctions between desiccant cooling systems and traditional air conditioners that use vapour compression are briefly outlined in Table 9.1.

Figure 9.4 compares the distribution of the annual COP for liquid and solid desiccant systems based on the total working hours of each system during different months of the year. The results show that the performance obtained in case of the liquid desiccant cooling system is subpar when compared to that of the solid desiccant cooling system [50, 51]. It was also claimed that solid desiccant devices have the advantage of requiring less water and energy. Yet, according to the numerous researchers in this field, the method's limited accuracy and better usefulness within a constrained set of parameters for typical climatic zones renders the prediction models unsuitable for a variety of climatic conditions and circumstances.

Table 9.2 depicts how the moisture removal capacity loss while working as a dehumidification system can be compared within the various types of desiccants used in aliquid dehumidifier having different chemical compositions and reactivation rates in the same type of system having capacity losses for 60% relative humidity after approximately 5000 hrs. cycle time [52–54].

TABLE 9.1
Comparison between VCR-Based Conventional Cooling and Desiccant Cooling

Parameter	VCR-based traditional cooling	Desiccant cooling
Operational cooling cost	High	Saves around 30–42%
Indoor air quality	Average	High
Effect on environment	Harmful	Eco-friendly
Energy source	High-grade electricity	Low-grade energy like industrial waste heat or renewable solar power
Moisture removal capacity	Average	High

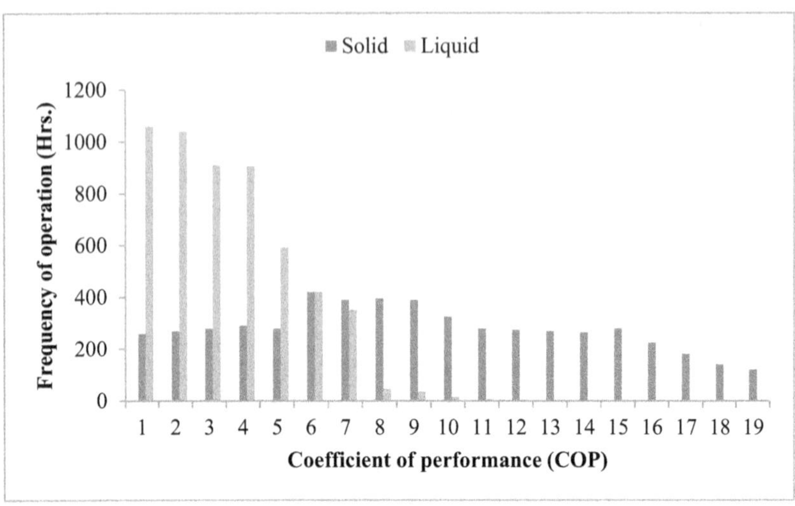

FIGURE 9.4 Performance comparison between solid and liquid desiccant cooling.

TABLE 9.2
Comparison of Different Desiccant Materials Based on Capacity Losses

Desiccant material	Capacity losses (%)
Alumina	41
Molecular sieve	30
LCIX type 1M desiccant	12
Alumina 13X	36
Silica	63
Alumina silica 13X	50

The substantial use of desiccant wheels in industrial air conditioning units has already begun to appear in the HVAC market recently using sensible cooling methods. Especially the countries having hot and humid climates in particular have already begun employing more cost-effective and environmentally friendly cooling methods as opposed to the vapour-compression-based traditional cooling systems. A comparison studyand economic analysis of a desiccant-assisted dehumidification and cooling system and a vapour compression system reveals that the performance of the hybrid desiccant cooling system was ameliorated by using solar thermal energy and ventilated air for cooling. This innovative technology has been successfully applied in specialised sectors as a consequence of innovation and developments in the previous few years. To reach the general market for air conditioning, however, more efficiency and reliability improvements, size and cost reductions, and increased building industry technological adoption are all necessary [55–57]. Given the potential of the desiccant cooling method, investments in more research and development of innovative desiccant materials thatcan get successfully regenerate at lower temperature near-ambient conditions, components with simpler designs, and sustainable cooling systems are necessary and justifiable.

9.5 OPPORTUNITIES AND FUTURE SCOPE

Desiccant-assisted dehumidification and cooling air conditioning systems are particularly appealing for areas with high latent loads and low sensible heat ratios. The desiccant subsystem may control the latent load when used in conjunction with a traditional vapour compression refrigeration cycle, and the vapour compression refrigeration cycle can control the sensible load. Desiccant substances are either liquid (absorbers) or solid sorbents (rotary desiccant wheel), both of which need to be heated again (regeneration by renewable energy sources). Desiccant cooling has the potential to boost system efficiency by decreasing the size (cooling capacity) of conventional vapour compression equipment and raising evaporation temperatures.

In order to predict the required cooling performance of the desiccant-based dehumidification and cooling systems, the use of advanced desiccant materials requires further research and innovation which aims to comprehend how material alterations and surface phenomena affect the sorption characteristics. The technological goal is to find a next-generation, economic, and low-cost advanced desiccant material that can be utilised to regenerate the water vapour sorption activity of the desiccant using solar radiation or heat from another low-cost primary waste energy source. This is a logical continuation of the basic research and innovation. Past reviews on desiccant materials and their relationship to low temperature regeneration have been employed thus far, and opportunities with its future scope are provided in this section. Naturally, a solid desiccant material can adsorb moisture to eliminate the latent load. For commercially used thermally regenerated desiccant cooling systems, when comparing with traditional VCR cooling, the coefficient of performance and initial investment cost are the primary concerns. By carrying out thermodynamic analysis, it is found that the adsorption characteristics of sorption materials are the prime concern which can beaffected critically by supply moisture and required reactivation temperature.

The ultimate aim is identifying innovative sorption materials with optimal performance by use of research and development. The reactivation temperature range used in different applications of desiccant-based dehumidification and cooling require appropriate energy sources to ameliorate the performance of the system. The cost, performance, and life cycle have a direct relationship with the enhanced reliability and sustainability of desired sorption properties for use in various applications of desiccant cooling. This innovative cooling technology not only reduces the running cost but also lowers the demand of electrical utilities to add peak-load generating capacity. The selection of proper desiccant material type mainly depends on a favourable performance-to-cost ratio. Widely available different varieties of desiccant materials can be certified, refined, and modified according to isotherm shape, rate of sorption, and chemical stability. The alteration in production methods with innovative desiccant materials could have a profound impact on substantially lowering the desiccant wheel manufacturing cost and weight. Model analysis for predicting the achievable dehumidification and cooling performance of different desiccant

TABLE 9.3
Comparison of Different Desiccant Materials Based on Regeneration Temperature

Type of solid desiccant material	Regeneration temperature(°C)
Activated alumina	120–260
Zeolite molecular sieve	175–315
Super polymer	45–75
SAPO-34/FAPO-34	50–70
Silica gel	60–100
MOF-801	65–95
Fluorite	180–200
Alumina gel (H-151)	180–440

materials have been documented extensively over many years. This is due to the fact that the moisture sorption characteristics of any desiccant materials play a crucial role in performance evaluation of the overall system.

The desiccant regeneration (Table 9.3) using freely available renewable solar energy can increase system economics by reducing operating expenses [58–61]. Particularly in India, where there is an abundance of solar radiation, these systems are thought to have a better applicability than the other major renewable energy sources. Due to the usage of solar energy, which is necessary to increase the system's cooling capacity in various HVAC applications, often in subtropical environments (almost in range 26–37°C DBT, 64–84% RH), the system's performance is improved. Like most revolutionary breakthroughs, solar power-based renewable cooling technology may have a high initial cost, but over time, it can result in considerable savings. In the past, the regeneration process might have been finished with an electric heater. On the other hand, using low-grade energy is thought to be a practical and helpful replacement. One example is the recovery of exhaust heat from various process industries to produce warm or hot water. Additional examples are waste process heat from industrial processes, which is mostly available in the process, and publicly accessible, renewable, enormous amounts of sun radiation. For instance, the regenerator and dehumidifier in a liquid desiccant-based dehumidification and indoor air cooling system might have identical schematic arrangements as previously demonstrated, but their internal flow processes might be in conflict. Before entering the regenerator, the diluted liquid desiccant solution warms up there. It is crucial to use intercooler to transfer heat from hot fluid to cold fluid before travelling between the absorber and regenerator column [62–63]. If the main heat source is unable to provide enough heat for regeneration, an ancillary heater is also necessary. Convective-type renewable solar collector regenerators can be further divided into those that use liquid or air as their operating fluid. The solar collector regenerator that uses air is typically the most popular because it is currently proven to be the most effective. In prior articles, diversified renewable energy sources to supply desiccant reactivation energy have been mentioned, including solar energy.

A variety of solar collectors can be used all year long to catch solar radiation, which is more prevalent in tropical nations. A desiccant-based hybrid cooling system that uses direct sunlight for reactivation heat and a solid desiccant-assisted vapour compression has been carefully researched. By adopting double glazing or effective shading techniques for desiccant dehumidification, solar radiation exposure to the large surface area of a building can be decreased. The use of photovoltaic panel with plants on a rooftop below the panel makes a symbiosis between a green roof and renewable energy, also called a bio-solar roof. Depending on the concentration ratio and many operating factors like the dry bulb temperatures that can be reached enhanced performance of the system, numerous solar radiation collection devices are used to collect reactivation energy. There still has to be more in-depth research on how to best optimise the many system settings for a changing environment in terms of variable outdoor temperature and humidity. The consumption of CFC-based refrigerants, which are predominantly employed in conventional HVAC systems andalso majorly contribute to global warming, can be decreased with desiccant-based green air conditioning. By installing natural ventilation systems in buildings and using double glazing to limit maximum solar heat received through the use of efficient shading mechanisms, energy consumption can be greatly lowered with more contemporary passive desiccant cooling techniques. Green roofing is a more contemporary technique for providing effective roof insulation. Passive cooling has been shown to reduce energy consumption in residential buildings by up to 28%. According to the previously mentioned literature review, numerous studies have been conducted to improve the functionality of desiccant-assisted cooling technology, which is very efficient and reasonably priced to operate hybrid cooling by comparatively cheaper solar-powered renewable heat sources for desiccant reactivation and has a reliable desiccant system. A recent study reviews research developments in parametric optimisation, system performance improvement, VCR integration, and comparison of solid and liquid desiccant systems. Techniques for evaluation are also covered. Despite rigorous study conducted earlier by many investigators in the field of energy utilisation in desiccant regeneration, cooling capability, and enhancement in overall system performance, further investigations are still necessary to ameliorate the scope of desiccant cooling in modern HVAC by means of energy savings and cost reductions.

9.6 CONCLUSIONS

Desiccant-assisted dehumidification and cooling is a cutting-edge space cooling technique for the built environment and a promising replacement for conventional vapour compression-based cooling, particularly in hot and humid climates, for effectively controlling indoor moisture. The desiccant-based hybrid cooling systems have a high potential to showenergetically better performance than the conventional cooling and be environment friendly. Moreover, they are costeffective for indoor cooling in regions where the use of thermal energy is more economical than the high-grade electrical power or freely available thermal energy as an option, that is, solar energy or industrial waste heat. Thus, rising electrical energy prices, environmental concerns, and regulatory requirements should all have an impact on future desiccant cooling needs and the desire for desiccant material that regenerates close to ambient conditions.

REFERENCES

[1] P.J. Banks. 1972. Coupled equilibrium heat and single adsorbate transfer in fluid flow through porous media—I, characteristic potentials and specific capacity ratios. Chem Eng Sci. 27:1143–55. https://doi.org/10.1016/0009–2509(72)80025–3.

[2] W.M. Worek, Z. Lavan. 1982. Performance of a cross-cooled desiccant dehumidif ier prototype. J Solar Energy Eng. 104:187–96. https://doi.org/doi.org/10.1115/1.3266301.

[3] S. Jain, P.L. Dhar. 1995. Evaluation of solid-desiccant-based evaporative cooling cycles for typical hot and humid climates. Int J Ref. 18:287–296. https://doi.org/10.1016/0140-7007(95)00016-5.

[4] D.G. Waugaman, A. Kini, C.F. Kettleborough. 1993. A review of desiccant cooling systems. J Energy Resour Tech. 115(1):1–8. https://doi.org/10.1115/1.2905965.

[5] B.S. Davanagere, S.A. Sherif, D.Y. Goswami. 1999. A feasibility study of solar desiccant air conditioning system- Part I: Psychrometrics and analysis of the conditioned zone. Int J Energy Res. 23:7–21. https://doi.org/10.1002/(SICI)1099–114X(199901)23:1<7:: AID-ER439>3.0.CO; 2-U.

[6] H.M. Henning, T. Erpenbeck, C. Hindenburg, I.S. Santamaria. 2001. The potential of solar energy use in desiccant cooling cycles. IntJ Refrig. 24(3):220–229. https://doi.org/10.1016/S0140-7007(00)00024-4.

[7] S.P. Halliday, C.B. Beggs, P.A. Sleigh. 2002. The use of solar desiccant cooling in the UK: a feasibility study. ApplTherm Eng. 22(12):1327–1338. https://doi.org/10.1016/S1359-4311(02)00052-2.

[8] M. Kanoğlu, M.O. Çarpınlıoğlu, M. Yıldırım, 2004. Energy and exergy analyses of an experimental open-cycle desiccant cooling system. Appl Therm Eng. 24(5–6):919–932. https://doi.org/10.1016/j.applthermaleng.2003.10.003.

[9] J.R. Camargo, C.D. Ebinuma, 2005. An evaporative and desiccant cooling system for air conditioning in humid climates. J Braz Soc.27: 243–247. https://doi.org/10.1590/S1678-58782005000300005.

[10] K. Daou, R.Z. Wang, Z.Z. Xia. 2006. Desiccant cooling air conditioning: a review. Renew Sustain Energy Rev. 10:55–77. https://doi.org/10.1016/j.rser.2004.09.010.

[11] C.X. Jia, Y.J. Dai, J.Y. Wu, R.Z. Wang. 2007. Use of compound desiccant to develop high performance desiccant cooling system. Int J Refrig.30(2):345–353. https://doi.org/10.1016/j.ijrefrig.2006.04.001.

[12] T.S. Ge, Y. Li, R.Z.Wang, Y.J. Dai.2009. Experimental study on a two-stage rotary desiccant cooling system. Int J Refrig.32(3):498–508. https://doi.org/10.1016/j.ijrefrig.2008.07.001.

[13] T.S. Ge, F. Ziegler, R.Z. Wang, H. Wang. 2010. Performance comparison between a solar driven rotary desiccant cooling system and conventional vapor compression system (performance study of desiccant cooling). Appl Therm Eng. 30(6–7):724–731. https://doi.org/10.1016/j.applthermaleng.2009.12.002.

[14] N. Enteria, K. Mizutani. 2011. The role of the thermally activated desiccant cooling technologies in the issue of energy and environment. RenewSustain Energy Rev. 15(4):2095–2122. https://doi.org/10.1016/j.rser.2011.01.013.

[15] Y. Ma, L. Guan. 2015. Performance analysis of solar desiccant-evaporative cooling for a commercial building under different Australian climates. Procedia. 121:528–535. https://doi.org/10.1016/j.proeng.2015.08.1024.

[16] D.B. Jani, M. Mishra, P.K. Sahoo. 2016. Solid desiccant air conditioning–A state of the art review. RenewSustain Energy Rev. 60:1451–1469. https://doi.org/10.1016/j.rser.2016.03.031.

[17] M. Sahlot, S.B. Riffat. 2016. Desiccant cooling systems: a review. Int J Low-Carbon Technol. 11(4):489–505. https://doi.org/10.1093/ijlct/ctv032.

[18] D.B. Jani, M. Mishra, P.K. Sahoo. 2017. Application of artificial neural network for predicting performance of solid desiccant cooling systems–A review. RenewSustain Energy Rev. 80: 352–366. https://doi.org/10.1016/j.rser.2017.05.169.

[19] R. Narayanan, E. Halawa, S. Jain. 2018. Performance characteristics of solid desiccant evaporative cooling systems in Australian Climate Zone of Brisbane. 11(10): 2574. https://doi.org/10.3390/en11102574.

[20] D.B. Jani, M. Mishra, P.K. Sahoo. 2015. Experimental investigations on hybrid solid desiccant–vapor compression air-conditioning system for Indian climate. In The proceedings of the 24th IIR international congress of refrigeration, Yokohama, Japan, Aug (pp. 16–22).

[21] K.K. Bhabhor, D.B. Jani. 2021. Performance analysis of desiccant dehumidifier with different channel geometry using CFD. J BuildEng. 103–021. https://doi.org/10.1016/j.jobe.2021.103021.

[22] D.B. Jani. 2022. Experimental assessment of rotary solid desiccant dehumidifier assisted hybrid cooling system. Int J Energy Resour Appl. 1:5–13. https://doi.org/10.56896/IJERA.2022.1.1.002.

[23] U. Eicker, D. Schneider, J. Schumacher, T. Ge, Y. Dai. 2010. Operational experiences with solar air collector driven desiccant cooling systems. Appl Energy. 87(12):3735–3747. https://doi.org/10.1016/j.apenergy.2010.06.022.

[24] D.B. Jani, M. Mishra, P.K. Sahoo. 2016. Exergy analysis of solid desiccant-vapour compression hybrid air conditioning system. International Journal of Exergy. 20(4):517–535. https://doi.org/10.1504/IJEX.2016.078106.

[25] A. Gagliano, F. Patania, F. Nocera, A. Galesi. 2014. Performance assessment of a solar assisted desiccant cooling system. Thermal Sci. 18(2):563–576. https://doi.org/10.2298/TSCI120526067G.

[26] M.J. Dadi, D.B. Jani. 2021. Experimental investigation of solid desiccant wheel in hot and humid weather of India. Int J Ambient Energy. 5983–5991. https://doi.org/10.1080/01430750.2021.1999326.

[27] D. B. Jani, M. Mishra, P.K. Sahoo. 2016. Performance prediction of rotary solid desiccant dehumidifier in hybrid air-conditioning system using artificial neural network. Appl Therm Eng. 98:1091–1103. https://doi.org/10.1016/j.applthermaleng.2015.12.112.

[28] L. Merabti, M. Merzouk, N.K. Merzouk, et al. 2017. Performance study of solar driven solid desiccant cooling system under Algerian coastal climate. Int. J. Hydrogen Energy. 42:28997–29005. https://doi.org/10.1016/j.ijhydene.2017.08.067.

[29] D.B. Jani, M. Mishra, P.K. Sahoo. 2018. Solar assisted solid desiccant—vapor compression hybrid air-conditioning system. Appl Solar Energy. 233–250. https://link.springer.com/chapter/10.1007/978-981-10-7206-2_12.

[30] M. Dadi, D.B. Jani. 2019. TRNSYS simulation of an evacuated tube solar collector and parabolic trough solar collector for hot climate of Ahmedabad. Soc Sci Res Network3367102. http://dx.doi.org/10.2139/ssrn.3367102.

[31] K.F. Fong, C.K. Lee, T.T. Chow, A.M.L. Fong. 2011. Investigation on solar hybrid desiccant cooling system for commercial premises with high latent cooling load in subtropical Hong Kong. ApplTherm Eng. 31(16):3393–3401. https://doi.org/10.1016/j.applthermaleng.2011.06.024.

[32] D.B. Jani. 2022. TRNSYS simulation of desiccant powered evaporative cooling systems in hot and humid climate. J Mech Eng. 1(1):1–6. https://web.archive.org/web/20220818202721id_/www.prakashjgec.com/_files/ugd/4247de_18ff01569155445c8b1060dd65e999be.pdf

[33] H. Caliskan, H. Hong, J.K. Jang. 2019. Thermodynamic assessments of the novel cascade air cooling system including solar heating and desiccant cooling units. Energy Convers Manag. 199:112013. https://doi.org/10.1016/j.enconman.2019.112013.

[34] K.K. Bhabhor, D.B. Jani. 2019. Progressive development in solid desiccant cooling: A review. Int J Ambient Energy. 43:992–1015. https://doi.org/10.1080/01430750.2019.1681293.

[35] A. Khalid, M. Mahmood, M. Asif, T. Muneer. 2009. Solar assisted, pre-cooled hybrid desiccant cooling system for Pakistan. RenewEnergy. 34(1):151–157. https://doi.org/10.1016/j.renene.2008.02.031.

[36] K.F. Fong, T.T. Chow, C.K. Lee, et al. 2010. Advancement of solar desiccant cooling system for building use in subtropical Hong Kong. Energy Build.42:2386–2399.https://doi.org/10.1016/j.enbuild.2010.08.008.

[37] D.B. Jani. 2019. An overview on desiccant assisted evaporative cooling in hot and humid climates. Algerian J Eng Technol. 1:1–7. http://dx.doi.org/10.5281/zenodo.3515477.

[38] K.F. Fong, C.K. Lee. 2020. Solar desiccant cooling system for hot and humid region–A new perspective and investigation. Solar Energy. 195:677–684. https://doi.org/10.1016/j.solener.2019.12.009.

[39] D. B. Jani. 2019. An overview on desiccant assisted sustainable environmental cooling technology. Int J Environ Plan Manage. 5(4):59–65. http://dx.doi.org/10.5281/zenodo.3515477.

[40] Q. Cui, H. Chen, G. Tao, H. Yao. 2005. Performance study of new adsorbent for solid desiccant cooling. Energy. 30(2–4):273–279. https://doi.org/10.1016/j.energy.2004.05.006.

[41] D.B. Jani. 2011. Desiccant cooling as an alternative to traditional air conditioners in green cooling technology. Instant J Mechanical Engineering. 1(1):1–13. www.raftpubs.com/ijme-mechanical-engineering/articles/ijme_raft1001.php.

[42] P. Patanwar, S.K. Shukla. 2014. Mathematical modelling and experimental investigation of the hybrid desiccant cooling system. Int J Sustainable Energy.3:103–111. https://doi.org/10.1080/14786451.2012.751915.

[43] D. B. Jani, M. Mishra, P.K. Sahoo. 2017. A critical review on solid desiccant-based hybrid cooling systems. Int J Air-cond Refrig. 25(03):1730002. https://doi.org/10.1142/S2010132517300026.

[44] R. Tu, Y. Hwang. 2018. Efficient configurations for desiccant wheel cooling systems using different heat sources for regeneration. Int J Refrig. 86:14–27. https://doi.org/10.1016/j.ijrefrig.2017.12.001.

[45] D.B. Jani, M. Mishra, P.K. Sahoo. 2018. Investigations on effect of operational conditions on performance of solid desiccant based hybrid cooling system in hot and humid climate. Therm SciEng Prog. 7:76–86. https://doi.org/10.1016/j.tsep.2018.05.005.

[46] Alahmer, A., Alsaqoor, S., Borowski, G. 2019. Effect of parameters on moisture removal capacity in the desiccant cooling systems. Case StudTherm Eng. 13:100364. https://doi.org/10.1016/j.csite.2018.11.015.

[47] D.B. Jani 2020. Solar assisted sustainable built environment: A review. JEnvironSoil Sci. 4(4):41–47. https://doi.org/10.32474/OAJESS.2020.04.000195.

[48] J. Guo, S. Lin, J.I. Bilbao, S.D. White, A.B. Sproul. 2017. A review of photovoltaic thermal (PV/T) heat utilisation with low temperature desiccant cooling and dehumidification. RenewSustain Energy Rev. 67:1–14. https://doi.org/10.1016/j.rser.2016.08.056.

[49] M. Dadi, D.B. Jani. 2019. Solar energy as a regeneration heat source in hybrid solid desiccant – vapor compression cooling system – a review. J Emerg Technol Innov Res. 6(5):421–425. www.semanticscholar.org/paper/Solar-Energy-as-a-Regeneration-Heat-Source-in-Solid-Dadi-Jani/b810b0ab34ff9218392006f507a3091c4c1cf964.

[50] Y.J. Dai, R.Z. Wang, Y.X. Xu. 2002. Study of a solar powered solid adsorption–desiccant cooling system used for grain storage. Renew Energy. 25(3):417–430. https://doi.org/10.1016/S0960-1481(01)00076-3.

[51] D.B. Jani. 2021. Use of artificial neural network in performance prediction of solid desiccant powered vapor compression air conditioning systems. Instant J Mech Eng. 3(3): 5–19. www.raftpubs.com/ijme-mechanical-engineering/articles/ijme_raft1005.php.

[52] K.S. Rambhad, P.V. Walke, D.J. Tidke. 2016. Solid desiccant dehumidification and regeneration methods: A review. Renew Sustain Energy Rev.59:73–83. https://doi.org/10.1016/j.rser.2015.12.264.

[53] D.B. Jani, K. Bhabhor, M. Dadi, S. Doshi, P.V. Jotaniya, H. Ravat, K. Bhatt. 2020. A review on use of TRNSYS as simulation tool in performance prediction of desiccant cooling cycle. J Therm Anal Calorim. 140(5):2011–2031. https://doi.org/10.1007/s10973-019-08968-1.

[54] A.S. Farooq, A.W. Badar, M.B. Sajid, M. Fatima, A. Zahra, M.S. Siddiqui. 2020. Dynamic simulation and parametric analysis of solar assisted desiccant cooling system with three configuration schemes. Solar Energy. 197:22–37. https://doi.org/10.1016/j.solener.2019.12.076.

[55] D.B. Jani, M. Mishra, P.K. Sahoo. 2018. Performance analysis of a solid desiccant assisted hybrid space cooling system using TRNSYS. J Build Eng. 19:26–35. https://doi.org/10.1016/j.jobe.2018.04.016.

[56] J.R. Mehta, S.M. Prasad, I.S. Khatri. 2016). Investigations on a porous rotating media liquid-air contacting device suitable for various alternative cooling technologies. Energy Procedia. 95:302–307. https://doi.org/10.1016/j.egypro.2016.09.008.

[57] D.B. Jani. 2022. A review on progressive development in desiccant materials. Mater Int. 2(2):73–82. https://doi.org/10.33263/Materials22.073082.

[58] S.P. Halliday, C.B. Beggs, P.A. Sleigh. 2022. The use of solar desiccant cooling in the UK: A feasibility study. Appl Therm Eng. 22(12):1327–1338. https://doi.org/10.1016/S1359-4311(02)00052-2.

[59] A.K. Verma, L. Yadav. 2023. Investigation of angular aspects of a novel four-sector rotary dehumidifier configuration for HVAC applications. JBuild Eng. 106490. https://doi.org/10.1016/j.jobe.2023.106490.

[60] A.K. Verma, L. Yadav, N. Kumar, A. Yadav. 2023. Mathematical investigation of different parameters of the passive desiccant wheel. Numer Heat Transfer, Pt A: Appl. 1–20. https://doi.org/10.1080/10407782.2023.2172491.

[61] A.L.K. Bharathi, S.I. Hussain, S. Kalaiselvam. 2023. Performance evaluation of various fiber paper matrix desiccant wheels coated with nano-adsorbent for energy efficient dehumidification. Int J Refrig. 146:416–429. https://doi.org/10.1016/j.ijrefrig.2022.12.002.

[62] T. Hussain. 2023. Optimization and comparative performance analysis of conventional and desiccant air conditioning systems regenerated by two different modes for hot and humid climates: experimental investigation. Energy Built Environ. 4(3):281–296. https://doi.org/10.1016/j.enbenv.2022.01.003.

[63] A. Pacak, A. Jurga, B. Kaźmierczak, D. Pandelidis. 2023. Experimental verification of the effect of air pre-cooling in dew point evaporative cooler on the performance of a solid desiccant dehumidifier. Int Commun Heat Mass Transf. 142:106651. https://doi.org/10.1016/j.icheatmasstransfer.2023.106651.

10 Resource-Efficient Urban Systems Aimed at Facing Urban Heat Islands (UHIs) and Local Climate Change

Ajay Tripathi and Govind Sahu

10.1 INTRODUCTION

Urbanization is a major scientific, humanitarian, and policy challenge of the 21st century. Currently, more than 50% of the world's population lives in cities, and the population of cities is increasing continuously [1, 2]. Scientific questions have emerged regarding the reasons for and pace of urbanization in many cities and the influence of urbanization on the local, regional, and large-scale environment. In particular, the effects of land cover and land use change (LCLUC) associated with urbanization on the health and daily life of a city's inhabitants are of great concern. Due to the inherent uniqueness of individual urbanizing cities with specific geography, climate, and socioeconomic status, the effect of urbanization needs to be analyzed for all cities.

According to the World Population Prospects 2019, currently, 18.6% of the world's urban population lives in China, followed by 17.7% in India and 2.3% in Nigeria [3]. China has received considerably more attention in the literature than both India and Nigeria with regard to the UHI effect. However, the prospects report a faster population growth rate per year for Nigeria (2.6%) and India (1.02%) than China (0.43%) [3]. Given that the percentage of the global world population for India is within one percentage point of China and the country has a faster growth rate, it is likely that the population of India will exceed that of China in the upcoming years (around 2027). Therefore, it is necessary to improve our understanding of UHIs in India. An agricultural economy and predominantly rural population have quickly industrialized in India during the past few decades due to fast urban expansion [4].

An area, region, or volume where the temperature is greater than that of its surroundings is known as a heat island. This common definition can apply to the environment at any scale, for example, from a few micrometers to many kilometers of area. In other words, a heat island may be created in the form of hot air bubbles on a scale of a few millimeters surrounded by cooler air or in the form of a high-pressure

DOI: 10.1201/9781003415695-10

weather system covered with a relatively cold mass of air on a many-kilometer scale. Moreover, UHI can be considered a more widespread or universal phenomenon. This definition will therefore be constrained for this chapter, and authors have only discussed the UHI. Hence, a UHI is the temperature difference between a selected urban location and a non-urban area selected for reference. It should go without saying that a reported UHI's size and characteristics can be influenced the selection of urban sites or a particular set of locations.

The difference in heat fluxes between urban and rural areas due to more absorption and storage of sensible heat in urban areas as compared to rural areas may be considered a main cause of the urban heat island (UHI) effect. The UHI intensity reduces with increasing wind speed, and it increases as the city's population grows. The literature reveals that a reduction in green space and reduced wind velocity due to scarcity of urban ventilation, the release of anthropogenic heat, and increased absorption and storage of solar radiation due to the widespread use of construction materials are some of the main factors that increase the UHI effect. According to studies, the UHI intensity varies between cities and is greatly influenced by the land cover and land use, building materials used, etc. According to studies, the excessive use of construction materials greatly influenced the UHI impact in Manchester, United Kingdom. According to several UHI mitigation measures, urban greening can greatly reduce the intensity of UHI directly or indirectly. This can lead to reductions in mean radiant temperature by 4°C and global air temperature by 4.5°C. The use of cool materials to reflect heat to the atmosphere, providing sufficient ventilation for better wind flow around buildings, minimizing anthropogenic heat to reduce the amount of heat transfer to the environment, and urban vegetation (green facades, green roofs, vertical greeneries, and green pavements), can help in reducing the UHI effect [5–7].

Urbanization is an anthropogenic modification that results in changes in surface materials as a result of plant suppression, soil sealing, and albedo variation greatly affecting the energy balance of selected local areas, creating an urban heat island (UHI) [8]. This impact results from the slower cooling and the increased absorption of electromagnetic energy of urbanized surfaces relative to those in adjacent areas covered with plants. This local variation results in increasing surface temperature and sensible heat, whereas decreasing the latent heat and relative humidity. The major factors of UHI formation are as follows: (i) heat-absorbing building materials, (ii) human-caused heat generation, (iii) minimization of and changes in wind speed due to surface roughness, and (iv) increased solar radiation absorption from surfaces with lower albedo. This chapter focuses on the cause and effect of UHI, including the variation in temperature of the local environment compared to its surrounding rural area [9–12]. The most frequent effects of UHI effect include: (i) effect on local microclimate, (ii) discomfort associated with increased temperature, (iii) effects on human health, and (iv) variation in hydrological behavior, such as water mass displacement [13, 14]. Increased mortality can result from the combined effect of UHI and natural events like heat waves (HWs) [15].

Moreover, anthropogenic sources of generation such as artificial heating and cooling of buildings, industrial operations, and transportation contribute heat into the urban environment, resulting in an increase in UHIs [16], and their intensities increase with time. According to Santamouris 2015 et al. [17], the intensity of the UHI increases with an increase in population density due to a reduction in the green and blue covered areas. Due to the local thermal balance, urban heat islands can exist at any latitude and at any time of day. The magnitude of UHI increases in clear and calm weather, while breeze and precipitation show a very high impact on it. Infrastructure related to housing is related to extreme climate change. Practically, poor urban design can intensify the effects of climate change, whereas buildings that were built for certain thermal conditions may eventually need to be applied in drier and hotter areas [18]. Finally, the development and operation of cities show major impacts on energy demand and, consequently, greenhouse gas emissions. To overcome this issue, more and more studies are needed, and urbanization, compact city growth, and urban sprawl should be studied.

10.2 CONCEPT OF UHIs

The temperature at any location depends on the energy balance at the underlying surface. The basic form of the energy balance equation for the underlying surface is given by [19]:

$$\alpha_s Q_{direct} + \alpha_s Q_{diffuse} + \alpha_l Q_l + Q_{anthropogenic}$$
$$= \varepsilon\, \sigma\, (T_o)^4 + h_c(T_o - T_a) + (kd\, T_o/dz)_{z=0} + Lm_{ev}$$

where α_s represents the absorptivity for short-wave radiation and α_l represents absorptivity for long-wave radiation; Q_{direct} is direct and $Q_{diffuse}$ is diffuse heat flux for short-wave radiation; Q_l is long-wave incoming heat flux; $Q_{anthropogenic}$ is anthropogenically generated heat flux; and surface emissivity is denoted by ε, the Stefan–Boltzmann constant by σ, surface temperature by T_o, air temperature by T_a, convective heat transfer coefficient by h_c, thermal conductivity by k, latent heat of vaporization by L, and rate of evaporation by m_{ev}. As a result, the first two terms on the left-hand side of the statement stand for the total heat flow from short-wave radiation absorbed at the surface, whereas the third term denotes the heat flux from long-wave radiation. Anthropogenic heat flow into the surface is represented by the fourth term on the left-hand side of the expression.

Long-wave radiative flux from the surface is represented by the first term on the right-hand side of the expression, sensible heat flux by the second, heat conduction through the surface by the third, and latent heat flux by the last term. The next sections explore a few of the terms in this equation either directly or indirectly. The anthropogenic heat flow into the surface or near-surface air layer is the fourth term from the left and includes heat from automobiles, construction equipment, and other heat sources. The sensible heat flux is the first term on the right side of the equation, followed by the ground heat flux, latent heat flux, and long-wave radiative flux from the surface. The terms in this equation are explored in more detail in the following sections, both directly and indirectly.

10.3 UHIs AND GLOBAL WARMING

UHI is responsible for an increase in temperature in a small area and local climate change, whereas global warming is responsible for global climate change. UHI is increased by the human activities in the urban area. A city is warmed by pollution in addition to the heat produced by automobile use, home heating, and industrial operations. When cities expand, the UHI effect increases, causing an unnatural warming of cities or local areas. However, across the world, glaciers are melting as a result of global warming, affecting the lives of many people with floods, droughts, and a shortage of drinkable water. A lot of questions and conflicts surround this complicated problem. Global temperatures are now higher than they have been in at least the previous millennium, and they are increasing much more quickly than initially anticipated by scientists. Flooding along the coast, an increase in harsh weather, the transmission of infectious diseases, and massive extinctions are all anticipated effects. The Earth's temperature has been increasing continuously, and many researchers now think that this is caused by human activity related to emissions of greenhouse gases like carbon dioxide (CO_2) and nitrogen oxides (NO_x) that are responsible for this warming.

UHI increases the use of Air-Conditioners (AC) in homes and offices, causing Chlorofluorocarbon (CFC) emission and resulting in global warming. It is well known that organic halogen compounds reduce atmospheric ozone levels, which contributes to the ozone hole over the Antarctic. A recent study revealed how much these substances have also contributed to global warming [20–30]. According to scientific calculations, heat trapped by halogenated compounds was responsible for one-third of the world's warming in the second half of the 20th century. This were liable for more than 50% of the melting of Arctic sea ice during this time. According to estimates, 24% of greenhouse gas emissions from humans come from CFCs. Compared to the majority of other known chemicals, halogenated organic compounds capture a significantly greater amount of heat in the atmosphere. For instance, the global warming potential of dichlorodifluoromethane (CFC-12) is roughly 11,000 times more than that of carbon dioxide. This means that although though CFCs are present in the atmosphere in much lower amounts than other greenhouse gases, their influence may still be quite high. Excessive use of AC and other home appliances needs more electricity, hence requiring more burning of coal and more greenhouse gases like CO_2 emissions, causing global warming. As a result of increased UHI, the demand for electricity is increasing rapidly, resulting in the burning of fossil fuels (such as coal, gas, and oils), which has been producing more greenhouse gases. The greenhouse gases were in balance in the past. Fossil fuels evolve over millions of years, yet it only takes a few minutes for them to burn, sending huge amounts of CO_2 into the atmosphere. In the past, natural factors like volcanic eruptions and the quantity of phytoplankton in the sea have been used to explain variations in CO_2 levels. It expected that at the end of the next 100 years, the temperature of Earth will be increases by 8–10°C with significant warming caused by the emission of CO_2 due to human activity. The reported growth in anthropogenic GHG (greenhouse gas) emissions is highly probable to be to responsible for most of the observed rise in average global temperatures since the mid-20th century [20].

The inventor of the greenhouse theory (James Hansen, NASA) and the most recognized climatologist (Richard Lindzen, MIT) predict that even if no action is taken to limit greenhouse gases, the globe's temperature will rise by roughly 1°C in the next 50 to 100 years. According to a team of UN scientists, developing nations are two times more at risk from climate change than industrialized nations, and small island nations are three times more at threat. The United States ranks fourth globally in terms of CO_2 emissions. People, animals, birds, and habitats are all affected by the consequences of global warming. No continent has actually been spared. The Adelie penguin population in Antarctica has reduced by 33% over the past 25 years as a result of decreasing sea ice. Even a small increase in temperature of 1°C can have a negative impact on sea levels. Coastal cities like Mumbai, Kolkata, and Chennai, as well as almost 60 island countries, including Bangladesh and the Maldives, will be at risk from this rise in sea level. Desert expansion is a result of global warming in the places such as North America, South Africa, India, Mexico, etc. By 2050 AD, more than a million species of animals and plants may be lost due to global warming [20].

In addition to the problems mentioned earlier, a changing climate can also result in a number of other issues, including the depletion of bodies of water on the surface, a decline in the groundwater level, severe water scarcities, the desertification of large areas that were previously fertile and productive lands, a change in crop patterns, lower agricultural yields, the spread of diseases, a lack of food, and the proliferation of microbes.

10.4 CAUSES OF UHIs

Over 50% of people on the planet live in cities [1–2]. Urbanization causes significant changes in land cover and land use as cities spread out further. Additionally, as urbanization progresses, so does the opportunity for vegetation to survive and thrive and the accessibility for people to see and enjoy nature [21]. Therefore, it is important to comprehend how urbanization may affect the environment as well as any resulting human effects, particularly in developing areas where urbanization rates are high and generally understudied [4, 22].

The urban heat island effect, also known as UHI, is a well-known phenomenon where metropolitan regions typically have greater surface and air temperatures than neighboring suburban and non-urban areas. The land surface characteristics, regional climate, season, time of day, city's geographic position, size of the city in terms of both area and population, and anthropogenic heat are some of the variables that affect UHI intensity [2]. For instance, Park [23] and Oke [24] show how UHI intensity rises in correlation with the city population. It was demonstrated by Ichinose et al. [25] that anthropogenic heat is seasonal and that the intensity of UHI is higher in the winter than in the summer. In addition, cities typically have more buildings, asphalt walkways, and roads than non-urban environments, as well as less vegetation, higher heat capacities, and more of these features. These characteristics of urban surfaces cause them to store more heat during the day and heat up more at night than natural surfaces [26]. Additionally, urban surfaces have larger Bowen ratios (i.e., the ratio of sensible to latent heat fluxes) and lower evapotranspiration rates, which aid in warming them up more quickly during the day than suburban and non-urban surfaces [27]. This daytime warming effect, however, might be lessened since increasing urban surface roughness enhances planetary boundary layer (PBL) mixing, which is most intense during the day [28]. As a result, there is less of a warming contrast between urban and non-urban surfaces [29, 30]. This aids in the upward heat transfer of more surface heat. Due to the changes in the aforementioned controlling factors between cities, previous estimations show that UHI intensity is largely geographically dependent and varies greatly among locations [2]. Peng et al. [31], examined 419 cities using satellite data and discovered that for 64% of the cities, the global surface UHI intensity was highest during the day. This finding illustrates that on an annual scale, surface UHI intensity is stronger in daylight. However, 36% of the cities that had a greater UHI at night were primarily concentrated in developing regions, such as western and southern Asia and northern Africa, where urbanization is frequently understudied [4]. Therefore, research should focus on urbanizing cities inside these growing areas to ascertain why they show a higher UHI intensity at night rather than during the day.

Further amplifying the UHI effect is growing urbanization, which also reduces vegetation density and spatial extent and lowers evapotranspiration [31]. On the other hand, if cities seek to maintain or enhance the amount of green space, the increased vegetation density and spatial extent generate increased evaporative cooling during the day [31], which lowers urban temperatures and thus lowers the daytime UHI [32]. To assess if a city's UHI has a warming or cooling effect, the greenness of the vegetation must be measured. The availability and energy absorption of leaf pigments like chlorophyll are frequently assessed using satellite-based vegetation indices, such as the normalized difference vegetation index (NDVI), which is also a good indicator of an urban climate [33, 34].

The difference in air temperature between two meteorological stations, one in an urban area and one adjacent in a non-urban area, has been used to measure UHI intensity [34]. In contrast, satellite data have recently employed LST to characterize spatiotemporal structures of UHI with adequate precision to distinguish between urban centers and non-urban surroundings and give global coverage at a high resolution [31]. This is due to a low density of weather stations over many locations. LST has been demonstrated to be comparable to air temperature in both urban and rural settings. Both datasets indicate UHI consequences, although it has been demonstrated that the UHI intensity measured using LST is greater in magnitude than the UHI intensity evaluated using air temperature [35]. UHI effects have been thoroughly investigated in several cities utilizing satellite products, including Houston, among others, in terms of their size concerning the elements stated in [2]. By 2030, the number of megacities—cities with more than 10 million inhabitants—is expected to rise from 18 to 43, with the majority of them located in developing nations like India, China, and Nigeria. Between 2018 and 2050, these three nations are anticipated to boost the urban population by 35% [3]. Focusing on India, severe urban expansion in recent decades has changed a nation with a predominately rural population and an agricultural-based economy into one with a rapidly industrializing society and major urbanization [4].

The increase in greenhouse gas emission and urbanization are linked to alterations in energy fluxes that, once they take place, may alter the microclimate and degrade the thermal environment. They also plays a role in climate change. Urbanization also results in an overabundance of buildings, many of which are constructed from materials with poor thermal properties. According to various studies, the degree of urbanization is related to the long-term trend in surface air temperature in urban centers [36–37]. A typical UHI occurs when there is any alteration in the energy balance of an urban area, resulting in a slower rate of cooling as compared to rural areas (for example, after sunset). However, a UHI can happen whenever there is a lot of heat stored in the urban canopy layer or any other heat sources (like anthropogenic sources). Heat islands can be created by a single causal factor or, more frequently, by the interaction of numerous factors. The main causes of UHI are as follows:

10.4.1 Thermal Capacity and Urban Geometry

Generally, high-heat-capacity materials are available in urban locations (e.g., brick, stone, concrete, asphalt, and pavement). Also, the surface-to-area ratio (SAR) for urban areas is significantly greater than that of rural areas, where SAR represents the ratio of the total surface area exposed to the sun and the total horizontal area. SARs in rural regions often hover around 1. Whereas the SAR in urban centers is found to be ≥5, it is about 2 or 3 in residential neighborhoods. Solar radiation is absorbed and stored more effectively in urban areas due to the interaction of greater SAR and higher heat capacity. The additional heat that has been trapped in various urban building materials and structures is released, heating the air and causing a UHI, As beams A and B shown in Figure 10.1.

10.4.2 Sky View Factor (SVF)

A UHI, especially of the nocturnal variety, is facilitated by the reduced SVF in urban regions as compared to that in open surroundings. This is because urban barriers prevent urban surfaces from radiatively cooling the walls, structures, etc. The beam at point B in Figure 10.1 is subjected to a number of reflections, resulting in a reduction in intensity and consequently warming the surfaces, and hence, air comes across it. The taller the buildings that surround the urban "canyon," the more reflections there may be. Variations in rate of cooling between the urban and surrounding rural area cause the generation of a UHI.

10.4.3 Albedo and Effective Albedo

In comparison to vegetation or bare land, the albedo of roads, pavement, building materials, and other urban constructions is smaller as compared to rural areas. However, the opposite is valid in some circumstances. According to the Taha [19] the value of urban albedo can be more than 0.20, whereas the surrounding grassland-covered area has an albedo of 0.15. A UHI generally comes into the picture when the value of urban albedo is found to be lower as compared to the surrounding rural areas. Apart from these various surfaces having low albedos (for example streets, pavement, walls, roofs, buildings, etc.), urban areas also have lower effective albedos than surrounding rural areas. Geometry (i.e., urban canyons) increases the solar radiation reflection, and the absorption of photons is increased by canyon surfaces instead of escaping back. The resulting decrease in effective albedo is shown by beam A in Figure 10.1. The beam becomes weaker and also some of the photons cannot meet the canyon floor due to successive reflections.

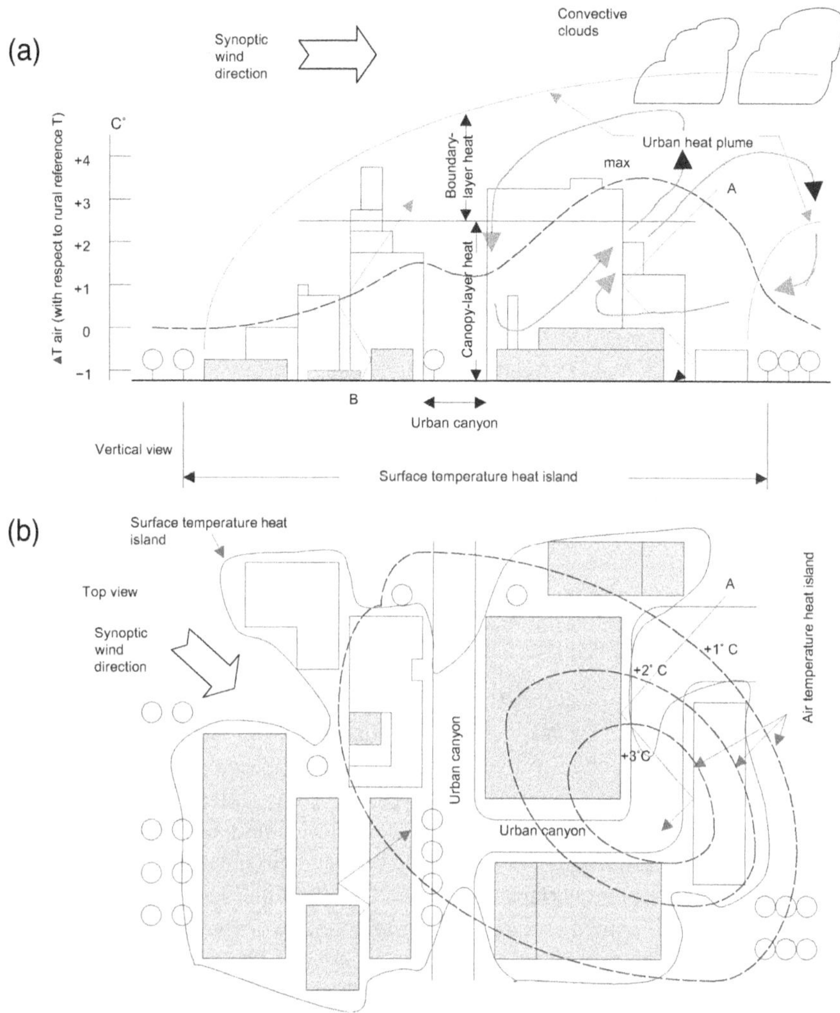

FIGURE 10.1 Schematic diagram of an urban heat island (UHI). (a) Front view of an urban area. (b) Top view of same area. (Regenerated from Figure 1 of Taha H. 2004 [19]).

10.4.4 Bowen Ratio

Urban locations have a less dense vegetative cover than rural regions, which changes how solar radiation is split into latent and sensible heat fluxes (among other things). Because of this, air temperatures are higher in metropolitan locations where the value of the Bowen ratio (β) is high. β is smaller in suburban or rural locations, where there is often more vegetation, evaporative cooling, and evaporation, causing lower air temperature. In metropolitan settings, the magnitude of β is found to be approximately 4 or 5, whereas, for vegetated canopies, typical values range from 0.8 to 1.5. For instance, β is roughly 0.1 for oceans and roughly 0.2 for tropical forests.

10.4.5 Anthropogenic Heat

Urban areas alter the energy balance both "passively and actively." A passive effect is caused by urban geometry, heat capacity, scarcity of covered grassland, and lower effective albedo, whereas the active effect considers the removal of extra heat to the air, including the release of heat due to heating or cooling purposes, heat release by furnaces, and anthropogenic or manmade heat. The most common sources of anthropogenic heat are industry, power plants, refineries and processing facilities, chimneys and stacks, and HVAC in automobiles and buildings. The value of anthropogenic heat may be calculated from the energy consumption for a particular purpose since its direct measurement is very difficult. Any of the previously mentioned causal elements, or, more frequently, different combinations of them, can result in a UHI. Because of this, the characteristics of UHIs and their scope might vary greatly depending on the place or time.

10.5 TECHNIQUES TO MEASURE CONTROLLING FACTORS OF UHIs

It is critical to quantify the correlations between these elements and determine which best causes urban heat in a metropolis to comprehend the mechanics of UHI production and create effective mitigation plans. After physical assessments of what components may be most important for a particular city's UHI, a range of conventional regression approaches and machine learning techniques can be employed to quantify the top controlling factors. Multiple linear regression (MLR), which is used to statistically predict one variable (predictand) from numerous other variables (predictors), is one of the most straightforward techniques. Some studies have utilized MLR to identify the most effective controls on urban heat for different cities, such Kolokotroni and Giridharan [38] for London, England, and Kim and Baik [2] for Seoul, South Korea. Kolokotroni and Giridharan [38] have examined leading controls of urban heat using machine learning techniques. These researchers discovered that machine learning approaches outperformed traditional MLR analysis by having lower error values and higher explained variance values when compared to their MLR results. These researchers employed a type of regression tree modeling (decision trees), which includes nonlinear interactions between predictor variables and does not make assumptions about the relationship between each predictor and the predictand [39]. The random forest (RF) model is capable of producing thousands of regression trees with the use of computers [40].

Each regression tree in the RF is created using a resampled version of the initial training dataset, and each branch in the developing process evaluates whether to use a new random subset of input variables [41, 42]. Together, these sources of randomness ensure the independence of each tree. Every city is different in how environmental factors may interact and influence urban heat due to their varying magnitudes and importance within the city's local climate. Although these studies have used a variety of methods to determine controls of urban heat but they compared their results with available data in literature. For instance, a particular collection of environmental elements might be more important in one city than another. As a result, it's possible that the conclusions from these earlier studies don't apply to other places. In reality, different studies that looked at different places came up with varied findings in terms of the decisive component. A study that examines the mechanisms that regulate urban heat in an Indian city is still lacking.

10.6 IMPACT OF UHIs

UHI has the potential to change the local meteorology, rates of air pollution emission, pollutant transport, and photochemistry. Naturally, they have an impact on energy usage.

10.6.1 Impact of UHIs on Local Climate

The development of a circulation or wind pattern is often a convective cell around the UHI zone. The convective cell for urban areas may be asymmetric in nature and blows downwind, which mainly depends upon the relative strength of the background winds and UHI winds represented by the thick arrow line in Figure 10.1. In general, a city's UHI circulation is stronger the larger and denser it is in terms of population. The increase in convection activity inside heat plumes caused by UHIs consequently increases the clouds and rain, as can be inferred from Figure 10.1. Of course, UHIs only result in increased convective clouds or precipitation when favorable conditions such as moisture, temperature condensation, and wind speed occur. According to Taha [19], the long-term effect of a UHI can result in increased cloudiness, rainfall, and thunderstorms by 8%, 14%, and 15% respectively.

10.6.2 Impact of UHIs on Ambient Temperature

The most noticeable impact of UHI is elevated ambient temperatures: UHIs typically cause increases in temperature of around 2°C on average and up to as high as 8°C. A potential UHI of 3.5°C downwind of the densest metropolitan area is shown in Figure 10.1. From the figure, it is clear that the "observed" UHI intensity can greatly depend upon the urban and reference rural areas.

10.6.3 Impact of UHIs on Pollution

When pollutants are released into the urban atmosphere, the boundary layer (BL) allows for their dispersion and mixing, which leads to lower apparent concentrations than in BLs where pollutant mixing is not allowed. This occurrence heavily depends on stability, the depth of the BL, and the amount and intensity of convection.

Increased emission of air pollutants that are temperature-dependent; for example, the temperature has a significant impact on the emission of hydrocarbons, such as isoprene emissions from plants. Additionally, the majority of fugitive hydrocarbon emissions are temperature dependent, including those from fuel tanks, automobiles, motor engines having a hot soak, and losses due to evaporation. The particulate matter emission also depends on the temperature. Higher precursor emissions typically result in worsening air quality, a faster rate of smog development, and less visibility.

10.6.4 Impact of UHIs on Photochemistry

Photochemical reactions are generally strongly associated with temperature in the photochemical formation of smog. As a result, a UHI may catalyze the creation of urban ozone, which is a serious health concern in large cities. There may appear to be a conflicting effect because greater temperatures can promote ozone generation while also increasing the mixing of contaminants in the BL. The majority of data and observations, however, point to a negative influence on air quality as the overall result of a UHI due to variations in the concentration of ozone.

10.6.5 Energy Impact of UHIs

UHIs often have a summertime increase in cooling loads and a wintertime fall in heating loads. The final result, of course, depends on the overall meteorological parameters and the changes in local climate such as available sunshine, wind patterns, topographically induced flows, precipitation, etc. Also, it is influenced by regional energy-specific factors such as population density, the predominant energy-using industries, building types, distribution, age, general building envelope features, HVAC equipment saturation, and regional fuel and power prices. But generally speaking, a UHI has more of an effect on cooling loads as compared to heating loads. Also, the latter effect is typically desired, and the former effect is unwanted.

Although building attributes, occupancy schedules, thermal integrity, and preferences for psychometric comfort may be difficult considerations inside the building, the energy requirements due to UHIs are relatively easy to comprehend. The UHI generally has a substantial impact on residential or small nonresidential buildings. The UHI is less sensitive in larger buildings, since internal loads (heat liberation from occupants and appliances) are more significant than heat interaction with the surroundings.

Finally, UHIs also have a secondary impact on other energy-related factors. For example, warmer weather increases the requirement for watering (for urban plants, parks, and nearby crops), which increases the demand for water pumping and distribution. A higher need for air conditioning in motor vehicles is also brought on by higher air temperatures. More energy is needed because of the higher air conditioning load (increased fuel consumption). Health and thermal comfort are other aspects of UHIs that also involve indirect energy usage. In the summer, UHIs rise, leading to an increase in heat-related hospital admissions and a severe worry for healthcare. These factors, although difficult to measure, involve higher energy use in medical institutions.

10.6.6 Energy Impact of UHIs on Local Climate Change

The quantity of energy used by different urban infrastructures, heating and cooling, and means of transportation may be greatly affected by UHIs. The implications of

UHIs that have received the greatest attention include a significant increase in the peak and global electricity consumption for HVAC systems as well as a significant decrease in their efficiency.

The amount of urban warming, the regional climate, and the properties of HVAC systems and buildings influence the increase in power and energy caused by UHIs. Using climate data from 30 urban and semi-urban areas of Athens, Greece, Santamouris et al. [43] assessed the effect of UHI on building energy use and discovered that the cooling load in the city center was twice as high as that in the surrounding area. Moreover, the peak electrical consumption triples, while the heating load in urban areas may drop by 30% to 50%. The COP of refrigeration systems is reduced by up to one-fourth as a result of a UHI, which also affects the performance of electrical and mechanical equipment. Santamouris et al. [44] studied the variation in the heating and cooling load of a particular building with time, which turns into regional and global climate change from 1970 to 2010. According to the data, the combined effects of local and global climate change resulted in an average 23% increase in cooling demand and a corresponding average 19% decrease in heating demand. According to a balance, annually, an 11% increase in energy demand was observed due to heating and cooling load.

10.6.7 Impact of UHIs on Health, Comfort, and Economy

Due to the harmful effects that urban overheating caused by UHI has on inhabitants' health, perceptions of indoor–outdoor comfort, and financial well-being, the urban environment has been modified in a significant way from several angles [45]. Particularly, low-income and vulnerable populations residing in crowded urban settings are those who are most sensitive to this environmental shift [46]. Sakka et al. (46) demonstrated that harmful interior thermal overheating was induced, particularly in low-tech structures exposed to a hot thermal environment for a prolonged time, as observed in southeastern Europe in 2007. In more detail, the researchers discovered that the average indoor temperature across all of the residential units under observation was nearly 4.2°C higher than the average indoor temperature recorded in Athens during typical summertime conditions, that is, before and after the time of the heat wave that was exacerbated by the UHI. Interior thermal field measurements showed how the UHI-emphasized heat wave exposed people to hot spells reaching 33°C for six straight days, which caused issues with interior comfort as well as an increase in the probability of non-negligible health conditions for the most susceptible families [46]. According to consistent preliminary results obtained using thermal energy models created within the framework of LUCID projects, the number of indoor overheating hours (for more than 28°C working temperature) tends to rise as one approaches another European megacity affected by UHI, namely the greater London area [47].

In reality, it has been shown that the UHI-heated outdoor urban environment has a significant impact on the temperature within buildings, particularly those with poor building envelopes and no active cooling HVAC systems [46]. Low-income families are also more susceptible to summer heat wave events and UHI in general due to their generally low awareness and education levels regarding the proper usage of passive cooling measures to improve indoor thermal conditions. According to this perspective, significant research contributions have concentrated on identifying the impact of UHI on the population of energy-poor people, with the core of the analysis centered on households' inability to provide the necessary environmental active systems, which are intended to address their basic needs for cooling and heating. According to data, persons who are energy inefficient are accustomed to living in houses with poor-performance envelopes, which are mostly caused by inadequate thermal insulation, ineffective shading mechanisms, and high rate of infiltration. Hence, low-income families are made even more sensitive to the UHI phenomena in congested cities around the world due to the energy poverty and a lack of readily available significant passive solutions. Additionally, the energy-poor population in Europe spends 30% more on house operating costs than the average European, or almost 40% of their income [37].

In hot weather, mortality rises, especially among the most susceptible, such as the elderly. Different countries, cities, and even the same area from year to year have varying degrees of connection between daily outdoor temperature and health outcomes [47]. In general, persons often exhibit an optimum temperature at which the death rate is lowest; temperatures over this temperature cause mortality rates to increase. In particular, it has been calculated that for every degree of temperature increase above a (locally specified) cut-off point, mortality for populations in the European Union will rise by 1% to 4% [48]. Important UK studies have revealed that the UHI phenomenon is predicted to make extreme urban overheating episodes worse in urban locations. For instance, an excess of 17% of deaths in England and Wales were attributed to the 2003 heat wave throughout the summer [49].

According to Taha [39], urban air pollution concentrations rise in warming cities. They may also rise during heat waves, which could have serious effects on mortality, as it did in Paris in the summer of 2003. This is because solar radiation and hot temperatures encourage the synthesis of volatile organic molecules (VOCs) like ozone. According to current air pollution control measures, ground-level ozone and fine particles are expected to cause 311,000 premature deaths by 2030.

10.7 MITIGATING THE URBAN HEAT ISLAND

In the last three decades, there has been a significant amount of research and development in the areas of understanding urban heat islands and their effects on the environment and health, developing mitigation strategies, and putting policies and programs in place to cool urban heat islands. The research initially concentrated on identifying why cities are generally warmer than their surrounding suburbs, noting that the urban fabric is responsible for these higher temperatures, and examining how potential changes to the urban fabric (such as altering the surface of the city and introducing vegetation) might impact the amount of energy used by buildings in cities during the summer. A study found that shade trees and cool roofs directly lower the amount of energy needed to cool buildings and that a city may be cooled by a few degrees by combining cool pavements, cool roofs, and urban vegetation [50]. Cooling cities (lower ambient temperatures) reduces the amount of energy used to cool buildings, improves the comfort of walking outdoors, and slows down the photochemical

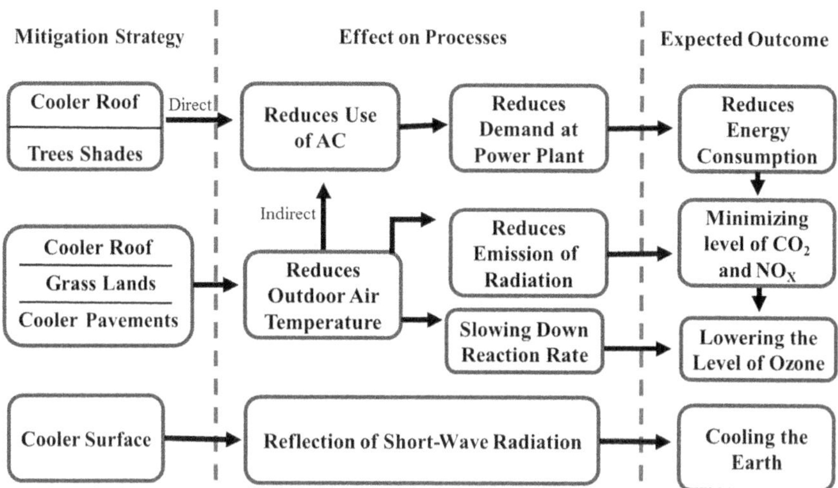

FIGURE 10.2 Effect of a mitigation strategy on the process parameters and its expected outcomes.

reactions that cause pollution (see Figure 10.2). With less energy being used for air conditioning, cool roofs and shaded areas result in less fuel consumption, air pollution, and greenhouse gas emissions. Urban greenery, cool pavement, and cool roofs keep the city cool and stop smog from forming. Cool surfaces such as roofs and pavements cause a negative radiation force by reflecting short-wave radiation to space. These first results spark a wave of investigation and public interest in reducing urban heat islands.

10.7.1 Development of Reflective Materials

Materials utilized in the urban fabric have a key influence on the thermal balance of the city by absorbing incident solar and infrared radiation and releasing some of the stored heat through convective and radiative processes in the atmosphere. As a result, ambient temperatures rise. Therefore, using "cool materials" in building envelopes and open spaces that can reflect a sizable portion of solar radiation and dissipate the heat they have absorbed through radiation helps to raise the urban albedo, maintain cooler surface temperatures, and thus present an effective way to reduce urban heat islands. The past few years have seen a lot of activity in the field of solar reflective materials, which has resulted in the creation of novel materials and methods that exhibit increased radiative qualities and better thermal characteristics, offering substantial promise for heat island abatement.

10.7.2 Development of Cool Roof Technologies

Due to their high solar reflectance, or albedo, cool roofs are solutions that are highly reflective of solar radiation. They can stop both individual buildings and huge cities from overheating. Since the 1980s in California and other US states, their potential has been quantitatively studied in response to both the UHI impact and the need to lower energy and peak power for air conditioning [51]. Numerous studies have since been conducted, showing the value of establishing sponsored incentive programs, product labeling, and standards to encourage the use of high-albedo materials for buildings, as well as the benefits of gradually increasing a city's albedo by selecting high-albedo surfaces to replace darker materials during routine maintenance of roofs [52]. Surveys of cool roofing materials for the United States were conducted, and significant reductions in peak power and cooling energy were demonstrated when researchers started concentrating on the resilience of high-albedo roof coatings, as reported by Akbari et al. [50]. Then, cities in the warm region of the United States started to take action to include cool roofs under the new ASHRAE construction standards and to incorporate cool surfaces as tradeable smog-offset credits in Los Angeles.

There are two distinct families of cool roofing solutions: cool white technologies, which are more frequently used for flat roof coverings, and cool color technologies, which are used for sloped roofs. The latter ones have a large capacity for reflection in the near-infrared (NIR, 0.7–2.5 mm), where more than 50% of solar energy falls but is invisible to the human eye, and a reflection spectrum in the visible range (0.4–0.7 mm), as needed to get a desired color.

There are many different types of cool white roofing materials, including built-up roofing, pre-painted metal roofs, field-applied coatings (paints, fluid-applied membranes, etc.), reinforced bitumen sheets made of modified bitumen (elastomeric or elastomeric), single-ply sheets and membranes (thermoset or thermoplastic), and tiles (ceramic, concrete, etc.). Non-metallic materials often exhibit thermal emittance in the range of 80% to 95% and initial solar reflectance as high as 80–85%. Its white color is the result of a large capacity for reflection in the visible region, but to produce a high value of reflectance in the near-infrared, a similar capacity is required. The use of white pigments like titanium dioxide (TiO_2) allow cool roof technologies to achieve a high reflectivity over the entire sunlight spectrum. The pigments can be spread in inorganic matrices like ceramic tiles and coatings as well as organic matrices like acrylic or bituminous binders. It's interesting to note that a high solar reflectance may be caused by the mean of several quite distinct reflection spectra that have been balanced by the solar irradiance spectrum. On the other hand, due to material degradation (chemical and physical), biological development, and—most importantly—soiling brought on by pollutant deposition, it is exceedingly difficult to maintain the initial reflectance value. A layer of particulate or other atmospheric suspensions may have a significant impact on the reflective performance since the most superficial matter has the highest reflectivity. As a result, aged solar reflectance ratings that were achieved through natural exposure to three different climatic conditions for at least three years are also offered within the context of the CRRC rating program. A technique for accelerated aging has recently been devised that can shorten the three-year natural aging process caused by soiling to three days of laboratory testing [53].

While biocides can be used to prevent biological growth, which is a problem in humid climates because of the cool roofs' low surface temperatures and the persistent condensation of atmospheric moisture, the development of matrices and white pigments that are chemically and physically stable allows for preventing reflectance degradation, such as that which is connected to surface yellowing. To lessen soiling, surface treatments that are extremely hydrophilic or hydrophobic can be used. You can also control the surface's porosity and roughness or employ self-cleaning coatings based on photocatalysis [54]. The most important requirements to prevent the deterioration of cool white roofs, however, are to choose a slope that ensures storm water outflow and to stay away from areas of stagnation.

Cool color technologies are being created for roof surfaces whose color is restricted by vegetation, such as tile roofs in the middle of old towns, or possibly by architectural choices. Solutions can be based on the choice of suitable pigments with selective reflection or by a successful method based on a multilayered coating in which a white, highly reflective basecoat is applied to a substrate and is followed by a topcoat with selective transparency. Yet the existence of structures that are near one another and form a so-called urban canyon might lead to radiation entering the canyon and being reflected back and forth multiple times before failing to reemerge. Roofs, facades, and other building surfaces not facing the sky can all use cool color technologies.

10.7.3 Development of Cool Pavement Technologies

The urban climate is greatly influenced by pavement, which also significantly contributes to the growth of the urban heat island phenomenon. Asphalt or concrete pavers cover a sizable portion of urban areas in the majority of developed nations. According to studies conducted in the United States, paved surfaces cover between 29% and 39% of metropolitan areas as viewed from above the urban canopy [50].

The amount of short-wave radiation received, long-wave radiation emitted and absorbed, convective heat transfer, gains and losses to the ground through conduction, and evaporation and condensation processes all contribute to the thermal balance of paved surfaces. According to Asaeda et al. [55], the amount of infrared radiation that pavements generate in Tokyo, Japan, is almost equal to half of the pace at which the city's business sectors consume energy.

Pavements' surface temperatures must drop to reduce the sensible heat they emit into the sky. The following methods can be used to accomplish this [14]:

a) Improving the reflectivity of pavement. Pavement surfaces with increased surface albedo reduce absorbed solar radiation and lower surface temperature. A variety of techniques have been put forth, both for the creation of brand-new paving surfaces and for the renovation of already-existing pavement. Resin-based pavement, utilizing light aggregates in asphaltic and concrete pavements, adding fly ash and slag cement as additives, using chip or sand sealing, using colorless reflecting binders, using sand/shot blasting, and abrading binder surfaces are some of the more well-liked techniques. Asphalt pavement can be modified by adding high-albedo materials, such as light-colored aggregates, color pigments, or sealants. In order to increase albedo after installation, asphalt pavement can also be maintained with treatments like chip seals (high-albedo aggregate used to resurface asphalt roads and highways), white topping (a layer of concrete applied over existing asphalt that is >10 cm thick or 10 cm thin), and micro surfacing (a thin sealing layer). To increase the albedo of paved surfaces, new technologies have been proposed. These include the use of white high-reflective paints on the pavement, infrared reflective color paints on the surface of the paved materials, heat reflective paints to cover the aggregates, and color-changing paints on the surface of the paved zones.

b) Increasing the permeability of the pavements to hasten the evaporation processes. This is also known as pervious, porous, or water-retaining pavement. Vegetative pavements, which comprise grass pavers and concrete grid surfaces that permit grass to grow in the intervals between them, show a lower surface temperature due to increased evapotranspiration. Examples of non-vegetative porous pavements include porous and pervious concrete, porous or rubberized asphalt, permeable interlocking concrete, etc. High-performance permeable pavement technologies used today include adding bottom ash and peat moss to old concrete, adding fine-texture old mortar to pervious concrete, adding water-holding fillers made of steel byproducts, adding fine blast furnace powder to water-retentive asphaltic pavements, and adding fly ash with narrow particle distributions to ceramic tiles.
c) Enhancing the pavements' perceived thermal capacity. This is mostly accomplished by incorporating phase change elements within the mass of the paving surfaces. The sensible heat discharged to the atmosphere can be reduced during the day by using PCMs to achieve a lower surface temperature.
d) Incorporating mechanical devices that disperse heat from the concrete or asphalt and lower the surface temperature of the pavements, such as tubes that circulate water.
e) Effectively shading the pavement surfaces to reduce the amount of solar radiation absorbed.

All of the aforementioned technologies are frequently utilized in large-scale projects, and the majority of them have a very high potential for reducing the amount of pavement, reducing the amount of sensible heat discharged into the atmosphere, and lessening the effects of UHI in urban areas. Santamouris provides detailed information on the current applications [14].

10.7.4 Greening the Urban Environment, the Impact of Trees in the City

Urban vegetation can improve the microclimate through processes like shade, evapotranspiration, controlling airflow, and heat exchange. Many studies have emphasized the value of urban parks as well as the strategic positioning of plants in landscape design. Urban parks significantly lower the average temperature in the nearby urban areas, while also reducing noise pollution, preventing soil erosion, stabilizing the soil, and calming visitors.

In cities, parks have a significant potential for mitigation. The precise impact on a city's climate quality is determined by several intricate regional and local factors, including the park's size and structure, the weather in the area, the plants used, how frequently they are watered, the thermal balance around the park, and the thermal characteristics of the entire city. The majority of current studies seeking to determine the ideal park size concluded that the larger the park, the higher its potential for mitigating. Yet research on the cooling potential of small and medium-sized urban parks found that they have a big impact on the climate in the area.

Because of the substantial pressure differential caused by the temperature difference between the park and the nearby urban areas, cool air can be transported from the park to the cities. According to Spronken-Smith [56], the impacted urban zone is only extended to a radius around the park's breadth. The size of the park has a significant impact on the extent of the affected urban zone. According to other experimental research, depending on the wind direction and speed as well as the thermal balance of the surrounding areas, the length of the affected urban area may be increased by 200 to 1000 m from the park limits. In reality, urban areas with a large anthropogenic heat output show very little influence from urban parks.

The next difficulty is to apply this information to industry practice now that the advantages of greenery are widely understood. The approaches for evaluating landscape plans in terms of plant cooling potential have not yet been fully defined, even though there are already frameworks in place to objectify landscape planning procedures like maintainability and irrigation. The introduction of other types of flora, such as vertical and rooftop greenery, exacerbates the difficulty of visualizing the thermal impact of any landscape design.

Scientists can now comprehend the role that vegetation plays in landscape design because of recent advances in modeling the outside thermal environment. Climatic maps can be used as a visualization tool at all scales. Several layers of spatial data can be analyzed simultaneously using the geographical information systems (GIS) platform, and climatic mapping has grown in importance in studies of the outdoor climate. Particularly, the mapping of green spaces has made substantial use of GIS, offering the opportunity to suggest landscape solutions through GIS mapping techniques.

Planning a landscape should take into account the objective selection and arrangement of plants. The importance of context and locality has been brought to light by the inclusion of thermal simulations in the investigations. The positioning of plants is substantially impacted by solar exposure, which dispels the widespread misconception that plants can improve the environment by uniformly lowering the temperature. Urban greenery must be utilized to its full potential as an ecosystem resource and in landscape planning to more fully exploit the cooling advantages of greenery as a method of reducing urban heat.

10.7.5 Actual Development of Green Roof Technologies

There are two types of green roofs: extensive and intensive. The old approach is characterized by a low-level planting of sedum, brush, or lawn and a thin coating of the growing media. This solution uses little water and needs no general upkeep. Intense green roofs are gardens with a thicker growing media, allowing rooted plants like shrubs and trees to develop and live; they are distinguished by higher water requirements and sufficient structural supports. In some instances, a different type, the semi-intensive green roof, is also utilized [57].

While a cool roof is an easy building solution to install for both new and existing buildings, a green roof is a more complicated technology made up of several layers, including, in the more general case, from bottom to top: the structural substrate of the roof, the waterproofing membrane, the root barrier, the thermal insulation layers (which are occasionally present or are sometimes included in the structural substrate of the roof), the aeration, and drainage systems.

Although the technology began to acquire popularity in the second half of the 20th century due to the increased thermal insulation offered by the systems, green roofs have been known and used since ancient times. In addition to serving this purpose, green roofing has recently emerged as a critical building technology. Stormwater management, water runoff control, improved urban air quality, an extension of roof life, greater architectural quality, and biodiversity are all advantages of green roofs.

The lack of green space in urban contexts is one of the most significant causes of the urban heat island phenomenon, and green roofs have been cited as a vital method to reduce outdoor air temperature [58]. The effectiveness and potential of green roofs as a type of mitigation depend on several (micro)climatic factors, including air temperature, solar radiation, ambient humidity, precipitation, and wind speed.

While several researchers have shown how green roofs can enhance the thermal and energy efficiency of built environments, both in the heating and cooling seasons, the same cannot be said for studies devoted to the calculation of the reduction of urban heat islands. In the latter situation, two methodologies are essentially used: evaluation of the latent and sensible heat flux from green roofs, and evaluation of the urban heat island reduction by mesoscale simulation assessments.

The city of Chicago, Illinois, in the United States, which is heavily engaged in mitigation efforts and where 50,000 m^2 of green roofs were put in 2008, was the subject of a mesoscale study by Smith and Roeber [59] that calculated the potential for mitigation. The latent phenomena were ignored since the study was conducted under the assumption that an extensive green roof would function similarly to a conventional roof with 0.8 solar reflectance. As a result of the green roofs replacing the old ones, the results showed a 2–3 K reduction in air temperature between 7:00 and 11:00 pm.

Real-world studies demonstrate the green roof technology's capacity to reduce the sensible and latent heat discharged into the urban environment, hence minimizing the urban heat island effect. To accurately forecast the drop in urban temperature as a function of climate, green roof characteristics, the area of a city covered by green roofs, and the distance between the roof and the street level, more research on the subject is still needed. To order to fully utilize the potential of the technology and benefit from market penetration to lower installation costs and increase the technology's cost-effectiveness, experimental, numerical, and theoretical studies should address these concerns in the upcoming year.

10.7.6 Mitigation of UHI Effects to Save Energy

At the same scales where the UHI affects energy (such as building and urban scales), these scales can also be used to address its mitigation (reversing or offsetting its impacts).

Hence, the objective at the building scale is to chill the building envelope, the nearby ambient air, or both. Instead, it's intended to keep the envelope and surrounding air from heating up as much as they otherwise would. According to current thought, high-albedo (reflective) materials should be used to cool the building envelope to lower the quantity of solar radiative flux absorbed and subsequently the amount of heat carried through the structure. The increased vegetative cover encourages evaporative cooling to cool the ambient air surrounding a building. The shade provided by the tree canopy also benefits the building's exterior by greatly reducing solar heat gain. More urban vegetation, especially of the evergreen variety, has energy benefits as well as provides wind protection. Even though it simply has a passing connection to heat islands, this has a considerable impact on energy. For instance, trees protect buildings from chilly winds in winter, preventing colder air from entering inner rooms and lowering the demand for heating energy, especially at higher latitudes. This effect is only significant for one- or two-story buildings because these types of buildings can benefit from wind protection provided by trees. Tree canopies at ground level do not provide shade for upper floors in taller buildings. Further advantages of urban greenery include bettering air quality by capturing and storing airborne contaminants. Also, it has aesthetic traits and values.

At the urban scale, the fundamental idea is to use high-albedo materials on built-up surfaces such as sidewalks, roads, parking lots, pavement, and roofs in addition to structures (such as roofs). A reforestation initiative in which numerous trees are planted all across urban areas would also be part of the approach. The results of such widespread adoption of reforestation and urban albedo initiatives would be to lessen, entirely offset, or occasionally more than balance the urban heat island. Calculations demonstrate that enhanced urban albedo can cause the regional air temperature to drop by 4°C or more at specific locations and times. These simulation experiments also demonstrate that a 30% increase in vegetative canopy cover can reduce air temperature by at least 3°C. Mesoscale meteorological modeling suggests that a drop in air temperature of 1–2°C is more prevalent in metropolitan areas. Hence, widespread use of high-albedo materials and urban forestry can reduce UHI in the majority of the cities previously addressed. The utility-scale effects of such initiatives would be savings of 4–10% from the summer afternoon peak electric demand for every 2°C drop in the UHI.

10.7.7 Other Mitigation Technologies

Many mitigation strategies have been created, refined, and successfully tried in actual projects to lessen the impact of heat islands. In addition to using cool materials and vegetation, they have also been used to dissipate surplus heat into low-temperature environmental heat sinks such as the ground, water, and surrounding air.

A fundamental and crucial strategy for enhancing the urban microclimate is the shading of outdoor areas. The thermal comfort conditions can be improved while the radiative temperature is significantly reduced with the help of effective shading systems. By allowing the air to flow and move through them, the use of proper permeable materials in open spaces can significantly lower both the ambient and radiative temperatures. Shade-giving furnishings and constructions like tents, pergolas, canopies, and other similar items can provide shade for the urban environment. Urban surfaces that are shaded receive less solar radiation, which results in cooler temperatures. As a result, the temperature in open areas is reduced. Several academics have already examined the effectiveness of shade in enhancing the urban microclimate. An unprotected courtyard with a mid-afternoon maximum air temperature of 34°C was compared to a comparable courtyard treated with a fabric shade mesh by Shashua-Bar et al. [60]. There was a minimum 1.5°C drop in air temperature. An important factor in urban environments that affects thermal environments and long-term thermal comfort was examined in another study carried out outdoors on a university campus in central Taiwan [61]. This study conducted several field experiments to examine the outdoor thermal conditions on urban streets in central Taiwan. Ten years of meteorological data were used to forecast long-term thermal comfort using the Rayman model. According to analytical findings, summertime temperatures in partly shaded places are often very hot, especially at noon. Yet extremely darkened areas typically have low physiologically comparable temperatures (PET revealed that scarcely shaded areas had high SVFs in the summer while highly shaded areas had low SVFs in the winter). Several shading levels and a variety of shading forms were suggested for outdoor area design. In terms of heat sinks, it is generally known that the ground temperature is rather stable and low throughout the year at a depth of about 2.5 to 3 m. By using ground-to-air heat exchangers—tubes buried in the ground—it is possible to benefit from the ground's warmth. The air enters the tubes, travels through the earth, and then leaves at a reduced temperature. This system may be referred to as "earth tubes" or "ground-coupled air heat exchangers," depending on the context. Some researchers have used the concept of earth cooling to transfer extra heat from the outdoors to the ground.

Furthermore, due to its thermal and optical qualities, water exhibits several benefits as a natural cooling method to cool the urban environment. Since water has a high specific heat that is on the order of four times that of typical urban materials, its thermal inertia is also four times greater. The thermal inertia has a dual impact, delaying and buffering the maximum temperature. Latent heat is a significant factor in the evaporation of water. When a drop of water evaporates, energy is released into the air and water around it, making them colder as a result. The low reflectivity of water results in reduced sun reflection on nearby surfaces, preventing their warming in the process.

Due to the increased evaporation throughout the day, water bodies have been widely employed to control the temperature of metropolitan areas. For example, some researchers have noted large drops in air temperature near metropolitan rivers and lakes. Due to river proximity, Hathway and Sharples [62] discovered considerable drops in urban air temperatures. They also claim that the cooling power is at its peak in early spring and begins to decline in June. When used in an urban setting, each of the aforementioned mitigation strategies reduces the heat island effect effectively. But depending on the exact site and weather patterns, combining some or all of the tactics will offer several advantages.

10.8 CONCLUSIONS

The most well-known effect of climate change that poses serious energy and environmental issues to cities is urban heat islands. Important research has been done to create effective mitigation solutions that can lower surface and ambient temperatures in urban areas to balance the effects of the phenomenon. Urban heat island mitigation reduces energy use and energy expenditure, enhances ambient conditions and air quality in cities, and contributes to the fight against global warming. Many implementation solutions are available and in use, and numerous administrative and scientific initiatives assist in the use of UHI mitigation measures.

To encourage cities to commit to building implementation programs, share their expertise with other cities, and create platforms for shared implementation plans, one example is the "100 Cool Cities" campaign. To measure the solar reflectance of roofing and paving materials, standards have been devised. Labeling groups like the European Cool Roof Council (ECRC 2015) and the Cool Roofs Rating Council (CRRC 2015) use these criteria to create programs and label products. To assist producers in evaluating the old optical performance of recently invented materials, an accelerated aging standard has been created. The solar reflectance of roofing materials has been incorporated into the building's minimum specific requirements by updating the building energy codes and standards. When it comes to promoting and implementing cool roofs, these guidelines are quite successful. Global Cool Cities Alliance (GCCA 2015), a non-profit group, was established to spread awareness of the cool city initiative worldwide. GCCA is now involved in assisting numerous nations with the implementation of cool pavement and roofing in their cities.

Although cool pavement and cool roofs are regarded as general technologies, city-specific programs must be created to take advantage of local resources and implementation techniques. Cities should start initiatives to:

- Calculate the effects of the measures on energy and air quality and perform a thorough analysis;
- Create and maintain an accurate land use/land cover database;
- Create specialized implementation plans (for roofs, paving, and trees);
- Synchronize efforts with regional and national organizations;
- Create regional energy standards, codes, and guidelines;
- Create a feedback system to enhance individualized programs;
- Create demonstration initiatives to increase public awareness.

REFERENCES

1. Grimm NB, Faeth SH, Golubiewski NE, Redman CL, Wu J, Bai X, Briggs JM. Global change and the ecology of cities. Science. 2008; 319(5864): 756–760. https://doi.org/10.1126/science.1150195
2. Kim YH, Baik JJ. Spatial and temporal structure of the urban heat island in Seoul. J Appl Meteorol Climatol. 2005; 44(5): 591–605. https://doi.org/10.1175/JAM2226.1
3. World Population Prospects 2019: United Nation. https://population.un.org/wpp/Publications/
4. Baud ISA, De Wit J (Eds.). New forms of urban governance in India: Shifts, models, networks and contestations. SAGE Publications India, 2009.
5. Bokaie M, Zarkesh MK, Arasteh PD, Hosseini A. Assessment of urban heat island based on the relationship between land surface temperature and land use/land cover in Tehran. Sustain. Cities Soc. 2016; 23: 94–104. https://doi.org/10.1016/j.scs.2016.03.009
6. Wang Q, Wang Y, Fan Y, Hang J, Li, Y. Urban heat island circulations of an idealized circular city as affected by background wind speed. Build Environ. 2019; 148: 433–447. https://doi.org/10.1016/j.buildenv.2018.11.024

7. Kotharkar R, Surawar M. Land use, land cover, and population density impact on the formation of canopy urban heat islands through traverse survey in the Nagpur urban area, India. J Urban Plan Dev. 2016; 142(1): 04015003. https://doi.org/10.1061/(ASCE)UP.1943-5444.0000277
8. Imhoff ML, Zhang P, Wolfe RE, Bounoua L. Remote sensing of the urban heat island effect across biomes in the continental USA. Remote Sens Environ. 2010; 114(3): 504–513. https://doi.org/10.1016/j.rse.2009.10.008
9. Grimmond CSB. Progress in measuring and observing the urban atmosphere. Theor. Appl. Climatol. 2006; 84: 3–22. doi: 10.1007/s00704-005-0140-5
10. Arnfield AJ. Two decades of urban climate research: A review of turbulence, exchanges of energy and water, and the urban heat island. Int J Climatol. 2003; 23(1): 1–26. https://doi.org/10.1002/joc.859
11. Gawuc L. Relationship between Surface Urban Heat Island intensity and sensible heat flux retrieved from meteorological parameters observed by road weather stations in urban area. In EGU General Assembly Conference, held 23–28 April, 2017 in Vienna, Austria. 2017. P. 16820.
12. Stewart ID, Oke TR. Local climate zones for urban temperature studies. Bull Am Meteorol Soc. 2012; 93(12): 1879–1900. https://doi.org/10.1175/BAMS-D-11-00019.1
13. Rosenzweig C, Solecki WD, Hammer SA, Mehrotra S (Eds.). Climate change and cities: First assessment report of the urban climate change research network. Cambridge University Press, 2011.
14. Santamouris, M. Environmental design of urban buildings: an integrated approach. Routledge, 2013.
15. Li D, Bou-Zeid E. Synergistic interactions between urban heat islands and heat waves: The impact in cities is larger than the sum of its parts. J Appl Meteorol Climatol. 2013; 52(9): 2051–2064. https://doi.org/10.1175/JAMC-D-13-02.1
16. Wilby RL. A review of climate change impacts on the built environment. Built Environ. 2007; 33(1): 31–45. https://doi.org/10.2148/benv.33.1.31
17. Kolokotsa D, Santamouris M. Review of the indoor environmental quality and energy consumption studies for low income households in Europe. Sci. Total Environ. 2015; 536: 316–330. https://doi.org/10.1016/j.scitotenv.2015.07.073
18. Menne B. Protecting health in Europe from climate change. World Health Organization, 2008.
19. Taha H. Heat islands and energy. Encyclopedia of Energy. 2004; 3: 133–143.
20. Goel A, Bhatt R. Causes and consequences of Global Warming. Int J Life Sci Biotechnol Pharma Res. 2012; 1(1): 27–31.
21. Andersson E. Urban landscapes and sustainable cities. Ecol Soc. 2006; 11(1):34. doi: www.jstor.org/stable/26267821
22. Mohajerani A, Bakaric J, Jeffrey-Bailey T. The urban heat island effect, its causes, and mitigation, with reference to the thermal properties of asphalt concrete. J Environ Manag. 2017; 197: 522–538.
23. Park HS. Features of the heat island in Seoul and its surrounding cities. Atmos. Environ. (1967). 1986; 20(10): 1859–1866. https://doi.org/10.1016/0004-6981(86)90326-4
24. Oke TR. City size and the urban heat island. Atmos. Environ. (1967). 1973; 7(8): 769–779. https://doi.org/10.1016/0004-6981(73)90140-6
25. Ichinose T, Shimodozono K, Hanaki K. Impact of anthropogenic heat on urban climate in Tokyo. Atmos Environ. 1999; 33(24–25): 3897–3909. https://doi.org/10.1016/S1352-2310(99)00132-6
26. Kim YH, Baik JJ. Maximum urban heat island intensity in Seoul. J. Appl. Meteorol. 2002; 41(6): 651–659. https://doi.org/10.1175/1520-0450(2002)041%3C0651:MUHIII%3E2.0.CO;2
27. Taha H. Urban climates and heat islands: Albedo, evapotranspiration, and anthropogenic heat. Energy Build. 1997; 25(2): 99–103. https://doi.org/10.1016/S0378-7788(96)00999-1

28. Garratt JR. The atmospheric boundary layer. Earth Sci Rev. 1994; 37(1–2): 89–134. https://doi.org/10.1016/0012-8252(94)90026-4
29. Xia G, Zhou L, Freedman JM, Roy SB, Harris RA, Cervarich MC. A case study of effects of atmospheric boundary layer turbulence, wind speed, and stability on wind farm induced temperature changes using observations from a field campaign. Clim. Dyn. 2016; 46: 2179–2196. doi: 10.1007/s00382-015-2696-9
30. Miao S, Chen F, LeMone MA, Tewari M, Li Q, Wang Y. An observational and modeling study of characteristics of urban heat island and boundary layer structures in Beijing. J Appl Meteorol Climatol. 2009; 48(3): 484–501. https://doi.org/10.1175/2008JAMC1909.1
31. Peng S, Piao S, Ciais P, Friedlingstein P, Ottle C, Breon FM, Myneni RB. Surface urban heat island across 419 global big cities. Environ. Sci. Technol. 2012; 46(2): 696–703. https://doi.org/10.1021/es2030438
32. Weng Q, Lu D, Schubring J. Estimation of land surface temperature–vegetation abundance relationship for urban heat island studies. Remote Sens Environ. 2004; 89(4): 467–483. https://doi.org/10.1016/j.rse.2003.11.005
33. Xia G, Zhou L, Freedman JM, Roy SB, Harris RA, Cervarich MC. A case study of effects of atmospheric boundary layer turbulence, wind speed, and stability on wind farm induced temperature changes using observations from a field campaign. Clim. Dyn. 2016; 46: 2179–2196. doi:10.1007/s00382-015-2696-9
34. Gallo KP, Owen TW. Satellite-based adjustments for the urban heat island temperature bias. J Appl Meteorol Climatol. 1999; 38(6): 806–813. https://doi.org/10.1175/1520-0450(1999)038%3C0806:SBAFTU%3E2.0.CO;2
35. Anniballe R, Bonafoni S, Pichierri M. Spatial and temporal trends of the surface and air heat island over Milan using MODIS data. Remote Sens Environ. 2014; 150: 163–171. https://doi.org/10.1016/j.rse.2014.05.005
36. Stone B, Hess JJ, Frumkin H. Urban form and extreme heat events: Are sprawling cities more vulnerable to climate change than compact cities. Environ. Health Perspect. 2010; 118(10): 1425–1428. https://doi.org/10.1289/ehp.0901879
37. Kolokotroni M, Davies M, Croxford B, Bhuiyan S, Mavrogianni A. A validated methodology for the prediction of heating and cooling energy demand for buildings within the Urban Heat Island: Case-study of London. Sol Energy. 2010; 84(12): 2246–2255. https://doi.org/10.1016/j.solener.2010.08.002
38. Kolokotroni M, Giridharan R. Urban heat island intensity in London: An investigation of the impact of physical characteristics on changes in outdoor air temperature during summer. Sol Energy. 2008; 82(11): 986–998. https://doi.org/10.1016/j.solener.2008.05.004
39. Vittinghoff E, Glidden DV, Shiboski SC, McCulloch CE. Regression methods in biostatistics: linear, logistic, survival, and repeated measures models. Springer, 2006.
40. Breiman L. Classification and regression trees. Routledge, 2017. https://doi.org/10.1201/9781315139470
41. Breiman, L. Random forests. Mach. Learn. 2001; 45: 5–32. https://doi.org/10.1023/A:1010933404324
42. Sussman HS. The Urban Heat Island of Bengaluru, India: Characteristics, Trends, and Mechanisms. State University of New York, 2022.
43. Littlefair P, Santamouris M, Alvarez S, Dupagne A, Hall D, Teller J, Papanikolaou N. Environmental site layout planning: solar access, microclimate and passive cooling in urban areas. CRC, 2000.
44. Mastrapostoli E, Karlessi T, Pantazaras A, Kolokotsa D, Gobakis K, Santamouris M. On the cooling potential of cool roofs in cold climates: Use of cool fluorocarbon coatings to enhance the optical properties and the energy performance of industrial buildings. Energy Build. 2014; 69: 417–425. https://doi.org/10.1016/j.enbuild.2013.10.024
45. Kolokotsa D, Santamouris M. Review of the indoor environmental quality and energy consumption studies for low income households in Europe. Sci. Total Environ. 2015; 536: 316–330. https://doi.org/10.1016/j.scitotenv.2015.07.073

46. Sakka A, Santamouris M, Livada I, Nicol F, Wilson M. On the thermal performance of low-income housing during heat waves. Energy Build. 2012; 49: 69–77. https://doi.org/10.1016/j.enbuild.2012.01.023
47. Kolokotroni M, Davies M, Croxford B, Bhuiyan S, Mavrogianni A. A validated methodology for the prediction of heating and cooling energy demand for buildings within the Urban Heat Island: Case-study of London. Sol Energy. 2010; 84(12): 2246–2255. https://doi.org/10.1016/j.solener.2010.08.002
48. World Health Organization. WHO guidelines for indoor air quality: Selected pollutants. World Health Organization, Regional Office for Europe, 2010.
49. Mavrogianni A, Davies M, Batty M, Belcher SE, Bohnenstengel SI, Carruthers D, Ye Z. The comfort, energy and health implications of London's urban heat island. Build Serv Eng Res Technol. 2011; 32(1): 35–52. https://doi.org/10.1177/0143624410394530
50. Akbari H. Cooling our communities. A guidebook on tree planting and light-colored surfacing. 2009. https://escholarship.org/uc/item/98z8p10x
51. Taha H, Akbari H, Rosenfeld A, Huang J. Residential cooling loads and the urban heat island—the effects of albedo. Build Environ. 1988; 23(4): 271–283. https://doi.org/10.1016/0360-1323(88)90033-9
52. Synnefa A, Santamouris M, Akbari H. Estimating the effect of using cool coatings on energy loads and thermal comfort in residential buildings in various climatic conditions. Energy Build. 2007; 39(11): 1167–1174. https://doi.org/10.1016/j.enbuild.2007.01.004
53. Sleiman M, Kirchstetter TW, Berdahl P, Gilbert HE, Quelen S, Marlot L, Destaillats H. Soiling of building envelope surfaces and its effect on solar reflectance–Part II: Development of an accelerated aging method for roofing materials. Sol. Energy Mater Sol. Cells. 2014; 122: 271–281. https://doi.org/10.1016/j.solmat.2013.11.028
54. Diamanti MV, Paolini R, Zinzi M, Ormellese M, Fiori MPG, Pedeferri M. Self-cleaning ability and cooling effect of TiO2-containing mortars. In Nanotech 2013: Technical Proceedings of the 2013 Nsti Nanotechnology Conference and Expo. 2013. 716–719.
55. Asaeda T, Ca VT, Wake A. Heat storage of pavement and its effect on the lower atmosphere. Atmos. Environ. 1996; 30(3): 413–427. https://doi.org/10.1016/1352-2310(94)00140-5
56. Spronken-Smith RA. Energetics and cooling in urban parks (Doctoral dissertation, University of British Columbia), 1994.
57. Theodosiou T. Green roofs in buildings: Thermal and environmental behaviour. Adv. Build. Energy Res. 2009; 3(1): 271–288. https://doi.org/10.3763/aber.2009.0311
58. Synnefa A, Karlessi T, Gaitani N, Santamouris M, Assimakopoulos DN, Papakatsikas C. Experimental testing of cool colored thin layer asphalt and estimation of its potential to improve the urban microclimate. Build Environ. 2011; 46(1): 38–44. https://doi.org/10.1016/j.buildenv.2010.06.014
59. Smith KR, Roebber PJ. Green roof mitigation potential for a proxy future climate scenario in Chicago, Illinois. J Appl Meteorol Climatol. 2011; 50(3): 507–522. https://doi.org/10.1175/2010JAMC2337.1
60. Shashua-Bar L, Pearlmutter D, Erell E. The influence of trees and grass on outdoor thermal comfort in a hot-arid environment. Int J Climatol. 2011; 31(10): 1498–1506. https://doi.org/10.1002/joc.2177
61. Hwang RL, Lin TP, Matzarakis A. Seasonal effects of urban street shading on long-term outdoor thermal comfort. Build Environ. 2011; 46(4): 863–870. https://doi.org/10.1016/j.buildenv.2010.10.017
62. Hathway A, Sharples S. Urban river microclimates. In PLEA 2011–Architecture and Sustainable Development, Conference Proceedings of the 27th International Conference on Passive and Low Energy Architecture. 2011. 183–187.

11 Well-Being, Thermal Comfort, and Environmental Liveability
Adaptation Studies

Saurabh Pathak, Ashish Panat, and R. K. Khandal

11.1 INTRODUCTION

Generally, a comfortable building envelope consists of proper lighting, air ventilation, and thermal comfort. Human comfort can be described as the mental state of communicating happiness with the surroundings in terms of the five basic parameters: thermal, auditory, visual, indoor air, and spatial. Beyond the design criteria for healthy buildings, the connection between architecture and health has traditionally received little consideration. The development of sustainable buildings at the turn of this century was initiated to efficiently meet these requirements while using the least amount of energy. Academic research is still being done on the effectiveness of high-tech automation systems in real-world settings. In this chapter, the human body is considered a system or occupant, and living space is referred to as the surrounding; apart from this, the region that does not affect the system is considered an environment. Various case studies have been discussed to emulate the well-being zone for a human being in a sustainable building. In a building, human comfort is achieved by the adaptation ability of a system relative to the surrounding conditions. Based on conditions, a unique thermal sensation is experienced by the system relative to an age group. It is more crucial to include a wide variety of qualitative and quantitative health factors when discussing well-being in buildings than to concentrate on a few specifically specified criteria. To improve human well-being, the building design paradigm needs to shift from apparent parameters like temperature and humidity to more rational approaches that take care of the health of occupants in a building envelope. Conventional architecture design, which emphasizes the aesthetic look of the building, has to move on to modern green building design that supports natural lighting, ventilation, and living beings.

In recent years, evaluating building performance has become a significant research field. Some studies have used building evaluation methods to provide normative standards for comfort levels. One of a building's key attributes is thermal comfort. In order to comprehend the thermal preconditions provided in spaces better,

DOI: 10.1201/9781003415695-11

researchers work on thermal physiological models. The quality of the interior thermal temperature has a significant impact on user satisfaction within a building. This suggests that architects and interior designers should have information about what occupants anticipate from the indoor environment.

11.2 THERMAL COMFORT

The World Health Organization defines health not as the absence of ill-health but as "*a state of complete physical, mental and social well-being.*" This condition, also known as a neutral condition, differs from person to person. The concept of wellness has evolved, and it now encompasses knowledge of the connections between social, psychological, and medical variables. The chemical energy found in the food that a living human body consumes can be coupled to a heat engine, which uses that energy to produce work and heat (Schoen et al., 2013).

11.2.1 Metabolism

Metabolism is the process through which the chemical energy present in food is transformed into heat and work. The metabolic rate is the rate at which chemical energy is transformed into heat and work. The thermal efficiency of a person can be defined as the ratio of usable labour output to energy input, much like a heat engine. For brief periods, a person's thermal efficiency can range from 0% to as high as 15–20% (Parsons, 2014g).

11.2.2 Neutral Condition

Associated with the human body, two different temperatures are significant.

11.2.2.1 Skin Temperature

The outermost surface of the human body, or skin temperature (T_{skin}), is measured using a standard thermometer. A value of approximately 33.7°C is what is seen for a neutral state (Parsons, 2014c).

11.2.2.2 Core Temperature

The term "core temperature" refers to the internal body temperature (T_{core}) of a person as measured by a conventional thermometer. It is measured at around 36.8°C in neutral conditions (Parsons, 2014c).

Figure 11.1 depicts the state of the human body concerning core temperature variation; on a typical basis, the neutral condition core temperature of the human body is measured to be 36.8°C. When the body's core temperature is between 35°C and 36.8°C or 36.8°C and 39°C, the body experiences some minor pain. The efficiency of the human body suffers greatly if its core temperature is between 31°C and 35°C or 39°C and 43°C. Human bodies become fatal if their core temperatures fall below 31°C or increase above 43°C ("Heat Stress," 2014; Parsons, 2014b).

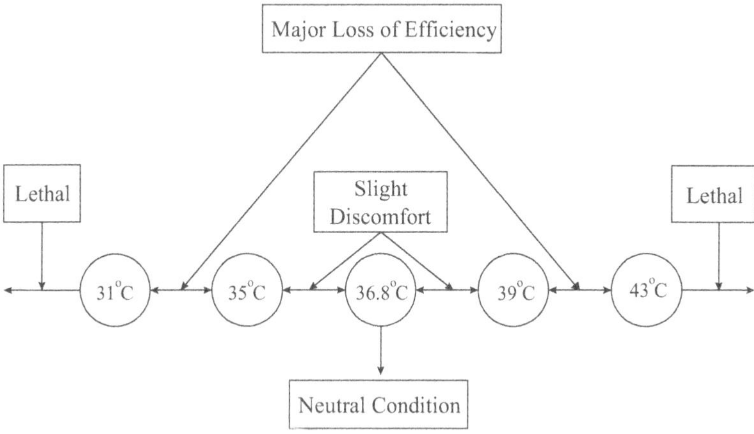

FIGURE 11.1 Condition of human body with variation in core temperature.

11.3 HEAT BALANCE EQUATION FOR A HUMAN BODY

The thermal energy balance between the human body and the thermal environment around it determines the body's temperature.

The heat balance equation 11.1 for the human body is (Parsons, 2014i):

$$Q_{gen} = Q_{sk} + Q_{res} + Q_{st} \qquad 11.1$$

Where:

Q_{gen} = Rate at which heat is generated in the body (can be calculated using equation 11.2) (Parsons, 2014h).

Q_{sk} = Total heat transfer rate from the skin.

Q_{res} = Heat transfer rate due to respiration.

Q_{st} = Rate at which heat is stored inside the body.

$$Q_{gen} = M(1-n) \approx \mathrm{M} \qquad 11.2$$

M = Metabolism rate

The thermal efficiency range of the human body is approximately zero (see equation 11.3), although it may rise up to 15% for short period of time (Parsons, 2014h).

$$\eta = \text{Thermal efficiency} \approx 0 \qquad 11.3$$

The activity determines the metabolic rate, expressed in "MET" units. A MET is determined to be equal to roughly 58.2 W/m^2, which is the sedentary person's metabolism rate per unit area. Table 11.1 displays the metabolic rate for a few specific activities.

TABLE 11.1
Activity Average Metabolic Rate (MET) (Fletcher et al., 2020)

Sr. no	Activity	Metabolism rate (MET)
01	Pilates	1.9
02	Basketball	4.0
03	Boxing	3.9
04	Circuits	3.2
05	Teaching	1.9
06	Fitness class	4.7
07	Netball	4.0
08	Rugby	4.0
09	Spinning	4.7
10	Table tennis	2.3
11	Testing	2.0

Given that the metabolic rate is determined by the amount of body surface area (naked), to get the total metabolic rate, it is critical to measure the area. The surface area of the human body is calculated using the Du-Bois equation 11.4 (Špelić et al., 2019) with the assumption that it is a cylinder with uniform heat generation and dissipation.

$$A_{hb} = 0.202m^{0.425}h^{0.725} \tag{11.4}$$

Where

A_{hb}= Surface area of human body (in m^2).

m = Mass of human body.

h = Height of human body.

As the Du-Bois equation considers the area of the naked body, a correction factor is needed to account for clothes. The surface area of the human body covered by clothing divided by the surface area uncovered may be used to define the correction factor. Hence, the heat produced may be easily determined from the metabolism rate and the surface area.

The amount of heat transfer rate by skin can be calculated using equation 11.5.

$$Q_{sk} = \pm Q_{conv} \pm Q_{rad} + Q_{evp} \tag{11.5}$$

Where

Q_{conv} = Heat transfer rate due to convection.

Q_{rad} = Heat transfer rate due to radiation.

Q_{evp} = Heat transfer rate due to evaporation.

Convection, radiation, and evaporation are the leading causes of heat transfer from the human body to the environment or vice versa through the skin of the human body. Although heat transfer from the environment to the human body is regarded

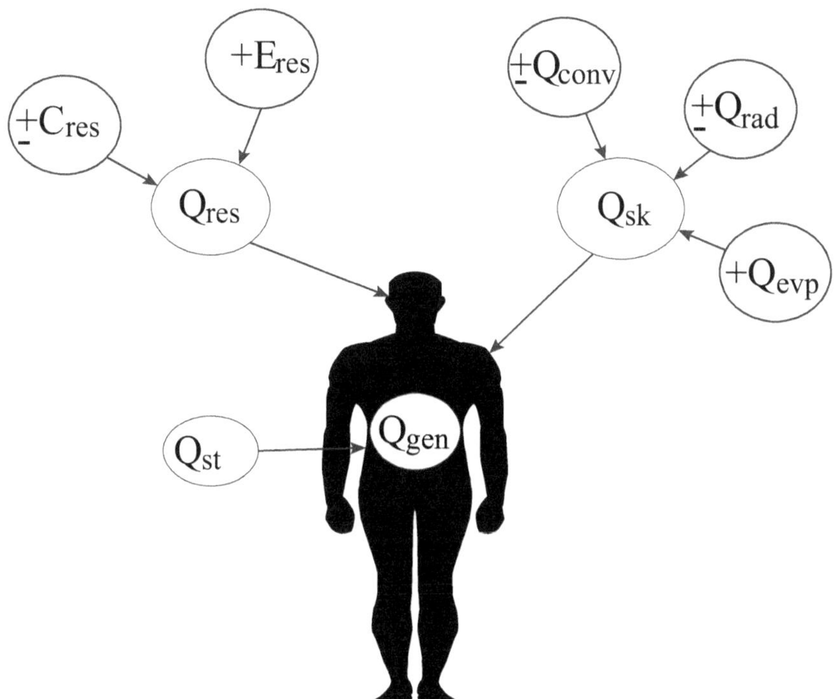

FIGURE 11.2 Heat interaction between the body and environment.

as unfavourable, the positive symbol used in the equation for the heat transfer rate via skin shows that heat is transferred from the human body to the environment. Evaporative heat transfer is always positive, unlike sensible heat transfer through convection and radiation, which can be either positive or negative. As shown in the illustration in Figure 11.2, if the ambient temperature is lower than the body's skin temperature, the body will lose heat by convection and radiation. Under the same situation, if moisture is on the skin, heat will be transferred from the human body to the environment owing to evaporation. Suppose the ambient temperature is higher than the body's skin temperature. In that case, the body will radiate and convect heat into the air, which is why a negative sign is employed (Parsons, 2014k).

The heat transfer rate due to respiration can be evaluated by equation 11.6.

$$Q_{res} = \pm C_{res} + E_{res} \quad 11.6$$

C_{res} = Dry heat loss or gain by the human body due to respiration.
E_{res} = Evaporative heat loss by the human body due to respiration.

The rate of heat storage in the body should be zero at the neutral condition for human comfort or survival. At neutral temperatures, a sedentary person loses around 40% of heat by evaporation, 30% through convection, and 30% through radiation. However, other factors could cause this ratio to be altered.

11.4 THERMOREGULATORY SYSTEM

11.4.1 Case 1. When the Environment Is Colder than the Neutral Zone

In this situation, heat loss from the body is more significant than heat input, causing the body's temperature to drop. The thermoregulatory system then tries to maintain the body's temperature.

11.4.1.1 Zone of Vasomotor Regulation against Cold (Vasoconstriction)

In this condition, blood vessels close to the skin constrict, limiting blood flow and heat transfer to the skin's immediate outer surface. When the outer skin tissue functions as an insulator, the body loses less heat and stays warmer.

11.4.1.2 Zone of Metabolism Regulation

Further environmental temperature drops mean vasomotor control will not offer a sufficient defence. Hence, to offset the increased losses, body heat synthesis is boosted by increased spontaneous activity and shivering. This effect can be seen most of the time throughout the winter season.

11.4.1.3 Zone of Inevitable Body Cooling

The body cannot prevent tissue cooling if the ambient temperature drops. The situation is quite dangerous, as the body temperature drops in these circumstances, necessitating immediate medical intervention. If emergency care is not sought and the body's core temperature falls below 31°C, death results.

11.4.2 Case 2. When the Environment Is Hotter than the Neutral Zone

In this situation, more heat is being added to the body than is being lost, which causes the body temperature to rise. To counteract this, the thermoregulatory system enters the picture.

11.4.2.1 Zone of Vasomotor Regulation against Heat (Vasodilation)

This causes the blood vessels next to the skin to enlarge, ultimately increasing blood flow and heat transfer to the skin's immediate outer surface. The increase in surface skin temperature raises the temperature at which heat can be transferred via convection and radiation. As a result, there is a slower rate of heat transfer from the environment to the body.

11.4.2.2 Zone of Evaporative Regulation

Sweat glands will grow more if the temperature in the environment rises further. The sweat glands become very active, producing copious amounts of perspiration on the skin's surface. An increase in evaporation can stop a rise in body temperature if the air's humidity and velocity allow it (Parsons, 2014a, 2014e).

11.4.2.3 Zone of Inevitable Body Heating

If the ambient temperature rises higher, so will the body temperature, creating an unavoidable zone of body heating. Because of this, considerable medical attention is needed in this area where the body temperature starts to rise. When the body's core temperature rises above 43°C without receiving medical assistance, the effect becomes fatal.

Following are the ill effects due to an increase in body temperature.

- Heat exhaustion
- Heat cramp
- Heat stroke

So, even while the human body has a regulatory mechanism that becomes inefficient beyond a certain point, ensuring the environment is conducive to a comfortable and secure way of life is still essential (Parsons, 2014a, 2014e, 2014f).

11.5 FACTORS AFFECTING THERMAL COMFORT

11.5.1 Physiological Factors

The physiological factor is one of the fundamental aspects that affect how quickly the human body burns calories. Age, exercise level, sex, and physical health of the individual make up the physiological parameters. The rate of metabolism is the key concern among all the activities. The physiological variables cannot be disregarded because the metabolic rate is a significant component in heat formation inside the body.

11.5.2 Insulating Factors due to Clothing

All of the scenarios discussed earlier are based on the heat balance equation for the human body, which also includes the nude body for calculating purposes. So, it is necessary to cause a change in the way that clothing transfers heat. The choice of clothes significantly impacts how quickly heat leaves the human body. "Clo" is the name of the unit used to measure the resistance provided by clothing.

$$\mathbf{1\ Clo = 0.155\ m^2k/W}$$

Table 11.2 represents typical "Clo" values for several apparel categories. For instance, the "Clo" value of a standard business suit is 1, whereas the "Clo" value of a pair of shorts is roughly 0.5. Therefore, the more clothing a person wears, the more resistance the clothing provides. Sometimes the unit log is used to measure the resistance of clothing since it is easier to measure it when there are fewer items on the human body.

11.5.3 Environmental Factors

Environmental factors have a significant impact on all the factors influencing thermal comfort. The dry bulb temperature (DBT), relative humidity (RH), air motion or air velocity, and ambient surface temperature are the environmental elements. The DBT has an impact on evaporation- and convection-based heat transfer. Air velocity impacts both convective and evaporative heat transfer, relative humidity (RH) affects

TABLE 11.2
Insulation Value for Different Clothes (*Clo—Clothing and Thermal Insulation*, n.d.)

Clothing		Insulation	
		I_{cr} *Clo*	m^2K/W
Nude		0	0
Underwear—pants	Pantyhose	0.02	0.003
	Panties	0.03	0.005
	Briefs	0.04	0.006
	Pants 1/2 long legs made of wool	0.06	0.009
	Pants long legs	0.1	0.016
Underwear—shirts	Bra	0.01	0.002
	Shirt sleeveless	0.06	0.009
	T-shirt	0.09	0.014
	Shirt with long sleeves	0.12	0.019
	Half-slip in nylon	0.14	0.022
Shirts	Tube top	0.06	0.009
	Short sleeve	0.09	0.029
	Light blouse with long sleeves	0.15	0.023
	Light shirt with long sleeves	0.2	0.031
	Normal with long sleeves	0.25	0.039
	Flannel shirt with long sleeves	0.3	0.047
	Long sleeves with turtleneck blouse	0.34	0.053
Trousers	Shorts	0.06	0.009
	Walking shorts	0.11	0.017
	Light trousers	0.2	0.031
	Normal trousers	0.25	0.039
	Flannel trousers	0.28	0.043
	Overalls	0.28	0.043
Coveralls	Daily wear, belted	0.49	0.076
	Work	0.5	0.078
Highly-insulating coveralls	Multi-component with filling	1.03	0.16
	Fibre-pelt	1.13	0.175
Sweaters	Sleeveless vest	0.12	0.019
	Thin sweater	0.2	0.031
	Long thin sleeves with turtleneck	0.26	0.04
	Thick sweater	0.35	0.054
	Long thick sleeves with turtleneck	0.37	0.057
Jacket	Vest	0.13	0.02
	Light summer jacket	0.25	0.039
	Smock	0.3	0.047
	Jacket	0.35	0.054

(Continued)

TABLE 11.2 (Continued)
Insulation Value for Different Clothes (*Clo—Clothing and Thermal Insulation*, n.d.)

Clothing		Insulation	
		I_{cl}, *Clo*	*m²K/W*
Coats, over-jackets, and over-trousers	Overalls multi-component	0.52	0.081
	Down jacket	0.55	0.085
	Coat	0.6	0.093
	Parka	0.7	0.109
Sundries	Socks	0.02	0.003
	Thin soled shoes	0.02	0.003
	Quilted fleece slippers	0.03	0.005
	Thick-soled shoes	0.04	0.006
	Thick ankle socks	0.05	0.008
	Boots	0.05	0.008
	Thick long socks	0.1	0.016
Skirts, dresses	Light skirt 15 cm. above knee	0.01	0.016
	Light skirt 15 cm. below knee	0.18	0.028
	Heavy skirt knee-length	0.25	0.039
	Light dress sleeveless	0.25	0.039
	Winter dress long sleeves	0.4	0.062
Sleepwear	Under shorts	0.1	0.016
	Short gown thin strap	0.15	0.023
	Long gown long sleeve	0.3	0.047
	Hospital gown	0.31	0.048
	Long pyjamas with long sleeve	0.5	0.078
	Body sleep with feet	0.72	0.112
Robes	Long sleeve, wrap, short	0.41	0.064
	Long sleeve, wrap, long	0.53	0.082

heat loss through evaporation, and surface temperature affects radiative heat transfer. In addition to the previously mentioned elements, such as draughts and asymmetrical cooling or heating, chilly or hot floors also impact thermal comfort. A comfortable air conditioning system's goal is to regulate the external elements. The standard air conditioning system does not influence the physiological and insulating elements, so comfortable conditions predominate in the occupied room.

Nonetheless, wearing the proper apparel may help to lower the air conditioning system's cost. So, one can only control the environmental component by installing an air conditioning system. This means that one can regulate the dry bulb temperature (DBT), moisture content, and air velocity inside the conditioned space. Age is one physiological element that cannot be controlled. It is unable to determine what the occupant's age should be. The action is likewise not under any oversight. There is no control over the clothing by an air conditioning system or its creator. However, one might lessen the need for an air conditioning system by changing or recommending appropriate attire.

11.6 INDOOR ENVIRONMENT QUALITY

Apart from aesthetic design and luxury lifestyle, we must push ourselves toward nature to live happily. As much as our work culture adheres to natural phenomena, it will help keep us healthy and happy and cause minor environmental damage. Hence, the indoor quality of building envelopes must be supported by proper lighting, comfortable temperature, pleasant sound, and sound design quality for the well-being of humans.

11.6.1 Light

Compared to electric light, natural light has many benefits such as versatility, efficiency, and the capacity to foster awareness of and connection to environmental conditions. In addition to providing a free source of light within a house, which contributes to an energy-saving plan, it will animate areas and add drama and variety. Also, the advantages for physical health are now widely recognized and help prevent seasonal affective disorder (SAD). Nonetheless, excessive lighting can be uncomfortable and interfere with sleep.

- In a building, a therapy room is designed called an Orient room which provides a dose of morning light to stimulate the **circadian rhythm**.
- Main habitable rooms should receive "good" daylight (above 3% average daylight factor), and a key family room should have access to direct sunlight for at least 2 hours per day.
- High head-height windows offer large sky views, which are crucial in densely populated areas and improve lighting in the space, increasing access to natural light.
- Particularly bedrooms should have excellent blackout alternatives to assist in restful sleep, such as thermal shutters (for cold months) and/or louvres that can be adjusted (for secure night-time ventilation in warm conditions).
- Individual control over the amount of daylight creates welcome chances for the occupant to modify circumstances to suit their usage habits, which increases their contentment with their surroundings.

11.6.2 Temperature

The thermal design plan should produce relaxing and energizing surroundings that can take advantage of the climatic conditions to increase energy efficiency. The system recognizes the ambient temperature in terms of air temperature and radiant conditions such as sunlight, air movement, and temperature distribution through surface materials (wood feels warm, stone feels cool).

- Use energy from the sun to create warm, bright areas to be in on winter days, such as sunrooms and window seats. Use heat storage materials to absorb and hold onto the heat.

- According to the principle of adaptive comfort, thermal conditions can change and fluctuate rather than being steady or "optimized." Success depends on occupant control and the design's capacity to change to accommodate users' changing demands and preferences.
- Design apertures that enable the development of secure night-time ventilation (for example, using louvered sections) and take advantage of stack and cross ventilation principles to cool a building down during hot spells.

11.6.3 Sound

Acoustic conditions can be exploited to generate chances to support user demands and preferences, just like other environmental design elements. Noise can be stressful, but it can also be beneficial to hear your surroundings and the sounds of nature. Similarly, while complete acoustic separation is rarely necessary, there are times and places where acoustic privacy is appreciated in the home.

- Establishing peaceful, quiet areas for reading and studying is crucial to promote learning behaviours.
- Acoustic separation in some settings facilitates activities like music and indoor exercise without bothering others.
- The design should include noise-attenuated air routes to fully utilize natural ventilation in an urban area, especially at night and when calm circumstances are desired for studying or sleeping.

11.6.4 Design Quality

- The colour of our surroundings, including interior walls, can influence how we learn and, in some settings, can be used to enhance learning.
- The shape of a location influences our perception of comfort and attractiveness. In recent studies, participants were more likely to assess environments as beautiful if curvilinear than rectilinear because curved forms are pleasant. As a result, blue, tall, curvy areas with views of the sky are more likely to be pleasant, social, and creative locations. On the other hand, red, rectilinear spaces with low ceilings are more likely to promote focus, concentration, and study.

Designing for well-being and health involves a wide range of options and standards. The idea is to create designs suitable for meeting quantitative health indicators while being adaptive to and integrated with a broader range of well-being-promoting ideas.

11.7 APPLICATION OF THE COMFORT EQUATION

For all possible combinations of the variables, comfort conditions can be determined using the comfort equation. It can also be demonstrated how different variables can be balanced against one another to create comfortable thermal conditions.

The effectiveness of a computer program used to forecast the equation's result was shown, and it began to provide answers to numerous queries regarding thermal comfort. A few examples of real-world situations for thermal comfort are as follows ("International Standards," 2014):

1. The comfort temperature for rest areas and a restaurant at one end of a swimming pool, where the air dry bulb temperature (DBT) equals the mean radiant temperature (DBT = T_{mrt}), as calculated by the comfort equation, is 29.1°C for naked (Clo = 0) and sedentary people (Q_g = 58 W m^2) with low relative air velocity (V = 0.2 m sl) and 70% relative humidity (RH = 70%).
2. A conference room should be comfortable at 23.3°C in the winter (Clo = 1.0; V = 0.1 m s; RH = 40%; DBT = T_{mrt}). When RH is 70% and Clo is 0.6, temp is 24.9°C in the summer.
3. It was discovered that 26°C is the ideal temperature in a clean environment with sedentary activity (v = 0.5 m s^{-1}; Clo= 0.75; RH = 50%; DBT = T_{mrt}).
4. A comfortable temperature in a store where customers are moving at a 1.5 km/h speed (V = 0.4 m/s; Clo = 1.0; RH = 50%) is DBT = T_{mrt} = 21.0°C.
5. Workers in slaughterhouses need garment insulation of 1.5 Clo (DBT = T_{mrt} = 8°C; V = 0.3 m/s; mild activity).
6. The air temperature required for comfort in a bus during the winter (DBT = T_{mrt} = 6°C) is 26°C (Clo = 1.0; V = 0.2 m s^{-1}; RH = 50%).
7. For comfort, a restaurant outside that uses infrared heaters to keep the air temperature above 12°C (V = 0.3 m/s; Clo = 1.5; RH = 50%) must maintain T_{mrt} = 40°C.
8. Sedentary passengers (Clo = 1.0; V = 0.2 m/s) should be in air temperatures of 20°C for comfort in a supersonic aircraft (DBT = T_{mrt} = 10°C).
9. Irradiant shields lower T_{mrt} in a foundry (DBT = 20°C; T_{mrt} = 50°C; RH = 50%, Clo = 0.5; light work (Q_g = 116 W m^2)) to 35°C. To maintain comfort, air velocity should be V = 1.5 m/s.

The quality and accuracy of the inputs will determine the outputs of all of the examples mentioned earlier and other applications. This will depend on measurement and estimation accuracy, as with other applications, and how the problem is formulated and interpreted. However, the flexibility of the comfort equation is evident. It represents a step forward in understanding and application and a source of guidance and inspiration for thermal comfort research.

11.8 COMFORT INDICES

Elements affecting human comfort include activity level, apparel, clothing temperature, relative humidity, air velocity, and ambient dry bulb temperature. There is no one particular mix that provides comfort because there are so many variables at play. There is a variety of conceivable combinations, all of which will provide an equivalent level of comfort. Hence, several comfort indices have been proposed to assess the efficiency of condition space. Direct and derived indices can be used to categorize these indices. How would anyone assess the qualities of an air-conditioned building, for instance, if there is a need to know if it is good or bad? The success of an air conditioning system, therefore, ultimately hinges on how comfortable it is inside. Whether it offers thermal comfort or not relies on several variables, as has been demonstrated. These aspects must be considered when evaluating an air-conditioned building's efficacy. In essence, the comfort indices were developed by space people to measure the effectiveness of conditions. Dry bulb temperature (DBT), humidity ratio air velocity, and mean radiant temperature (T_{mrt}) (found using equation 11.7) are direct indices.

$$T_{mrt} = [T_g^4 + CV^{1/2}(T_g - DBT)]^{1/4} \tag{11.7}$$

T_g = Globe temperature (°K)
DBT = Dry bulb temperature (°K)
V = velocity of air (m/s)
C = constant (0.247×10^9)

11.8.1 Globe Temperature (T_G)

This is the temperature that a thermocouple positioned in the middle of a hollow, 15.24 cm in diameter, black-painted cylinder kept in the conditioned space measures at a steady state. The thermocouple reading, which is expressed in Kelvin, is the outcome of a balance between convective and radiative heat exchanges between the environment and the earth. It is connected to either the ambient surface temperature or the mean radiant temperature.

Two or more direct indices are combined into a single element to create the derived indices. Below are a few of the significant derived indices:

1. Effective temperature (ET)
2. Operative temperature (T_{op})
3. Predicted mean vote (PMV)
4. Predicted percentage of dissatisfied (PPD)
5. Skin wettedness (w)

11.8.2 Effective Temperature (ET)

Dry bulb temperature (DBT) and relative humidity (RH) effects are combined into one quantity known as effective temperature. It is defined as the environment's temperature at 50% relative humidity (RH), at which the skin loses the same amount of heat overall as the actual environment. Several different variables, including exercise, clothing, air velocity, and T, or radiant temperature, influence the value. A standard effective temperature (Wang & Hong, 2020) is established for a clothing value of Clo = 0.6, an activity of 1 MET, an air velocity of 0.1 m/s, and the assumption that the mean radiant temperature is the same as the dry bulb temperature.

11.8.3 Operative Temperature (T_{OP})

The air dry bulb temperature (DBT) and mean radiant temperature are averaged and weighted to determine the operative temperature (T_{mrt}). So, it is possible to infer that it combines both the dry bulb temperature (DBT) and the ambient surface temperature. If h_r and h_c are nearly identical, "T_{op}" will give this (see equation 11.8). The operating temperature therefore approaches the arithmetic mean of the mean radiant temperature and the ambient dry bulb temperature.

$$T_{op} = \frac{h_r T_{mrt} + h_c T_{amb}}{h\mu + h_c} \approx \frac{T_{mrt} + T_{amb}}{2} \quad 11.8$$

h_r = Radiative heat transfer coefficient.
h_c = Conductive heat transfer coefficient.
T_{amb} = Dry bulb temperature of ambient air.

11.8.4 The Predicted Mean Vote (PMV)

The predicted mean vote (PMV) is a novel thermal sensation index that forecasts the mean sensation vote for a group of people for any given combination of air temperature, mean radiant temperature, air velocity, humidity, activity, and apparel on a standard scale. PMV ranges from −3 to +3 and from 0 to 3. That is, −3 means chilly, −2 means cool, −1 means slightly cool, 0 means neutral, +1 means slightly warm, +2 means warm, and +3 means hot. When the comfort equation is satisfied, it would be anticipated that PMV = 0 for a sizable population (neutral). Heat stored in the body, Q_{st}, will be 0 in comfortable conditions. At times, thermoregulatory reactions try to maintain a temperature balance, which causes skin temperatures and sweat production to depart from what is comfortable. Hence, the physiological stress on the body is connected to the heat load. With two levels of activity (1.0 and 1.2 MET), low air movement, 50% relative humidity, three levels of clothing (Clo = 0.5, 0.75, and 1.0), and air temperature = mean radiant temperature, PMV values for typical indoor conditions are shown in Table 11.3.

11.8.5 The Predicted Percentage of Dissatisfied (PPD)

The Predicted percentage of dissatisfied (PPD) gauges how many are "potential complainers" or are anticipated to be "decidedly dissatisfied" or "decidedly uncomfortable" and thus dissatisfied. There will be a range of levels of dissatisfaction among the group because people are all different. According to one study, 1,296 participants who wore light clothing and spent three hours at a specific ambient temperature showed positive effects (Clo = 0.6). Every 30 minutes for three hours, they asked subjects to rate their thermal sensations, and the mean of the most recent three votes for each subject was "reckoned" to the closest vote (integer). The distribution of the mean votes was then calculated for each ambient temperature between 18.9°C and 32.2°C. The unsatisfied were those who voted −3 (very cold), −2 (cool), +2 (warm), and +3 (warm) (hot). The proportion of people who reported being thermally unsatisfied across the range of ambient temperatures examined was plotted using probit analysis. One line was derived for warm dissatisfaction and the other for cold

TABLE 11.3
PMV Values for Certain MET Values of 1 and 1.2 with Given Conditions (Wilson & Sharples, 2015)

MET	1.0			1.2		
$T_a = T_{mrt}$/Clo	0.5	0.75	1	0.5	0.75	1
18	−3	−0.27	−1.58	−2.27	−1.42	−0.88
19	−2.88	−1.96	−1.30	−1.92	−1.19	−0.68
20	−2.51	−1.64	−1.02	−1.58	−0.92	−0.45
21	−2.13	−1.30	−0.74	−1.31	−0.68	−0.23
22	−1.75	−0.98	−0.46	−0.97	−0.40	0.00
23	−1.33	−0.66	−0.18	−0.69	−0.15	0.23
24	−0.95	−0.33	0.10	−0.36	0.12	0.46
25	−0.56	−0.01	0.38	−0.06	0.37	0.69
26	−0.18	0.31	0.66	0.26	0.64	0.91
27	0.20	0.64	0.95	0.57	0.91	1.15
28	0.59	0.96	1.24	0.88	1.18	1.37
29	0.98	1.31	1.54	1.22	1.46	1.63
30	1.37	1.65	1.83	1.51	1.70	1.83
31	1.80	1.99	2.13	1.87	2.01	2.11
32	2.21	2.34	2.43	2.17	2.29	2.36
33	2.62	2.69	2.74	2.54	2.57	2.60

dissatisfaction. The overall percentage of dissatisfaction as a function of the PMV was then presented in a semi-logarithmic plot and transformed from ambient temperature to the PMV values obtained from the controlled testing settings (see equation 11.9) (Wang & Hong, 2020).

$$PPD = 100 - 95\exp(-0.03353\ PMV^4 - 0.2179\ PMV^2) \qquad 11.9$$

11.8.6 Skin Wettedness (w)

The skin wettedness (w) is calculated as the difference between the actual rate of evaporation from the skin surface (Q_{sk}) and the highest rate of evaporation possible in that environment (Q_{max}). Hence, it can theoretically have a value between 0 and 1. When heat is lost through evaporation from the body (skin), the minimum value is typically regarded to be w = 0.06 for insensible sweat, and the highest value is w = 1.0 (influenced by humidity, air velocity, and clothing). Skin wettedness is a helpful indicator of heat exhaustion.

11.9 LOCAL THERMAL DISCOMFORT

11.9.1 "Neutral" but Uncomfortable

A person with a "neutral" overall thermal experience is presumed to enjoy neither being warmer nor cooler. If they want to be warmer, they are probably too cool; if they prefer to be more relaxed, they are probably too warm. Hence, a neutral experience denotes contentment, "no change," and the assumption that the person is experiencing "whole-body" or general thermal comfort. Nonetheless, there are situations in which a person can say that, while they feel neutral generally, their thermal experience fluctuates around their body, and they are not at ease. They might, for instance, be neutral overall but have uncomfortably warm hands and uncomfortably cold feet, or they might have a left side that is warm and a right side that is frigid because the person feels localized thermal discomfort; a "neutral" rating does not reflect comfort. The local thermal discomfort is determined by the sensation brought on by a change in skin temperature, either upward or downward. Hence, the body must be in thermal balance, mean skin temperature, sweating (wettedness), and local thermal discomfort are factors that can influence an individual's perception of thermal comfort, but their presence or absence alone does not determine whether a person is in thermal comfort.

11.9.2 Body Maps, Local Thermal Sensation, and Sweating

Sensors in the skin sense whether the body is too hot or cold and communicate the information to the brain via the spinal cord, and effector mechanisms from the hypothalamus or the spinal cord immediately cause a change in skin condition. The entire body experiences this, yet the "shortcut" through the spinal cord may result in a local response. The body will begin to sweat if necessary if it is hot or cold. It will boost blood flow to the skin if it is cold. There are nerve endings in the skin, some of which react to cold and others to heat (there are also others that respond to pressure and others that produce pain). Over the entire body's surface, they are dispersed variably. The thermal sensation is influenced by the area stimulated, the location and length of the stimulus, the pre-existing thermal state, the temperature, and the rate of temperature change.

Body maps display the distribution of physiological reactions such as skin temperature, perspiration, and thermal sensitivity over the body's surface. Designing apparel and settings may be a beneficial application for body mapping. It may be plausible to suppose that local temperature sensation and discomfort are connected to the density, distribution, and sensitivity of sensors and effectors around the body, which is relevant to local thermal discomfort. One intriguing observation was that as the stimulus level increasing, the variations in sensitivity across the body decrease. As a result, the sensitivity distribution may alter when a near-threshold or even moderate stimulus is applied when pain is considered.

11.9.3 Local Thermal Discomfort Caused by Draughts

A draught is described by ISO 7730 (2005) as "unwanted local cooling of the body caused by air movement." This term implies convective cooling, which may also refer to evaporative cooling if the skin is wet. Although the phrase "radiant draught" is occasionally used to express the sensation of skin chilling via thermal radiation loss to a cold surface such as a cold window, air movement also implies a slight amount of pressure on the skin. The definition is rigorously violated because there is no pressure on the skin brought on by air movement. However, a draught may result from a chilly surface from cold air falling to the floor and irritating the ankles and feet. Draught pain will be lessened for the same stimulus in those who are active and warmer than neutral (but not sweating) (*Developing an Adaptive Model of Thermal Comfort and Preference (Conference) | OSTI.GOV*, n.d.). People who are colder than average may find draughts to be more uncomfortable for the same local cooling. A cold person will try to maintain heat; therefore, more cooling will not be desired. This is in line with the findings of studies on thermal pleasure, which show that it is pleasant to give a warm stimulus to someone who is cold or a cold stimulus to someone who is warm, while it is unpleasant to give a cold stimulus to someone who is cold or a warm stimulus to someone who is hot.

Two other things to note about draughts: first, women generally have smaller hands and fingers than men, which makes them more susceptible to the cold. So, even when they wear clothing identical to that worn by men, it would be expected that women would be more susceptible to draughts than men. Females may experience a sense of draught more frequently than males because dress fashion in women frequently exposes the neck and legs (although leggings are sometimes worn). The final issue for further consideration in the following section is that people will leave a draught if they are in one and the circumstances, personal ability, opportunity, disposition, and context permit it. In this case, the draught could serve as a behavioural cue instead of a constant cause of local heat discomfort.

11.9.4 Local Thermal Discomfort Caused by Radiation

The mean radiant temperature (T_{mrt}), which is the net average of all radiation into and out of the body in all directions, is used to reflect the contribution of heat transfer via radiation to and from the body when forecasting overall (whole-body) thermal comfort. Some situations (especially when it comes to solar radiation) have radiation that is more prevalent in some directions than others. A person's posture and resulting "projected surface area" toward the directed radiation have an impact on this as well. Infrared radiation from and to hot and cold surfaces, in addition to solar radiation, facilitates radiation exchange and contributes to local thermal discomfort. When compared to more prolonged wavelength radiation for indoor and other lower-temperature surfaces, solar radiation has a relatively short wavelength. This is because the shorter the wavelength of radiation released, the hotter the surface is. The ratio of the difference between the fourth powers of two surfaces' absolute temperatures determines how much heat is transferred by radiation between them. There will be a net heat transfer from the body if the skin temperature is higher than the surface temperatures; conversely, if the skin temperature is lower than the ambient surface temperatures, there will be a net heat gain via radiation. This might have an impact on local thermal discomfort.

11.9.5 Local Thermal Discomfort Caused by Vertical Temperature Differences

ISO 7730 (2005) provides the following equation 11.10 for the percentage of dissatisfied due to temperature differences between head and feet (Parsons, 2014 (d)

$$PD = \frac{100}{1 + exp(5.76 - 0{\cdot}836\Delta t_a V)} \quad 11.10$$

Displacement ventilation occurs when cooled air is supplied to the floor through apertures in the floor or from vertical "bins" at floor level (sometimes with a chilled ceiling to keep moisture at bay). Heat is applied to the air by both humans and machines, including computers, causing it to rise and be "exhausted" at the ceiling (so that it is breathed only once, and any pollution is removed at the ceiling). Displacement ventilation was utilized in a field trial in a contemporary office in the metropolis of London. However, people reported headaches and tiredness in the late afternoon. During an inquiry, it was determined that the cause of the problem was a build-up of carbon dioxide (caused by people), since office workers' desks had blocked the fresh air entry ducts on the floor due to drought. It is necessary to conduct more research on vertical temperature differences because it is unclear if the discomfort is brought on by the difference itself or by the local air temperature and air movement.

11.9.6 Local Thermal Discomfort Caused by Warm and Cold Floors

Pressure is felt when the skin makes contact with surfaces that are skin temperature. A feeling of warmth will also be present if the material's surface temperature is higher than the skin's, and a feeling of coolness will also be present if it is lower. It's likely that there could be discomfort and that the skin will be harmed if the ensuing skin temperature spikes or plummets (e.g., due to burning or freezing). There will not be any damage between extremes (such as skin temperatures of 5°C and 40°C), but there might be a pain. The degree of the (effect) feeling will depend on the state of the skin, the thermal conductivity (k), specific heat (c), density (d), nature of the surface, thermal sensation already present, degree of contact, and duration of contact. Cold flooring, furniture such as including chairs, handrails, tools, immersion in liquids, certain food kinds, and obviously, other concrete examples of contact with surfaces of moderate temperature may cause discomfort. It is a well-known fact that even when two materials are at the same surface temperature, touching cold or hot metal will produce a more intense sensation than touching wood or plastic. The sensation is a result of how quickly heat is transferred from the material to the skin. It is possible to hypothesize a contact temperature (tc), which happens right on contact at the interface between the two surfaces, in a straightforward model where there is perfect contact between two materials (with infinite capacity so that it can be assumed that neither material will change temperature). Whether a person can adjust to lessen or remove the source of localized heat discomfort, lessen its consequences, or move away from it entirely will determine the degree of the effect. To prevent local thermal pain as well as to achieve general thermal comfort, an adaptive opportunity will be significant (Parsons, 2021 (a).

11.10 ADAPTIVE THERMAL COMFORT

Every living thing, from flowers and trees to reptiles, mammals, and fish, not to mention people, constantly interacts with the environment in which it is found and modifies its physical makeup and behaviour to reach and maintain an optimal state. This guarantees survival and is connected to what we refer to as "comfort" in individuals. The desire for comfort drives all human behaviour. It is applicable in all physical settings and is not limited to specific settings like types of buildings, vehicles, or the outdoors. It is a constant process and a fundamental aspect of human nature. It would be incorrect and would impede knowledge of the factors that offer thermal comfort to see it differently. Indeed, the consideration here is not just limited to temperature settings. This rule applies to reactions to acoustic, vibrational, visual, air quality, and other environmental factors that work together to provide an integrated response to the overall environment in which a person is immersed. The process through which people adjust to their surroundings, encompassing both physiological and behavioural responses, in order to feel comfortable is known as "adaptive thermal comfort." Long recognized as a means of achieving survival, the use of behaviour to regulate body temperature in response to thermal conditions is known as behavioural thermoregulation. The then-accepted heat balancing method of anticipating thermal comfort conditions can be partially adaptable; it does not compete with the adaptive method of thermal comfort but rather enhances it. There is no doubt that behavioural adaptation affects the conditions necessary for thermal comfort, but it is unclear whether physiological (as defined earlier) and psychological adaptation actually impact the requirements for thermal comfort around the world. So, it may be concluded that behavioural responses to thermal surroundings are virtually exclusively responsible for adaptive thermal comfort.

11.10.1 Biological Adaptation

Living things, which all display both physiological and behavioural reactions, are known to demonstrate biological adaptation. Although camels in the desert do not store water in their humps, they do exhibit survival strategies. For instance, they fluctuate their internal body temperature between high levels during the day and low levels at night when they are dehydrated (up to 30% of body weight). Within 15 minutes, they also exhibit quick and correct (not excessive) rehydration. The kangaroo rat, which lives in the desert as well, does not seem to consume any water at all and instead gets its water from the oxidation of "dry" food and any free water within it. When it is hot outside, sheep, goats, cats, and other animals cool their brains selectively by passing cool blood from their "wet" noses through a network (rete) of tiny blood capillaries in the carotid artery, which exchanges heat and lowers blood temperature before it reaches the brain. Animals with behavioural thermoregulation include birds that extend their wings to shed heat in the shade and gain heat in the sun.

11.10.2 Human Adaptation

The scientific literature initially described adaptation in humans as a set of idiosyncrasies of people from "remote regions" who primarily live outside. Fish filleters were reported to keep their hand blood flowing even in the chilly water so they could complete jobs without losing their physical dexterity. These modifications, together with acclimatization, which heightens perspiration following heat exposure, are referred to as habituation and frequently gradually go away after the exposure stops. Morphological (genetic) alterations occur. Since these parts have a high surface area-to-mass (volume) ratio, people from cold climates have smaller noses and ears to prevent heat loss, and fat people have more thermal insulation—a benefit in cold climates. Based on a psycho-physiological thermoregulation concept, adaptive thermal comfort offers a more dynamic interaction between the environment and the attainment of thermal comfort. Predicting that a large group of people will experience the mean temperature as "cold" is a valuable indicator of how the group—or even an individual—will react to a static environment. However, in terms of human response and thermal comfort, it should be viewed as a catalyst for change rather than as a prediction of the outcome. Humans do not like to be "cold;" thus, they will add more insulation to their clothing, relocate to a warmer setting, or use any other perceived possibilities to become comfortable. The PMV and SET indices are examples of non-adaptive thermal comfort techniques. They are valid but have limitations. The PMV number should therefore be viewed as the strength of the drive for a behavioural reaction rather than being thought of as a measure of thermal comfort.

11.10.3 Adaptive Models and Thermal Comfort

The PMV index can give thermal comfort ranges in a flexible manner. This, however, ignores the adaptive opportunity's dynamic nature in design. People have always been known to adapt (behave) in order to live and find comfort in their environment. Yet it wasn't until the early 1960s that adaptive techniques were first used for thermal comfort applications. For example, most recently, ISO 7730 (2005), ASHRAE 55–2017 with addendum 2019, and other indices were formalized into national and international standards for determining criteria for interior environments in buildings and the PMV comfort index. Nonetheless, research into adaptive thermal comfort, primarily through field investigations, persisted, Humphreys (1978) being particularly significant with his synthesis of global data. He demonstrated a connection between the temperatures within structures and the outside. Hence, for instance, people appeared to prefer warmer temperatures in hotter climates (for comfort) (*Developing an Adaptive Model of Thermal Comfort and Preference (Conference) | OSTI.GOV*, n.d.). It is frequently assumed that humans who live in hot areas have evolved to favour warmer environments for comfort. Thermal comfort is determined by a person's feeling of the combined impacts of air temperature, radiant temperature, air velocity, humidity, activity level, and clothing, which can change inside buildings regardless of the weather outside. The possibility of more adaptable and practical environmental design optimizes indoor conditions to enhance occupant comfort, productivity, and well-being, ultimately fostering healthier and more sustainable built environments.

11.10.4 Adaptive Thermal Comfort Regression Models

The ability to forecast comfortable temperatures from the weather gives building engineers the advantage of having access to global weather information. The equation can be used to determine the comfort temperature (T_{Com_f}), as shown in equation 11.11(Parsons, 2014d).

$$T_{Com_f} = 0.53(T_{OM}) + 13.8^{\circ}C \quad 11.11$$

Where (T_{OM})is the mean of the monthly mean of the daily maximum outdoor temperature and the monthly mean of the daily minimum outdoor temperature.

One of the earliest official attempts to include the adaptive thermal comfort paradigm into thermal comfort standards was made by the American Society for Heating, Refrigerating, and Air-Conditioning Engineers (ASHRAE, 1998). Richard De Dear and Gail Brager, who, along with others, carried out the massive global survey on behalf of ASHRAE, outlined a global database of thermal comfort field tests and how this was used to construct an adaptive model. The decision of which external environmental factors should be employed to forecast internal comfort temperatures presented several challenges. The sole reliance on air temperatures ignored significant contributions from solar radiation and other factors. An attempt to use new effective temperature (ET*) showed greater validity. However, assumptions had to be made because not all the variables' meteorological data were available in the form needed for ET*. The ASHRAE 55–2004 standard offered the most comfortable temperature, as presented in equation 11.12.

$$T_{Com_f} = 0.31(T_O) + 17.8^o C \quad 11.12$$

Where (T_O) is loosely defined as the prevailing mean outdoor temperature. European standard EN 15251 (2007) used data from European studies to provide the adaptive model in equation 11.13.

$$T_{Com_f} = 0.33(T_{rm}) + 18.8^o C \quad 11.13$$

Where T_{rm} is the exponentially weighted running mean of the outdoor temperature.

11.10.5 Adaptive Models that Modify the PMV Index

The PMV index could be changed to use an adaptive strategy. The PMV value might be multiplied by an anticipation factor, reducing the PMV severity and, consequently, the PPD if persons who did not expect any better would not be displeased with or complain about a thermal environment. To determine and specify the parameters for thermal comfort, however, would seem to be an improper strategy given that people would apparently expect nothing better. The most popular and practical adaptation strategy for the majority of situations is changing the insulation of garments. In order to provide a PMV value that takes into account adaptive behaviour, the proposed method that all adaptive opportunities might be incorporated into an equal adjustment to clothing value (I_{eq}). This would give effective clothing insulation to be employed in the PMV equation. The method's formulation limits the effective garment insulation to values between zero and twice that of the initial clothing before correction (I_{st}). Although this is a practical suggestion that can be used in many applications, the process is complicated by this restriction. According to the definition, equivalent clothing insulation is "the garment insulation that would give persons without adaptation the same level of thermal comfort as people who adapt to their environment." It has been determined that the mechanisms for adjusting to discomfort in heat and cold can differ. Following are the first (equation 11.14) and second (equation 11.15) equations for discomfort due to heat and cold, respectively.

$$I_{eq} = I_{st} - (I_{Aj} \times I_{st}) \qquad 11.14$$

$$I_{eq} = I_{st} + (I_{Aj} \times I_{st}) \qquad 11.15$$

where I_{Aj}'s adaptive opportunity value ranges from 0 (minimum) to 0.25 (low), 0.5 (middle), 0.75 (high), and 1 (maximum). As a result, wearing apparel decreases in the heat and increases in the cold (*Developing an Adaptive Model of Thermal Comfort and Preference (Conference) | OSTI.GOV*, n.d.).

The optimal indoor temperature for the summer and winter seasons is displayed on the ASHRAE comfort chart. Based on statistical sampling of a significant number of occupants with activity levels less than 1.2 MET. The directional radiation, the ASHRAE comfort chart, was created and examining the impacts of the adaptation on the circumstances to which a person or persons are subjected is a more casual and, thus, effective way to figure out how adaptive behaviour affects the environment. For instance, altering clothes to affect how well they insulate, slowing down activity levels, lowering metabolic heat production, turning on fans, raising air velocity, and employing other innovative strategies offer opportunities to tailor indoor environments dynamically, providing personalized comfort experiences while also promoting energy efficiency and environmental sustainability.

In comparison to the generic techniques mentioned earlier, this would offer a more trustworthy forecasting strategy. Due to various adaptive behaviours, there are various methods of adaptation and suggested temperature offsets for neutral temperatures (e.g., operating a desk fan at 2 m/s raises the neutral temperature by 2.8°C). This is desirable but challenging to accomplish in practice since each adaptive opportunity frequently involves a complex interplay with the thermal conditions encountered across time.

11.11 ASHRAE COMFORT CHART

Building on what has been accomplished while also eschewing approaches that do not incorporate adaptive thermal comfort is crucial for the advancement of our knowledge of thermal comfort and the design of thermal comfort. To define comfort levels, it is necessary to take into account how the seven elements—air temperature, radiant temperature, humidity, air velocity, clothing, activity, and adaptive opportunity—interact. Figure 11.3 illustrates how an adaptive strategy can be displayed on a psychrometric chart.

The ASHRAE comfort chart, which resembles a psychrometric chart is shown. Nevertheless, the operative temperature, not the dry bulb temperature, is shown on the x axis here. The humidity ratio is shown on the y axis, and the effective temperature lines are indicated by an angled line. A horizontal line at 2°C is the dew point line. The ASHRAE human comfort chart also displays a relative humidity line at 60%. Due to overlaid areas for the winter and summer zones, there are some areas that are being overlapped. A person wearing summer clothing will therefore feel

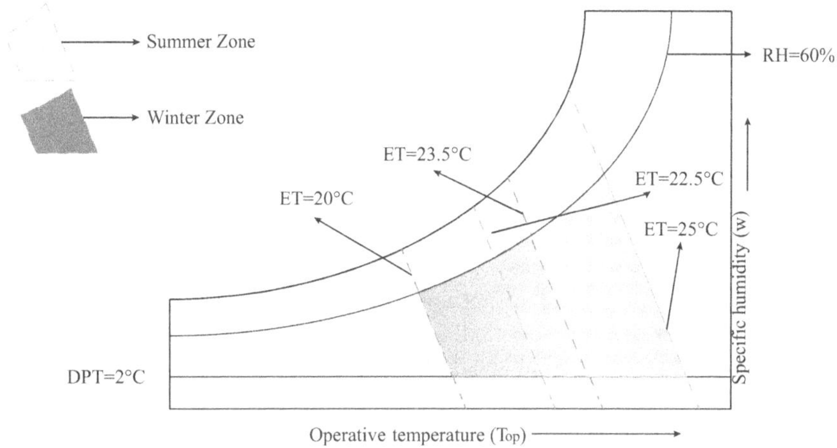

FIGURE 11.3 ASHRAE human comfort chart.

somewhat cooler, and a person wearing winter clothing will feel slightly warmer, if the surroundings (conditioned space) lie between overlapping zones.

The effective temperature line, constant relative humidity line, and dew point temperature line form the boundaries of the ASHRAE human comfort chart (Schoen et al., 2013). It is advised to keep indoor air at a maximum relative humidity of 60% and a minimum dew point temperature of 2°C. According to the ASHRAE human comfort chart, the ideal effective temperature range for summer and winter is 22°C to 25°C and 20°C to 23.5°C, respectively.

11.12 THERMAL ENVIRONMENT IN PLANES, TRAINS, AUTOMOBILES, OUTDOORS, IN SPACE, AND UNDER PRESSURE

Particularly when exposure to the temperature conditions varies with time and space, there are situations that deviate from what can be considered the "typical" steady-state indoor office environment. In addition to other environments like those in space, hypo- and hyperbaric environments, and outdoor environments, these also include vehicle environments, where people frequently occupy relatively small spaces with limited adaptive opportunity and are exposed to the changing external environment through windows and to significant contact with surrounding surfaces (Parsons, 2021b).

11.12.1 Thermal Environment in Automobiles

With gasoline-powered automobiles, there is, albeit at a cost, adequate energy for heating, ventilation, and air conditioning. The provision of thermal comfort in "green" cars powered, for instance, by electricity, when power constraints create a necessity for creative solutions, is a challenge for the future. The fundamentals of thermal comfort design will not alter. When given the same parameters, everyone will react similarly. Nonetheless, some places' specific needs could necessitate extra care. Cool air should be directed into the chest of drivers and passengers in cars for thermal comfort ("Automotive Ergonomics," 2016). One more value that is equivalent in temperature is defined as thermal comfort in vehicles.

11.12.1.1 Equivalent Temperature

The term "equivalent temperature" was used to describe the constant temperature of a hypothetical, still-air enclosure in which a person would exchange the same amount of dry heat through radiation and convection as they would in the actual environment. Hence, it defines equivalence using the conventional environment approach. Sweating and heat loss through evaporation, which are frequent sources of discomfort in cars, are, by definition, excluded. It is especially pertinent to the assessment of asymmetric situations, and although various techniques are acceptable, measurements are frequently conducted using thermal manikins.

The adoption of a "new material" that might efficiently heat seats in a consistent and easily adjustable manner is an effort to create a more individualized (adaptive) strategy for thermal comfort in motor vehicles (also energy efficient). (In reality, the substance is currently utilized in Grand Prix racing to heat race vehicle tires.) The discovery that people could stay comfortable at a surprising range of temperatures, with the driver's cool hands on the unheated steering wheel serving as a limiting factor and, eventually, everyone's feet being frigid. This approach has the benefit of allowing each passenger to regulate their own seat temperature. Fung and Parsons (1996), who carried out a number of climatic chamber studies on human subjects, have published thorough reports on the thermal characteristics of vehicle seats and their impact on thermal comfort.

11.12.2 Thermal Environment on Train Journeys

In order to travel by public transportation, one must leave their home or other bases, travel to a "station," frequently wait outside, board a vehicle (a bus, train, etc.), ride the vehicle, exit the vehicle, and then travel to their destination. The traveller will typically need to experience a variety of thermal settings because of this. Consider the entire train ride, for instance. There are various thermal states in which the traveller could arrive at the station. Just keep in mind to concentrate on thermal comfort while waiting for the train, throughout the voyage, and when getting off the train. There are several ways to get to and from a train station. Heat radiation and comfort have been investigated, with applications to both trains and automobiles. A similar procedure was employed to enhance train-related knowledge. Garment fit and color evaluations demonstrated that under identical thermal conditions, tight-fitting black clothing felt approximately one scale unit warmer than loose-fitting white clothing. An essential consideration for commuters and other train travellers is crowding. Other passengers' radiation input will increase in addition to the psychological stress, and convective and evaporative heat transfer will be limited.

11.12.3 Thermal Environment on Aeroplanes

Due to the hazardous outside environment, the interior conditions for passengers on aeroplanes are completely controlled. Passengers receive little direct solar exposure through the windows, and while they have few opportunities for adaptation, they can move around somewhat and typically control the air speed (jets) from above. The air pressure decreases as we ascend higher in the atmosphere, resulting in less oxygen available for breathing. In order to combat this in passenger aircraft, the cabin is "pressurized" to the equivalent of being at around 1,800 m (about 6,000 feet) above sea level. When compared to the identical thermal circumstances at sea level, heat transfer via convection is less efficient in this hypobaric environment, but heat transfer through evaporation is more effective.

11.12.4 Thermal Environment in Space Vehicles

The absence of gravity almost entirely in space is one of the critical distinctions between aircraft settings or structures on Earth and those in space. As a result, there will not be any evaporation in space due to natural convectional heat transfer. Body heat does not cause cooler, denser air to rise, and gravity does not cause warmer, less dense air to replace it when it does. This will rely on the local gravity levels for buildings, space stations, and other constructions on other worlds. According to research, factors influencing thermal comfort in space include air temperature, radiant temperature, humidity, air velocity, clothing insulation, activity level, and air pressure.

11.12.5 Outdoor Thermal Environment

Weather factors affect outdoor thermal environments, and solar radiation and relatively high and unstable air velocities will have a significant impact on thermal comfort. For instance, there is a lot of interest in creating relaxing, engaging, and comfortable outdoor spaces in urban areas. There is a class of outdoor settings that individuals seek out for enjoyment, like when sunbathing on the beach. These "unique" situations should be handled differently from "typical" thermal comfort circumstances. Architecture and environmental designers are very interested in thermal pleasure and delight.

11.12.6 Thermal Environment in Hyperbaric Environments

It is uncommon to think about discomfort in hyperbaric settings. Even slight variations in atmospheric or other air pressure will not significantly affect thermal comfort in hypobaric situations. In contrast, Fanger (1970) asserts that deep mines, pressurized rooms for medical treatment, and ocean-floor sea laboratories are all unique hyperbaric habitats with little impact from variations in atmospheric pressure. We can also include compressed air tunnels that are used to keep water from adjacent seas, lakes, or rivers at bay, as well as hyperbaric lifeboats where divers can wait for rescue if a "mother-ship" isn't accessible.

11.12.7 Thermal Environment on Mountains

As there is less atmosphere on top of mountains, people who climb them experience lower atmospheric pressures than they would at sea level. Nevertheless, pressure affects heat transfer between a person and the environment, decreasing convection and boosting evaporation. Air temperature, radiant temperature, air velocity, humidity, clothing, and activity continue to be critical factors. Mountaintops receive significantly more solar radiation, the air temperature often drops, the wind picks up, and clouds, rain, hail, and snow are frequent occurrences. Due to their low thermal inertia, tents are significantly impacted by ambient temperature. In hypobaric conditions, the amount of oxygen that is available to the body may decrease, which may cause discomfort.

11.13 THERMAL COMFORT AND SEX, AGE, AND FOR PEOPLE WITH DISABILITIES

It is impressive that, roughly speaking, all persons experience identical thermal comfort circumstances when we take into consideration their clothes, activity, exposure to air, radiant heat, humidity, and air velocity, as well as their mechanisms and ability for adaptation. Since everyone's thermoregulatory system functions essentially the same, healthy individuals will experience similar physiological reactions to heat and cold. Nonetheless, because every person in the world has distinctive qualities, it is not surprising that people react to temperature settings differently from one another.

11.13.1 Thermal Comfort and Sex

The comfort reactions of male and female participants when they are both wearing the same clothing insulation show relatively minor if any, differences. When men and women switched clothes, as a result, their "neutral" temperatures changed. Different comfort needs will result from the differences in clothes between men and women ("Thermal Comfort," 2014). A further conclusion is that women may be more sensitive outside of comfortable surroundings, but the menstrual cycle has no impact on the need for thermal comfort. According to studies, thermal sensitivity and comfort are the same under "neutral" and warm settings, but when environments turn cold, females express greater discontent than males.

11.13.2 Thermal Comfort and Age

If the thermal comfort conditions are defined in terms of the interaction of air temperature, radiant temperature, humidity, air velocity, clothing, and activity, it is discovered that sensitivity is inversely related to age. Yet the adaptive opportunity will vary, mainly because age will have an impact on previous experience as well as physical and mental capacity. New-borns have a high potential for heat loss due to evaporation right after delivery since they are moist. Infants typically have large head-to-body ratios, big whole-body surface area-to-mass (volume) ratios, high skin blood flow, immature thermoregulatory systems, and brown fat, all of which contribute to their high heat-loss capacities. All new-borns benefit from care to maintain a healthy body temperature, and cuddling with the mother will help. The ideal temperature for children between the ages of 9 and 11 is 24°C, while the ideal temperature for those between the ages of 11 and 17 is 24.3°C. When older people and college students are compared, it is discovered that there is little difference in their responses, with older people preferring slightly warmer air temperatures. Elderly people have lower metabolic rates than younger people, and it has been noted that, in reality, lifestyle and attire will vary. The ideal temperature for adults over 60 varies from person to person, but in a neutral environment, a temperature of about 24°C is advised.

11.13.3 People with Disabilities

The best place to start when designing an environment is by creating temperatures that are bearable for people without mental impairments. In order to preserve comfort, it may also be necessary for those without mental disorders to provide support. Since every person is different and has a wide range of traits, this will be tailored to their particular needs. It was discovered that each individual with a physical handicap has a variety of distinctive qualities and needs in addition to their straightforward categorization as having a disability. A person who has limited physical mobility and cannot move to escape a draught or modify clothing needs special consideration when designing their surroundings in order to ensure thermal comfort.

11.14 THERMAL COMFORT AND HUMAN PERFORMANCE

Considering extensive research into how the thermal environment affects people's capacity to execute tasks and jobs, it is acceptable to draw the conclusion that the best settings for human performance are those that also offer thermal comfort. This is in line with the definition of thermal comfort, which states that it is a state in which there is no physical strain on the body because no change is necessary. As a result, it does not require resources in terms of maintaining health, paying attention to a task, or losing the ability to perform tasks. There have been several studies on how the thermal environment affects human performance, and one helpful way to forecast outcomes is to think about the three main effects.

11.14.1 The HSDC Method

The HSDC stands for health, safety, distraction, and capacity. In order to forecast how the environment would affect human performance and production, a model of human performance has been developed. It entails multiplying the amount of worktime (e.g., over a shift) available for safe work (HS), the amount of time spent working without being distracted by the environment (D), and the amount of work or tasks that can be accomplished in comparison with the amount that can be accomplished under ideal conditions. In other words, HSDC represents the decrease in performance at a task or work when compared to optimum performance under comfortable conditions.

11.14.1.1 Thermal Comfort and Health and Safety (HS)

If people become so uncomfortable owing to thermal circumstances that they refuse to work or if restrictions lead to a cessation of work as thermal limitations are surpassed, then performance and productivity will decrease to zero in times when people are not working (e.g., for all or part of a day). In conditions that are mild and where thermal comfort—rather than health—is the primary concern, this is unlikely to happen. Nonetheless, this might happen in uncomfortable thermal settings (Parsons, 2014j).

11.14.1.2 Distraction (D)

Thermal distraction, which is defined as "a tendency for a person to attend to attend the thermal state (hot, uncomfortable, or cold) instead of accomplishing a task," is a sort of distraction. Therefore, if distracted, performance in the task will be zero during that time. There are many potential sources of distraction, such as paying attention to stimuli or tasks other than those that are being carried out; however, in terms of how the environment affects performance, it is the distraction brought on by environmental factors that is of interest; similarly, in terms of thermal comfort, it will be related to the level of discomfort, which may result in dissatisfaction. Studies say that one in five of the people dissatisfied will be inclined to do something about it and hence be distracted; we can predict the PPD from the thermal conditions and hence the percentage of time a person will be distracted. If the PMV is −2, cold, the PPD is 50%, so 10% of people will be distracted. If we assume that people are only distracted 10% of the time, we can say that performance in this cold environment will only be 90% of what would be in comfortable conditions, where there would be no distractions.

11.14.1.3 Capacity (C)

The quantity of work that can be accomplished by a typical human being in a thermally neutral environment is referred to as capacity. A task taxonomy is typically used to clarify what we mean by human performance as the first step in the process. The ability to complete a task is what is meant by "human performance." Inconsistency in outcomes is a regular observation for researchers looking at how the thermal environment affects capacity and, ultimately, performance at tasks, with the possible exception of the lack of manual dexterity brought on by chilly hands. Many tasks have been researched, and it is usually discovered that, under identical temperature and contextual conditions, performance varies depending on the study. In some studies, performance declines; in others, it improves; and in some, there is no difference. In mild conditions, there is a small decrease in the ability to accomplish tasks. Hence, the HSDC model would emphasize distraction as the main potential cause of performance loss across all situations. It makes sense to assume that task performance will be influenced by the thermoregulatory response. Speed of response may be accelerated by vasodilation, whereas manual dexterity may be slowed down by vasoconstriction. The grip will be hampered by sweaty hands. Therefore, it is logical to draw the conclusion that if moderate settings have an impact on one's ability to perform tasks, this influence hasn't yet been expressed consistently.

11.14.2 The ISO Standards Initiative

To study, how the physical environment affect the performance of people, ISO NWI 23454-1 (2019): "Human performance in physical environments: Part 1 — A performance framework" is adopted. The plan is to create two general standards that

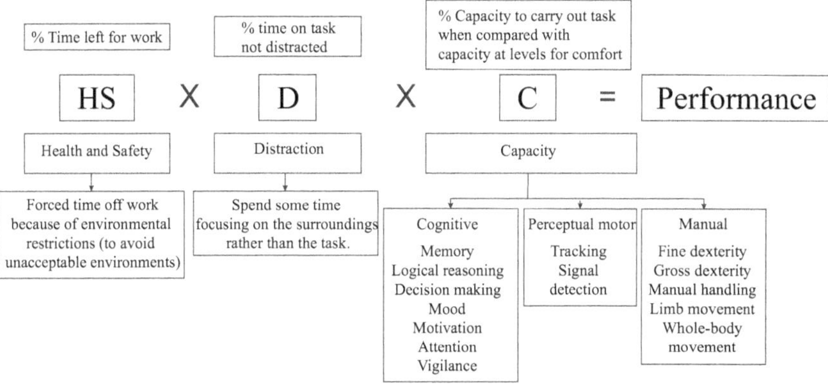

FIGURE 11.4 HSDC model for predicting human performance.

can be used to explain how certain environmental factors affect certain things (thermal, noise, light, etc.). The first generic standard will put forth a performance model for people. Figure 11.4 illustrates this. Additionally, it will outline a performance prediction technique based on the HSDC method ("International Standards," 2014). The second will outline standardized tests that can be used to assess how well the model's elements are performing. Human performance in physical environments: Part 2 — Measurements and methods for evaluating the effects of the physical environment on human performance (ISO DP 23454–2 (2019)).

11.15 INTERNATIONAL STANDARDS

The two most essential standards used internationally to specify thermal comfort conditions, primarily but not exclusively for buildings around the world, are ASHRAE 55, which is reviewed (but not necessarily revised) yearly, and ISO 7730 (2005), "Ergonomics of the Thermal Environment—Analytical determination and interpretation of Thermal Comfort Using Calculation of the PMV and PPD indices and Local Thermal Comfort Criteria." The ASHRAE 55–2017 "Thermal Environmental Conditions for Human Occupancy (ANSI/ASHRAE Certified)" standard is in effect for 2017. As an intermediate step, the addenda to the standard are published annually. The predicted mean vote (PMV)/predicted percentage of dissatisfied (PPD) indices are calculated using an updated computer program written in the JavaScript programming language in the addendum for 2019. ASHRAE 55 illustrates comfort limitations in terms of air temperature and humidity on a psychrometric chart. It also gives an adaptable strategy for obtaining thermal comfort. Comfort zones are shown on the psychrometric chart of operative temperature and dew point, which is connected to humidity. The zones of thermal comfort are provided as spaces on the psychrometric chart that are bordered by lines for light activity (see Figure 11.3). The International Organization for Standardization (ISO) 7730 (2005) is the international standard that specifies conditions for thermal comfort and the prediction of the degree of thermal discomfort. Other standard bodies use ISO 7730 (2005) as the basis for standards in specific areas of application, such as outdoors or indoors. ASHRAE provides the answer to frequently asked questions such as the recommended indoor temperature and humidity levels for homes are between 67°F (19.5°C) and 82°F (27.8°C), with standardizing the definition and computation of the PMV and PPD indices as well as techniques for predicting local thermal discomfort is one of its key purposes. It is a component of a set of international thermal comfort standards that also address subjective evaluation (ISO 10551, 2019), physical environments for people with special needs (ISO 28803, 2012), the thermal sensation caused by skin contact with surfaces of moderate temperature (ISO 13732 Part 2, 2001), environmental survey (ISO 28802, 2012), thermal comfort in vehicles (ISO 14505 Parts 1–4, 2006), and the practical design of environments for thermal comfort. The estimation of clothing's thermal properties, the measurement of the thermal environment, and the estimation of metabolic heat production based on activity type are all supported standards (ISO 9920, 2007; ISO 7726, 1998; ISO 8996, 2004). ISO 7730 (2005) discusses and defines how to calculate the PMV of a large group of persons exposed to a thermal environment defined by the air temperature, radiant temperature, humidity, air velocity, metabolic rate, and clothing. The PMV is a value on the thermal sensory scale: 3, hot; 2, warm; 1, slightly warm; 0, neutral; −1, slightly cool; −2, cool; −3, frigid. Thermal comfort is defined as having a neutral feeling (PMV = 0). The PMV value is used to calculate the PPD, which forecasts how many people will be unhappy with the circumstances.

11.15.1 Building Envelope for Comfort of Occupants

ASHRAE 700–2015 provides specific instruction to the builder to prepare buildings that provide comfort to occupants based on the construction and position of equipment inside the building. The intent of ASHRAE 700–2015 is to control the pollutant sources. According to it, in a building with natural draft combustion space or equipment used for heating water, all equipment which liberates heat (boilers, natural draft furnace, or water heater) should be kept outside the conditioned spaces where the outdoor air source is available and make sure the place is well sealed and separated from conditioned space. The air handling unit (AHU) associated with the same should be inside the conditioned space. It is mandatory to ensure that any direct heating equipment must be vented to the outside or open space which is direct to environment, and any equipment which uses solid fuel for combustion, like stoves or heaters, must be certified by the American Society for Testing and Materials (ASTM) or Environmental Protection Agency (EPA). For any structure with a common wall with the door between the garage and conditioned space, it is essential to ensure that the door is tightly sealed and gasketed and a minimum of 85% of wood materials which are used for floor, walls, roof sheathing, structural panels, custom woodwork, or component closet shelving need to be moisture resistant. The carpets should not be installed adjacent to bathing fixtures, and water closets following flooring types are recommended.

1. Ceramic tile flooring
2. Organic-free, mineral-based flooring
3. Clay masonry flooring

TABLE 11.4
VOC Content Limits for Architectural Coatings (Green & Corporation, 2013)

Coating category	Limit (g/L)
Flat coatings	50
Non-flat coatings	100
Non-flat high-gloss coatings	150
Aluminium roof coatings	400
Basement specially coatings	400
Bituminous roof coatings	50
Bituminous roof priming	350
Bond breakers	350
Concrete curing compounds	350
Concrete/masonry sealers	100
Driveway sealers	50
Dry fog coatings	150
Faux finishing coating	350
Fire-resistive coating	350
Floor coatings	100
Form-release compounds	250
Graphic arts coatings	500
High temperature coatings	420
Low solid coatings	250
Magnesite cement coatings	120
Mastic texture coatings	450
Metallic pigmented coatings	100
Multi-colour coating	500
Pre-treatment wash primers	250
Primers, sealers, and undercoatings	420
Reactive penetrating sealers	100
Recycled coatings	350

4. Concrete flooring
5. Metal flooring

A minimum of 10% of the interior wall surfaces are covered with wall coverings, and a minimum 85% of wall coverings follow the emission concentration limit. Table 11.4 shows the recommended volatile organic compound (VOC) content limits for architectural coatings.

Carbon monoxide alarms should be provided as per the standard (IRC section R315) for safety purposes. The prevalent source of pollutants is the entrance of the building, which can be controlled by using exterior grilles/interior grilles or mats. Smoking should be avoided in the living space, especially in multifamily buildings. For more pollutant control, spot ventilation should be used; clothes dryers are vented outdoors, and kitchen exhaust units are ducted outdoors and have a minimum ventilation rate of 47.2 L/s. The exhaust fan used in the building should be Energy Star.

The stack effect should be considered while designing fenestration in spaces. For building ventilation systems, the exhaust or supply fan should be in such standards to be used for a long time in running mode. For moisture management (vapour, rainwater, plumbing, HVAC), cold water pipes in unconditioned spaces are insulated to a minimum, and plumbing is installed in conditioned spaces. The central HVAC system should be used for additional control of the occupant environment. Practices to control ventilation, moisture, pollutant sources, and sanitation should protect indoor air quality (IAQ). Some innovative practices like a humidity monitoring system, which shows the temperature and relative humidity of conditioned space, and kitchen exhaust units having a capacity of more than 189 L/s should be installed.

REFERENCES

Automotive Ergonomics: Driver-Vehicle Interaction. Automot Ergon (2016). https://doi.org/10.1201/B13017

Developing an Adaptive Model of Thermal Comfort and Preference (Conference) | OSTI.GOV. (n.d.). Retrieved March 27, 2023, from www.osti.gov/biblio/653186

Fletcher, M. J., Glew, D. W., Hardy, A., & Gorse, C. (2020). A modified approach to metabolic rate determination for thermal comfort prediction during high metabolic rate activities. *Building and Environment*, *185*(July), 107302. https://doi.org/10.1016/j.buildenv.2020.107302

Heat Stress. (2014). *Human Thermal Environments*, 362–393. https://doi.org/10.1201/B16750-23

International Standards. (2014). *Human Thermal Environments*, 484–541. https://doi.org/10.1201/B16750-27

Parsons, K. (2014a). Human thermal environments: The effects of hot, moderate, and cold environments on human health, comfort, and performance, third edition. *Human Thermal Environments: The Effects of Hot, Moderate, and Cold Environments on Human Health, Comfort, and Performance, Third Edition*, 1–586. https://doi.org/10.1201/B16750/HUMAN-THERMAL-ENVIRONMENTS-KEN-PARSONS

Parsons, K. (2014b). Cold stress. *Human Thermal Environments*, 394–425. https://doi.org/10.1201/B16750-24

Parsons, K. (2014c). Human skin contact with hot, moderate and cold surfaces. *Human Thermal Environments*, 448–483. https://doi.org/10.1201/B16750-26

Parsons, K. (2014d). Human skin contact with hot, moderate and cold surfaces. *Human Thermal Environments*, 448–483. https://doi.org/10.1201/B16750-26/HUMAN-SKIN-CONTACT-HOT-MODERATE-COLD-SURFACES-KEN-PARSONS

Parsons, K. (2014e). Human thermal physiology and thermoregulation. *Human Thermal Environments*, 98–117. https://doi.org/10.1201/B16750-12

Parsons, K. (2014f). Human Thermal physiology and thermoregulation. *Human Thermal Environments*, 98–117. https://doi.org/10.1201/B16750-12/HUMAN-THERMAL-PHYSIOLOGY-THERMOREGULATION-KEN-PARSONS

Parsons, K. (2014g). Metabolic heat production. *Human Thermal Environments*, 226–249. https://doi.org/10.1201/B16750-17/METABOLIC-HEAT-PRODUCTION-KEN-PARSONS

Parsons, K. (2014h). Metabolic heat production. *Human Thermal Environments*, 226–249. https://doi.org/10.1201/B16750-17

Parsons, K. (2014i). The human heat balance equation and the thermal audit. *Human Thermal Environments*, 72–97. https://doi.org/10.1201/B16750-11

Parsons, K. (2014j). Thermal environments and human performance. *Human Thermal Environments*, 426–447. https://doi.org/10.1201/B16750-25/THERMAL-ENVIRONMENTS-HUMAN-PERFORMANCE-KEN-PARSONS

Parsons, K. (2014k). Thermal models. *Human Thermal Environments*, 200–225. https://doi.org/10.1201/B16750-16

Parsons, K. (2021a). Weather, climate change and energy use. *Human Thermal Environments*, 542–553. https://doi.org/10.1201/B16750-28

Parsons, K. (2021b). Weather, climate change and energy use. *Human Thermal Environments*, 542–553. https://doi.org/10.1201/B16750-28/WEATHER-CLIMATE-CHANGE-ENERGY-USE-KEN-PARSONS

Rawal, R., Manu, S., Shukla, Y., Thomas, L. E., De Dear, R. (2016). Clothing insulation as a behavioural adaptation for thermal comfort in Indian office buildings. *Proc - 9th Int Wind Conf 2016 Mak Comf Relev* 2016:403–15. https://doi.org/https://www.researchgate.net/publication/300477778

Schoen, L. J., Alspach, P. F., Arens, E. A., Aynsley, R. M., Bean, R., Eddy, J., Int-Hout, D., Khalil, E. E., Simmonds, P., Stoops, J. L., Turner, S. C., Walter, W. F., Hall, R. L., Anderson, J. R., Barnaby, C. S., Bruning, S. F., Clark, J. A., Emmerich, S. J., & Ferguson, J. M. (2013). *A—Ashare 55. 2013.* www.ashrae.org/technology.

Shah, K. W., Li, W. (2019). A Review on Catalytic Nanomaterials for Volatile Organic Compounds VOC Removal and Their Applications for Healthy Buildings. *Nanomater 2019*, Vol 9, Page 910 2019;9:910. https://doi.org/10.3390/NANO9060910

Špelić, I., Mihelic-Bogdanic, A., & Šajatovic, A. H. (2019). Standard methods for thermal comfort assessment of clothing. *Standard Methods for Thermal Comfort Assessment of Clothing*, 1–227. https://doi.org/10.1201/9780429422997/STANDARD-METHODS-THERMAL-COMFORT-ASSESSMENT-CLOTHING-IVANA-

Thermal Comfort. (2014). *Human Thermal Environments*, 296–329. https://doi.org/10.1201/B16750-20

Wang, Z., & Hong, T. (2020). Learning occupants' indoor comfort temperature through a Bayesian inference approach for office buildings in United States. *Renewable and Sustainable Energy Reviews*, *119*. https://doi.org/10.1016/J.RSER.2019.109593

Wilson, J. R., & Sharples, S. (2015). Evaluation of human work, Fourth Edition. *Evaluation of Human Work, Fourth Edition*, 1–984. https://doi.org/10.1201/B18362/EVALUATION-HUMAN-WORK-JOHN-WILSON-SARAH-SHARPLES

Index

U

V

W

For Product Safety Concerns and Information please contact our EU representative GPSR@taylorandfrancis.com
Taylor & Francis Verlag GmbH, Kaufingerstraße 24, 80331 München, Germany

www.ingramcontent.com/pod-product-compliance
Lightning Source LLC
LaVergne TN
LVHW010852110826
845149LV00005B/1391

* 9 7 8 1 0 3 2 5 4 2 1 2 6 *